Multisensory Flavor Perception

Related Titles

Flavour Science: Proceedings from XIII Weurman Flavour Research Symposium
(ISBN 978-0-12398-549-1)

Breakthrough Food Product Innovation Through Emotions Research
(ISBN 978-0-12387-712-3)

Product Experience
(ISBN 978-0-08045-089-6)

Woodhead Publishing Series in Food Science, Technology and Nutrition: Number 298

Multisensory Flavor Perception

From Fundamental Neuroscience Through to the Marketplace

Edited by

Betina Piqueras-Fiszman

Charles Spence

AMSTERDAM • BOSTON • HEIDELBERG • LONDON
NEW YORK • OXFORD • PARIS • SAN DIEGO
SAN FRANCISCO • SINGAPORE • SYDNEY • TOKYO
Woodhead Publishing is an imprint of Elsevier

Woodhead Publishing Limited is an imprint of Elsevier
The Officers' Mess Business Centre, Royston Road, Duxford, CB22 4QH, UK
50 Hampshire Street, 5th Floor, Cambridge, MA 02139, USA
The Boulevard, Langford Lane, Kidlington, OX5 1GB, UK

Copyright © 2016 Elsevier Ltd. All rights reserved.

No part of this publication may be reproduced or transmitted in any form or by any means, electronic or mechanical, including photocopying, recording, or any information storage and retrieval system, without permission in writing from the publisher. Details on how to seek permission, further information about the Publisher's permissions policies and our arrangements with organizations such as the Copyright Clearance Center and the Copyright Licensing Agency, can be found at our website: www.elsevier.com/permissions.

This book and the individual contributions contained in it are protected under copyright by the Publisher (other than as may be noted herein).

Notices
Knowledge and best practice in this field are constantly changing. As new research and experience broaden our understanding, changes in research methods, professional practices, or medical treatment may become necessary. Practitioners and researchers must always rely on their own experience and knowledge in evaluating and using any information, methods, compounds, or experiments described herein. In using such information or methods they should be mindful of their own safety and the safety of others, including parties for whom they have a professional responsibility.

To the fullest extent of the law, neither the Publisher nor the authors, contributors, or editors, assume any liability for any injury and/or damage to persons or property as a matter of products liability, negligence or otherwise, or from any use or operation of any methods, products, instructions, or ideas contained in the material herein.

British Library Cataloguing-in-Publication Data
A catalogue record for this book is available from the British Library

Library of Congress Cataloging-in-Publication Data
A catalog record for this book is available from the Library of Congress

ISBN: 978-0-08-100350-3 (print)
ISBN: 978-0-08-100351-0 (online)

For Information on all Woodhead Publishing
visit our website at http://elsevier.com/

Publisher: Nikki Levy
Acquisition Editor: Rob Sykes
Editorial Project Manager: Josh Bennett
Production Project Manager: Caroline Johnson
Designer: Ines Cruz

Typeset by MPS Limited, Chennai, India

Contents

List of contributors		ix
Foreword		xi
Woodhead Publishing Series in Food Science, Technology and Nutrition		xv

1 Introduction   **1**
Betina Piqueras-Fiszman and Charles Spence
1. The senses and multisensory integration   1
2. Expectations   4
3. On the hedonics and memorability of flavor   5
4. Living in different taste worlds   6
5. A closer look into our brain and bodily reactions   6
6. The impact of the senses in context: from the supermarket to disposal   7
References   8

2 Attention and Flavor Binding   **15**
Richard J. Stevenson
1. Introduction   15
2. Flavor binding   16
3. How might flavor binding occur?   22
4. Conclusions   30
References   31

3 Oral Referral   **37**
Juyun Lim
1. Introduction   37
2. Conceptual background   37
3. Oral referral   44
4. Conclusions   51
References   52

4 Oral-Somatosensory Contributions to Flavor Perception and the Appreciation of Food and Drink   **59**
Charles Spence and Betina Piqueras-Fiszman
1. Introduction   59
2. Multisensory flavor perception   60
3. Oral-somatosensation   62
4. Oral-somatosensation: Neural substrates   67
5. The role of oral-somatosensation in determining our food likes/dislikes   68
6. The role of oral-somatosensation in expected and experienced satiety   69

	7 On the contribution of somatosensory stimulation outside the mouth	70
	8 Conclusions	72
	References	73
5	**Sound: The Forgotten Flavor Sense**	**81**
	Charles Spence	
	1 Introduction	81
	2 The sounds of food and beverage preparation	82
	3 The sounds of packaging	84
	4 The sound of food and drink and our interaction with them	85
	5 Background noise	89
	6 Background music	93
	7 Sonic seasoning	96
	8 Conclusions	99
	References	100
6	**Food Color and Its Impact on Taste/Flavor Perception**	**107**
	Charles Spence and Betina Piqueras-Fiszman	
	1 Introduction	107
	2 Psychological effects of food color: setting sensory expectations	108
	3 Names, brands, and colors	115
	4 Psychological effects on food color on behavior	116
	5 Marketing color	119
	6 Individual differences in the psychological effects of color	119
	7 Conclusions	124
	References	125
7	**Multisensory Flavor Priming**	**133**
	Garmt Dijksterhuis	
	1 Introduction	133
	2 A taxonomy of primes	140
	3 Odor and flavor priming	143
	4 Bottom-up and top-down priming	145
	5 A suggestion for a conceptual framework for situated (flavor) perception and behavior	147
	6 Conclusions	148
	Acknowledgments	149
	Appendix	149
	References	151
8	**Flavor Liking**	**155**
	John Prescott	
	1 Development of liking	155
	2 Evaluative conditioning	157

	3	Models of evaluative learning	**160**
	4	Mechanisms of binding of odors and tastes	**161**
	5	Conclusions: The adaptive significance of flavors	**163**
	References		**163**

9   Flavor Memory — **169**
Jos Mojet and Ep Köster
1. Introduction — **169**
2. Different forms of memory — **170**
3. Flavor learning and memory over lifetime — **175**
4. Influence of flavor memory on eating behavior — **177**
5. How should real-world flavor memory be studied? — **178**
6. Conclusions — **180**
References — **180**

10   Individual Differences in Multisensory Flavor Perception — **185**
Cordelia A. Running and John E. Hayes
1. Introduction — **185**
2. Overview of genetics and molecular biology — **185**
3. Tactile and chemesthetic percepts — **188**
4. Odor — **190**
5. Taste — **192**
6. Functional outcome: food intake — **198**
7. Conclusions and future directions — **201**
Acknowledgments — **202**
References — **202**

11   Pleasure of Food in the Brain — **211**
Alexander Fjaeldstad, Tim J. van Hartevelt and Morten L. Kringelbach
1. Introduction — **211**
2. Brain principles of eating — **212**
3. Computational processing related to eating — **213**
4. Reward processing in the brain — **223**
5. Perspectives/challenges — **227**
References — **227**

12   The Neuroscience of Flavor — **235**
Charles Spence
1. Introduction — **235**
2. Flavor expectations and flavor experiences — **236**
3. Neural circuits underlying multisensory flavor perception — **237**
4. Branding and pricing — **241**
5. Conclusions — **243**
References — **244**

13	Responses of the Autonomic Nervous System to Flavors	249
	René A. de Wijk and Sanne Boesveldt	
	1 The problem	249
	2 The central and autonomic nervous systems	249
	3 Emotions and specificity of responses	251
	4 Measures of autonomic nervous system activity	253
	5 Patterns of ANS activity	253
	6 Concluding remarks	264
	References	265
14	Assessing the Influence of the Drinking Receptacle on the Perception of the Contents	269
	Charles Spence and Xiaoang Wan	
	1 The shape of the receptacle	270
	2 Sensation transference	277
	3 Drinking vessels and consumption behavior	284
	4 Implications for marketing practice	287
	5 Future research directions and conclusions	288
	References	290
15	The Roles of the Senses in Different Stages of Consumers' Interactions With Food Products	297
	Hendrik N.J. Schifferstein	
	1 Introduction	297
	2 Research approaches to the study of dynamic experiences	298
	3 Findings concerning the role of sensory perception in different interactions with food products	303
	4 Conclusions	307
	References	309
16	Sensory Branding: Using Brand, Pack, and Product Sensory Characteristics to Deliver a Compelling Brand Message	313
	David M.H. Thomson	
	1 Taking a multisensory approach to branding	313
	2 A quick word about mental constructions	315
	3 Sensory characteristics	315
	4 Pleasantness and reward	316
	5 Conceptual associations	319
	6 Linking conceptualization to reward	320
	7 The fundamental nature of brands	323
	8 Extending sensory branding to pack and product	324
	9 "The Matrix"	326
	10 Sensory signatures	328
	11 Consonance and "fit-to-brand"	329
	12 Conclusion	334
	References	334

Index 337

List of contributors

Sanne Boesveldt Division of Human Nutrition, Wageningen University, Wageningen, The Netherlands

René A. de Wijk Consumer Science & Health, WUR, Wageningen, The Netherlands

Garmt Dijksterhuis University College Roosevelt, University of Utrecht, Middelburg, The Netherlands; University of Copenhagen, Sensory and Consumer Science Section, Copenhagen, Denmark

Alexander Fjaeldstad Department of Psychiatry, University of Oxford, Oxford, United Kingdom; Department of Otorhinolaryngology, Aarhus University Hospital, Aarhus, Denmark

John E. Hayes Department of Food Science, College of Agricultural Sciences, The Pennsylvania State University, University Park, PA, United States

Ep Köster Helmholtz Institute, Psychology, University of Utrecht, Utrecht, The Netherlands

Morten L. Kringelbach Department of Psychiatry, University of Oxford, Oxford, United Kingdom; Center of Functionally Integrative Neuroscience (CFIN), Aarhus University, Aarhus, Denmark

Juyun Lim Department of Food Science and Technology, Oregon State University, Corvallis, OR, United States

Jos Mojet Food & Biobased Research, Wageningen University and Research Centre, Wageningen, The Netherlands

Betina Piqueras-Fiszman Marketing and Consumer Behavior Group, Wageningen University, Wageningen, The Netherlands

John Prescott TasteMatters Research & Consulting, Sydney, Australia

Cordelia A. Running Sensory Evaluation Center, College of Agricultural Sciences, The Pennsylvania State University, University Park, PA, United States

Hendrik N.J. Schifferstein Department of Industrial Design, Delft University of Technology, Delft, The Netherlands

Charles Spence Crossmodal Research Laboratory, Department of Experimental Psychology, University of Oxford, Oxford, United Kingdom

Richard J. Stevenson Department of Psychology, Macquarie University, Sydney, NSW, Australia

David M.H. Thomson MMR Research Worldwide, Wallingford, Oxfordshire, United Kingdom

Tim J. van Hartevelt Department of Psychiatry, University of Oxford, Oxford, United Kingdom; Center of Functionally Integrative Neuroscience (CFIN), Aarhus University, Aarhus, Denmark

Xiaoang Wan Department of Psychology, School of Social Sciences, Tsinghua University, Beijing, P.R. China

Foreword

Throughout the history of gastronomy, there has always been an understanding that food is both a human necessity as well as being a great source of pleasure, one that is of prime importance in celebrations and occasions the world over (Jones, 2008). Unfortunately, however, it is also true that in many countries both developed and developing, poor nutrition is one of the leading causes of health problems. Nevertheless, regardless of the impact it has on our health and well-being, eating, and particularly dining, is an activity that engages each and every one of our senses and can bring us great satisfaction.[1]

It was Nicolas Kurti, a physicist who was particularly fond of cooking (and of the science of cooking; eg, see Kurti & This-Benckhard, 1994a, 1994b), who once said *"I think it is a sad reflection on our civilization that while we can and do measure the temperature in the atmosphere of Venus we do not know what goes on inside our soufflés."* (McQuaid, 2015, p. 230). I would add to this the issue that given the importance of food in our lives, is it not vital that we should have a better understanding behind the psychology and neuroscience of eating? Fields of research that are starting to provide answers here are sometimes referred to as neurogastronomy (Shepherd, 2012; see also http://isneurogastronomy.com/about-us), or more inclusively, gastrophysics (Spence & Piqueras-Fiszman, 2014). Understanding the associations and perceptions we have with different foods and how this impacts our relationship and interactions with them has, I think, the potential to completely change the way in which we approach food design. Indeed, the last year or so has seen a veritable explosion of new courses, conferences, and journals on food design.

As chefs, we continuously look for ways in which to captivate and delight our guests with the dishes we present them with. Culinary science which has given rise to "molecular gastronomy," and "modernist cuisine" has for several decades provided chefs all over the world with inspiration for the plate. In my own work (Youssef, 2013), I have tried to translate the latest techniques from the high-end kitchen into the language of the home chef. Multisensory flavor perception is an extension of this in some ways—it is about taking an interest in the science of dining. This opens the doors for chefs to work with scientists, artists, designers, and musicians towards developing truly memorable multisensory experiences for diners which engage them on several levels and through all of their senses (see Spence & Piqueras-Fiszman, 2014, for a review). What I think we are seeing currently is a shift from the science

---

[1] The exception being the dine-in-the-dark restaurant, about which no one seems to have anything good to say (Spence & Piqueras-Fiszman, 2012, 2014).

of the kitchen, of new culinary techniques and materials (McGee, 1984/2004), to a growing realization of the importance of the new science of the table.

For my team and I, this book is one which we depend on as we develop all the sensory aspects of our experimental dining concepts. The research carried out reveals fascinating patterns in food perception and associations that may seem intuitive, but we never quite understand why. As a chef, I often find myself wondering how I will plate a dish to maximize its visual appeal, now it turns out there may indeed be several pieces of research which have identified "preferences" that we as humans have towards food presentation (see Michel, Woods, Neuhäeuser, Landgraf, & Spence, 2015; Spence, Piqueras-Fiszman, Michel, & Deroy, 2014). Then, of course, there is the fascinating world of scent and the impact aroma has on our perceptions of flavor, once again this is an area where till now intuition has been our main guide (see Spence & Youssef, 2015, for a review). Now equipped with a more informed understanding of the role aroma plays we have been able to incorporate unique aroma elements into our culinary creations. Finally, it is worth highlighting the impact of sound and texture on flavor, taste and mouthfeel. These are senses that are often overlooked in the context of food and dining (at least by chefs). However, the research summarized in this volume, together with our own experience of putting these ideas to the test, and ultimately into practice, shows just how important they are to creating enhanced multisensory dining experiences.

The chapters in this book look to change how we define "flavor"; taking it from the conventional two-dimensional definition of a combination of taste and smell, and broadening it to include all the other senses. This has the potential to bring great changes to the ways in which foods are designed by both chefs and large food manufacturers. Given all the current health concerns about our diet and nutritional intake, as well as concerns related to sustainability and food security—enhanced knowledge about multisensory flavor perception may hold the key to helping understand how we can potentially improve people's food choices by altering the sensory cues in food, packaging, and presentation, with the ultimate goal of making nutritious and sustainable foods (such as insects) more appealing to the population at large (see Deroy, Reade, & Spence, 2015). As a case in point, Menu 1 shows the current menu from Mexico a culinary dining concept in which four out of the seven dishes on the tasting menu contain insects. I have benefited greatly from the latest scientific insights in terms of thinking how best to put this dining concept together.

Mexico by Kitchen Theory 2015 Menu
The Holy Trinity – corn, beans, chilli
Nopal – nopal, oaxacan cheese, fresh tortilla, salsa, lime, coriander
Memories of Oaxaca – shellfish, octopus, corn, lime, epazote, coriander
El Chapulín Colorado – hearts of palm, octopus, cucumber, jalapeno, aguachile, avocado, tostada, chapulines (Mexican grasshopper)
An Offering for the Gods – venison, mole negro, pumpkin, burnt tortilla, spiced mealworm powder
Mezcal – orange, mezcal, tajin chili seasoning, coriander, ant salt
Vanilla and the bee – vanilla, cinnamon, camomile, honey, bee pollen

Vanilla and the Bee

Looking forward, I see a great deal of excitement for chefs working together with scientists, building on the culinary skills of the chef by rigorous experimentation. We are, I think, already starting to see the first wave of research papers coming out of this collaboration (eg, Spence, Shankar, & Blumenthal, 2011; Spence et al., 2015; Yeomans, Chambers, Blumenthal, & Blake, 2008; Youssef, Juravle, Youssef, Woods, & Spence, 2015).

Jozef Youssef,
Founder,
Kitchen Theory,
London

References

Deroy, O., Reade, B., & Spence, C. (2015). The insectivore's dilemma. *Food Quality and Preference*, *44*, 44–55.
Jones, M. (2008). *Feast: Why humans share food*. Oxford: Oxford University Press.
Kurti, N., & This-Benckhard, H. (1994a). Chemistry and physics in the kitchen. *Scientific American*, *270*(4), 66–71.
Kurti, N., & This-Benckhard, H. (1994b). The amateur scientist: The kitchen as a lab. *Scientific American*, *270*(4), 120–123.
McGee, H. (1984/2004). *On food and cooking: The science and lore of the kitchen* (rev. ed.). New York, NY: Scribner.
McQuaid, J. (2015). *Tasty: The art and science of what we eat*. New York, NY: Simon and Schuster. p. 230.

Michel, C., Woods, A. T., Neuhäeuser, M., Landgraf, A., & Spence, C. (2015). Orienting the plate: Online study assesses the importance of the orientation in the plating of food. *Food Quality and Preference*, *44*, 194–202.

Shepherd, G. M. (2012). *Neurogastronomy: How the brain creates flavor and why it matters*. New York, NY: Columbia University Press.

Spence, C., & Piqueras-Fiszman, B. (2012). Dining in the dark: Why, exactly, is the experience so popular? *The Psychologist*, *25*, 888–891.

Spence, C., & Piqueras-Fiszman, B. (2014). *The perfect meal: The multisensory science of food and dining*. Oxford: Wiley-Blackwell.

Spence, C., Piqueras-Fiszman, , Michel, C., & Deroy, O. (2014). Plating manifesto (II): The art and science of plating. *Flavour*, *3*, 4.

Spence, C., Shankar, M. U., & Blumenthal, H. (2011). "Sound bites": Auditory contributions to the perception and consumption of food and drink. In F. Bacci & D. Melcher (Eds.), *Art and the senses* (pp. 207–238). Oxford: Oxford University Press.

Spence, C., Wan, X., Woods, A., Velasco, C., Deng, J., Youssef, J., et al. (2015). On tasty colours and colourful tastes? Assessing, explaining, and utilizing crossmodal correspondences between colours and basic tastes. *Flavour*, *4*, 23.

Spence, C., & Youssef, J. (2015). Olfactory dining: Designing for the dominant sense. *Flavour*, *4*, 32.

Yeomans, M., Chambers, L., Blumenthal, H., & Blake, A. (2008). The role of expectancy in sensory and hedonic evaluation: The case of smoked salmon ice-cream. *Food Quality and Preference*, *19*, 565–573.

Youssef, J. (2013). *Molecular cooking at home: Taking culinary physics out of the lab and into your kitchen*. London: Quintet Publishing.

Youssef, J., Juravle, G., Youssef, L., Woods, A., & Spence, C. (2015). On the art and science of naming and plating food. *Flavour*, *4*, 27.

Woodhead Publishing Series in Food Science, Technology and Nutrition

1. **Chilled foods: A comprehensive guide**
   Edited by C. Dennis and M. Stringer

2. **Yoghurt: Science and technology**
   Y. Tamime and R. K. Robinson

3. **Food processing technology: Principles and practice**
   P. J. Fellows

4. **Bender's dictionary of nutrition and food technology Sixth edition**
   D. A. Bender

5. **Determination of veterinary residues in food**
   Edited by N. T. Crosby

6. **Food contaminants: Sources and surveillance**
   Edited by C. Creaser and R. Purchase

7. **Nitrates and nitrites in food and water**
   Edited by M. J. Hill

8. **Pesticide chemistry and bioscience: The food-environment challenge**
   Edited by G. T. Brooks and T. Roberts

9. **Pesticides: Developments, impacts and controls**
   Edited by G. A. Best and A. D. Ruthven

10. **Dietary fibre: Chemical and biological aspects**
    Edited by D. A. T. Southgate, K. W. Waldron, I. T. Johnson and G. R. Fenwick

11. **Vitamins and minerals in health and nutrition**
    M. Tolonen

12. **Technology of biscuits, crackers and cookies Second edition**
    D. Manley

13. **Instrumentation and sensors for the food industry**
    Edited by E. Kress-Rogers

14. **Food and cancer prevention: Chemical and biological aspects**
    Edited by K. W. Waldron, I. T. Johnson and G. R. Fenwick

15. **Food colloids: Proteins, lipids and polysaccharides**
    Edited by E. Dickinson and B. Bergenstahl

16 **Food emulsions and foams**
   Edited by E. Dickinson
17 **Maillard reactions in chemistry, food and health**
   Edited by T. P. Labuza, V. Monnier, J. Baynes and J. O'Brien
18 **The Maillard reaction in foods and medicine**
   Edited by J. O'Brien, H. E. Nursten, M. J. Crabbe and J. M. Ames
19 **Encapsulation and controlled release**
   Edited by D. R. Karsa and R. A. Stephenson
20 **Flavours and fragrances**
   Edited by A. D. Swift
21 **Feta and related cheeses**
   Edited by A. Y. Tamime and R. K. Robinson
22 **Biochemistry of milk products**
   Edited by A. T. Andrews and J. R. Varley
23 **Physical properties of foods and food processing systems**
   M. J. Lewis
24 **Food irradiation: A reference guide**
   V. M. Wilkinson and G. Gould
25 **Kent's technology of cereals: An introduction for students of food science and agriculture Fourth edition**
   N. L. Kent and A. D. Evers
26 **Biosensors for food analysis**
   Edited by A. O. Scott
27 **Separation processes in the food and biotechnology industries: Principles and applications**
   Edited by A. S. Grandison and M. J. Lewis
28 **Handbook of indices of food quality and authenticity**
   R. S. Singhal, P. K. Kulkarni and D. V. Rege
29 **Principles and practices for the safe processing of foods**
   D. A. Shapton and N. F. Shapton
30 **Biscuit, cookie and cracker manufacturing manuals Volume 1: Ingredients**
   D. Manley
31 **Biscuit, cookie and cracker manufacturing manuals Volume 2: Biscuit doughs**
   D. Manley
32 **Biscuit, cookie and cracker manufacturing manuals Volume 3: Biscuit dough piece forming**
   D. Manley

33 Biscuit, cookie and cracker manufacturing manuals Volume 4: Baking and cooling of biscuits
    D. Manley

34 Biscuit, cookie and cracker manufacturing manuals Volume 5: Secondary processing in biscuit manufacturing
    D. Manley

35 Biscuit, cookie and cracker manufacturing manuals Volume 6: Biscuit packaging and storage
    D. Manley

36 Practical dehydration Second edition
    M. Greensmith

37 Lawrie's meat science Sixth edition
    R. A. Lawrie

38 Yoghurt: Science and technology Second edition
    Y. Tamime and R. K. Robinson

39 New ingredients in food processing: Biochemistry and agriculture
    G. Linden and D. Lorient

40 Benders' dictionary of nutrition and food technology Seventh edition
    D. A. Bender and A. E. Bender

41 Technology of biscuits, crackers and cookies Third edition
    D. Manley

42 Food processing technology: Principles and practice Second edition
    P. J. Fellows

43 Managing frozen foods
    Edited by C. J. Kennedy

44 Handbook of hydrocolloids
    Edited by G. O. Phillips and P. A. Williams

45 Food labeling
    Edited by J. R. Blanchfield

46 Cereal biotechnology
    Edited by P. C. Morris and J. H. Bryce

47 Food intolerance and the food industry
    Edited by T. Dean

48 The stability and shelf-life of food
    Edited by D. Kilcast and P. Subramaniam

49 Functional foods: Concept to product
    Edited by G. R. Gibson and C. M. Williams

50 Chilled foods: A comprehensive guide Second edition
Edited by M. Stringer and C. Dennis

51 HACCP in the meat industry
Edited by M. Brown

52 Biscuit, cracker and cookie recipes for the food industry
D. Manley

53 Cereals processing technology
Edited by G. Owens

54 Baking problems solved
S. P. Cauvain and L. S. Young

55 Thermal technologies in food processing
Edited by P. Richardson

56 Frying: Improving quality
Edited by J. B. Rossell

57 Food chemical safety Volume 1: Contaminants
Edited by D. Watson

58 Making the most of HACCP: Learning from others' experience
Edited by T. Mayes and S. Mortimore

59 Food process modeling
Edited by L. M. M. Tijskens, M. L. A. T. M. Hertog and B. M. Nicolaï

60 EU food law: A practical guide
Edited by K. Goodburn

61 Extrusion cooking: Technologies and applications
Edited by R. Guy

62 Auditing in the food industry: From safety and quality to environmental and other audits
Edited by M. Dillon and C. Griffith

63 Handbook of herbs and spices Volume 1
Edited by K. V. Peter

64 Food product development: Maximising success
M. Earle, R. Earle and A. Anderson

65 Instrumentation and sensors for the food industry Second edition
Edited by E. Kress-Rogers and C. J. B. Brimelow

66 Food chemical safety Volume 2: Additives
Edited by D. Watson

67 Fruit and vegetable biotechnology
Edited by V. Valpuesta

68 Foodborne pathogens: Hazards, risk analysis and control
    Edited by C. de W. Blackburn and P. J. McClure

69 Meat refrigeration
    S. J. James and C. James

70 Lockhart and Wiseman's crop husbandry Eighth edition
    H. J. S. Finch, A. M. Samuel and G. P. F. Lane

71 Safety and quality issues in fish processing
    Edited by H. A. Bremner

72 Minimal processing technologies in the food industries
    Edited by T. Ohlsson and N. Bengtsson

73 Fruit and vegetable processing: Improving quality
    Edited by W. Jongen

74 The nutrition handbook for food processors
    Edited by C. J. K. Henry and C. Chapman

75 Colour in food: Improving quality
    Edited by D. MacDougall

76 Meat processing: Improving quality
    Edited by J. P. Kerry, J. F. Kerry and D. A. Ledward

77 Microbiological risk assessment in food processing
    Edited by M. Brown and M. Stringer

78 Performance functional foods
    Edited by D. Watson

79 Functional dairy products Volume 1
    Edited by T. Mattila-Sandholm and M. Saarela

80 Taints and off-flavours in foods
    Edited by B. Baigrie

81 Yeasts in food
    Edited by T. Boekhout and V. Robert

82 Phytochemical functional foods
    Edited by I. T. Johnson and G. Williamson

83 Novel food packaging techniques
    Edited by R. Ahvenainen

84 Detecting pathogens in food
    Edited by T. A. McMeekin

85 Natural antimicrobials for the minimal processing of foods
    Edited by S. Roller

86 **Texture in food Volume 1: Semi-solid foods**
 Edited by B. M. McKenna

87 **Dairy processing: Improving quality**
 Edited by G. Smit

88 **Hygiene in food processing: Principles and practice**
 Edited by H. L. M. Lelieveld, M. A. Mostert, B. White and J. Holah

89 **Rapid and on-line instrumentation for food quality assurance**
 Edited by I. Tothill

90 **Sausage manufacture: Principles and practice**
 E. Essien

91 **Environmentally-friendly food processing**
 Edited by B. Mattsson and U. Sonesson

92 **Bread making: Improving quality**
 Edited by S. P. Cauvain

93 **Food preservation techniques**
 Edited by P. Zeuthen and L. Bøgh-Sørensen

94 **Food authenticity and traceability**
 Edited by M. Lees

95 **Analytical methods for food additives**
 R. Wood, L. Foster, A. Damant and P. Key

96 **Handbook of herbs and spices Volume 2**
 Edited by K. V. Peter

97 **Texture in food Volume 2: Solid foods**
 Edited by D. Kilcast

98 **Proteins in food processing**
 Edited by R. Yada

99 **Detecting foreign bodies in food**
 Edited by M. Edwards

100 **Understanding and measuring the shelf-life of food**
 Edited by R. Steele

101 **Poultry meat processing and quality**
 Edited by G. Mead

102 **Functional foods, ageing and degenerative disease**
 Edited by C. Remacle and B. Reusens

103 **Mycotoxins in food: Detection and control**
 Edited by N. Magan and M. Olsen

104 **Improving the thermal processing of foods**
Edited by P. Richardson

105 **Pesticide, veterinary and other residues in food**
Edited by D. Watson

106 **Starch in food: Structure, functions and applications**
Edited by A.-C. Eliasson

107 **Functional foods, cardiovascular disease and diabetes**
Edited by A. Arnoldi

108 **Brewing: Science and practice**
D. E. Briggs, P. A. Brookes, R. Stevens and C. A. Boulton

109 **Using cereal science and technology for the benefit of consumers: Proceedings of the 12th International ICC Cereal and Bread Congress, 24–26th May, 2004, Harrogate, UK**
Edited by S. P. Cauvain, L. S. Young and S. Salmon

110 **Improving the safety of fresh meat**
Edited by J. Sofos

111 **Understanding pathogen behaviour: Virulence, stress response and resistance**
Edited by M. Griffiths

112 **The microwave processing of foods**
Edited by H. Schubert and M. Regier

113 **Food safety control in the poultry industry**
Edited by G. Mead

114 **Improving the safety of fresh fruit and vegetables**
Edited by W. Jongen

115 **Food, diet and obesity**
Edited by D. Mela

116 **Handbook of hygiene control in the food industry**
Edited by H. L. M. Lelieveld, M. A. Mostert and J. Holah

117 **Detecting allergens in food**
Edited by S. Koppelman and S. Hefle

118 **Improving the fat content of foods**
Edited by C. Williams and J. Buttriss

119 **Improving traceability in food processing and distribution**
Edited by I. Smith and A. Furness

120 **Flavour in food**
Edited by A. Voilley and P. Etievant

121 **The Chorleywood bread process**
   S. P. Cauvain and L. S. Young

122 **Food spoilage microorganisms**
   Edited by C. de W. Blackburn

123 **Emerging foodborne pathogens**
   Edited by Y. Motarjemi and M. Adams

124 **Benders' dictionary of nutrition and food technology Eighth edition**
   D. A. Bender

125 **Optimising sweet taste in foods**
   Edited by W. J. Spillane

126 **Brewing: New technologies**
   Edited by C. Bamforth

127 **Handbook of herbs and spices Volume 3**
   Edited by K. V. Peter

128 **Lawrie's meat science Seventh edition**
   R. A. Lawrie in collaboration with D. A. Ledward

129 **Modifying lipids for use in food**
   Edited by F. Gunstone

130 **Meat products handbook: Practical science and technology**
   G. Feiner

131 **Food consumption and disease risk: Consumer–pathogen interactions**
   Edited by M. Potter

132 **Acrylamide and other hazardous compounds in heat-treated foods**
   Edited by K. Skog and J. Alexander

133 **Managing allergens in food**
   Edited by C. Mills, H. Wichers and K. Hoffman-Sommergruber

134 **Microbiological analysis of red meat, poultry and eggs**
   Edited by G. Mead

135 **Maximising the value of marine by-products**
   Edited by F. Shahidi

136 **Chemical migration and food contact materials**
   Edited by K. Barnes, R. Sinclair and D. Watson

137 **Understanding consumers of food products**
   Edited by L. Frewer and H. van Trijp

138 **Reducing salt in foods: Practical strategies**
   Edited by D. Kilcast and F. Angus

139 **Modelling microorganisms in food**
   Edited by S. Brul, S. Van Gerwen and M. Zwietering

140 **Tamime and Robinson's yoghurt: Science and technology Third edition**
   Y. Tamime and R. K. Robinson

141 **Handbook of waste management and co-product recovery in food processing Volume 1**
   Edited by K. W. Waldron

142 **Improving the flavour of cheese**
   Edited by B. Weimer

143 **Novel food ingredients for weight control**
   Edited by C. J. K. Henry

144 **Consumer-led food product development**
   Edited by H. MacFie

145 **Functional dairy products Volume 2**
   Edited by M. Saarela

146 **Modifying flavour in food**
   Edited by A. J. Taylor and J. Hort

147 **Cheese problems solved**
   Edited by P. L. H. McSweeney

148 **Handbook of organic food safety and quality**
   Edited by J. Cooper, C. Leifert and U. Niggli

149 **Understanding and controlling the microstructure of complex foods**
   Edited by D. J. McClements

150 **Novel enzyme technology for food applications**
   Edited by R. Rastall

151 **Food preservation by pulsed electric fields: From research to application**
   Edited by H. L. M. Lelieveld and S. W. H. de Haan

152 **Technology of functional cereal products**
   Edited by B. R. Hamaker

153 **Case studies in food product development**
   Edited by M. Earle and R. Earle

154 **Delivery and controlled release of bioactives in foods and nutraceuticals**
   Edited by N. Garti

155 **Fruit and vegetable flavour: Recent advances and future prospects**
   Edited by B. Brückner and S. G. Wyllie

156 **Food fortification and supplementation: Technological, safety and regulatory aspects**
Edited by P. Berry Ottaway

157 **Improving the health-promoting properties of fruit and vegetable products**
Edited by F. A. Tomás-Barberán and M. I. Gil

158 **Improving seafood products for the consumer**
Edited by T. Børresen

159 **In-pack processed foods: Improving quality**
Edited by P. Richardson

160 **Handbook of water and energy management in food processing**
Edited by J. Klemeš, R.. Smith and J.-K. Kim

161 **Environmentally compatible food packaging**
Edited by E. Chiellini

162 **Improving farmed fish quality and safety**
Edited by Ø. Lie

163 **Carbohydrate-active enzymes**
Edited by K.-H. Park

164 **Chilled foods: A comprehensive guide Third edition**
Edited by M. Brown

165 **Food for the ageing population**
Edited by M. M. Raats, C. P. G. M. de Groot and W. A Van Staveren

166 **Improving the sensory and nutritional quality of fresh meat**
Edited by J. P. Kerry and D. A. Ledward

167 **Shellfish safety and quality**
Edited by S. E. Shumway and G. E. Rodrick

168 **Functional and speciality beverage technology**
Edited by P. Paquin

169 **Functional foods: Principles and technology**
M. Guo

170 **Endocrine-disrupting chemicals in food**
Edited by I. Shaw

171 **Meals in science and practice: Interdisciplinary research and business applications**
Edited by H. L. Meiselman

172 **Food constituents and oral health: Current status and future prospects**
Edited by M. Wilson

173 Handbook of hydrocolloids Second edition
Edited by G. O. Phillips and P. A. Williams

174 Food processing technology: Principles and practice Third edition
P. J. Fellows

175 Science and technology of enrobed and filled chocolate, confectionery and bakery products
Edited by G. Talbot

176 Foodborne pathogens: Hazards, risk analysis and control Second edition
Edited by C. de W. Blackburn and P. J. McClure

177 Designing functional foods: Measuring and controlling food structure breakdown and absorption
Edited by D. J. McClements and E. A. Decker

178 New technologies in aquaculture: Improving production efficiency, quality and environmental management
Edited by G. Burnell and G. Allan

179 More baking problems solved
S. P. Cauvain and L. S. Young

180 Soft drink and fruit juice problems solved
P. Ashurst and R. Hargitt

181 Biofilms in the food and beverage industries
Edited by P. M. Fratamico, B. A. Annous and N. W. Gunther

182 Dairy-derived ingredients: Food and neutraceutical uses
Edited by M. Corredig

183 Handbook of waste management and co-product recovery in food processing Volume 2
Edited by K. W. Waldron

184 Innovations in food labeling
Edited by J. Albert

185 Delivering performance in food supply chains
Edited by C. Mena and G. Stevens

186 Chemical deterioration and physical instability of food and beverages
Edited by L. H. Skibsted, J. Risbo and M. L. Andersen

187 Managing wine quality Volume 1: Viticulture and wine quality
Edited by A. G. Reynolds

188 Improving the safety and quality of milk Volume 1: Milk production and processing
Edited by M. Griffiths

189 **Improving the safety and quality of milk Volume 2: Improving quality in milk products**
Edited by M. Griffiths

190 **Cereal grains: Assessing and managing quality**
Edited by C. Wrigley and I. Batey

191 **Sensory analysis for food and beverage quality control: A practical guide**
Edited by D. Kilcast

192 **Managing wine quality Volume 2: Oenology and wine quality**
Edited by A. G. Reynolds

193 **Winemaking problems solved**
Edited by C. E. Butzke

194 **Environmental assessment and management in the food industry**
Edited by U. Sonesson, J. Berlin and F. Ziegler

195 **Consumer-driven innovation in food and personal care products**
Edited by S. R. Jaeger and H. MacFie

196 **Tracing pathogens in the food chain**
Edited by S. Brul, P. M. Fratamico and T. A. McMeekin

197 **Case studies in novel food processing technologies: Innovations in processing, packaging, and predictive modeling**
Edited by C. J. Doona, K. Kustin and F. E. Feeherry

198 **Freeze-drying of pharmaceutical and food products**
Edited by T.-C. Hua, B.-L. Liu and H. Zhang

199 **Oxidation in foods and beverages and antioxidant applications Volume 1: Understanding mechanisms of oxidation and antioxidant activity**
Edited by E. A. Decker, R. J. Elias and D. J. McClements

200 **Oxidation in foods and beverages and antioxidant applications Volume 2: Management in different industry sectors**
Edited by E. A. Decker, R. J. Elias and D. J. McClements

201 **Protective cultures, antimicrobial metabolites and bacteriophages for food and beverage biopreservation**
Edited by C. Lacroix

202 **Separation, extraction and concentration processes in the food, beverage and nutraceutical industries**
Edited by S. S. H. Rizvi

203 **Determining mycotoxins and mycotoxigenic fungi in food and feed**
Edited by S. De Saeger

204 **Developing children's food products**
Edited by D. Kilcast and F. Angus

205 **Functional foods: Concept to product Second edition**
Edited by M. Saarela

206 **Postharvest biology and technology of tropical and subtropical fruits Volume 1: Fundamental issues**
Edited by E. M. Yahia

207 **Postharvest biology and technology of tropical and subtropical fruits Volume 2: Açai to citrus**
Edited by E. M. Yahia

208 **Postharvest biology and technology of tropical and subtropical fruits Volume 3: Cocona to mango**
Edited by E. M. Yahia

209 **Postharvest biology and technology of tropical and subtropical fruits Volume 4: Mangosteen to white sapote**
Edited by E. M. Yahia

210 **Food and beverage stability and shelf life**
Edited by D. Kilcast and P. Subramaniam

211 **Processed meats: Improving safety, nutrition and quality**
Edited by J. P. Kerry and J. F. Kerry

212 **Food chain integrity: A holistic approach to food traceability, safety, quality and authenticity**
Edited by J. Hoorfar, K. Jordan, F. Butler and R. Prugger

213 **Improving the safety and quality of eggs and egg products Volume 1**
Edited by Y. Nys, M. Bain and F. Van Immerseel

214 **Improving the safety and quality of eggs and egg products Volume 2**
Edited by F. Van Immerseel, Y. Nys and M. Bain

215 **Animal feed contamination: Effects on livestock and food safety**
Edited by J. Fink-Gremmels

216 **Hygienic design of food factories**
Edited by J. Holah and H. L. M. Lelieveld

217 **Manley's technology of biscuits, crackers and cookies Fourth edition**
Edited by D. Manley

218 **Nanotechnology in the food, beverage and nutraceutical industries**
Edited by Q. Huang

219 **Rice quality: A guide to rice properties and analysis**
K. R. Bhattacharya

220 **Advances in meat, poultry and seafood packaging**
Edited by J. P. Kerry

221 **Reducing saturated fats in foods**
Edited by G. Talbot

222 **Handbook of food proteins**
Edited by G. O. Phillips and P. A. Williams

223 **Lifetime nutritional influences on cognition, behaviour and psychiatric illness**
Edited by D. Benton

224 **Food machinery for the production of cereal foods, snack foods and confectionery**
L.-M. Cheng

225 **Alcoholic beverages: Sensory evaluation and consumer research**
Edited by J. Piggott

226 **Extrusion problems solved: Food, pet food and feed**
M. N. Riaz and G. J. Rokey

227 **Handbook of herbs and spices Second edition Volume 1**
Edited by K. V. Peter

228 **Handbook of herbs and spices Second edition Volume 2**
Edited by K. V. Peter

229 **Breadmaking: Improving quality Second edition**
Edited by S. P. Cauvain

230 **Emerging food packaging technologies: Principles and practice**
Edited by K. L. Yam and D. S. Lee

231 **Infectious disease in aquaculture: Prevention and control**
Edited by B. Austin

232 **Diet, immunity and inflammation**
Edited by P. C. Calder and P. Yaqoob

233 **Natural food additives, ingredients and flavourings**
Edited by D. Baines and R. Seal

234 **Microbial decontamination in the food industry: Novel methods and applications**
Edited by A. Demirci and M.O. Ngadi

235 **Chemical contaminants and residues in foods**
Edited by D. Schrenk

236 **Robotics and automation in the food industry: Current and future technologies**
Edited by D. G. Caldwell

237 **Fibre-rich and wholegrain foods: Improving quality**
Edited by J. A. Delcour and K. Poutanen

238 **Computer vision technology in the food and beverage industries**
Edited by D.-W. Sun

239 **Encapsulation technologies and delivery systems for food ingredients and nutraceuticals**
Edited by N. Garti and D. J. McClements

240 **Case studies in food safety and authenticity**
Edited by J. Hoorfar

241 **Heat treatment for insect control: Developments and applications**
D. Hammond

242 **Advances in aquaculture hatchery technology**
Edited by G. Allan and G. Burnell

243 **Open innovation in the food and beverage industry**
Edited by M. Garcia Martinez

244 **Trends in packaging of food, beverages and other fast-moving consumer goods (FMCG)**
Edited by N. Farmer

245 **New analytical approaches for verifying the origin of food**
Edited by P. Brereton

246 **Microbial production of food ingredients, enzymes and nutraceuticals**
Edited by B. McNeil, D. Archer, I. Giavasis and L. Harvey

247 **Persistent organic pollutants and toxic metals in foods**
Edited by M. Rose and A. Fernandes

248 **Cereal grains for the food and beverage industries**
E. Arendt and E. Zannini

249 **Viruses in food and water: Risks, surveillance and control**
Edited by N. Cook

250 **Improving the safety and quality of nuts**
Edited by L. J. Harris

251 **Metabolomics in food and nutrition**
Edited by B. C. Weimer and C. Slupsky

252 **Food enrichment with omega-3 fatty acids**
Edited by C. Jacobsen, N. S. Nielsen, A. F. Horn and A.-D. M. Sørensen

253 **Instrumental assessment of food sensory quality: A practical guide**
Edited by D. Kilcast

254 **Food microstructures: Microscopy, measurement and modeling**
Edited by V. J. Morris and K. Groves

255 **Handbook of food powders: Processes and properties**
Edited by B. R. Bhandari, N. Bansal, M. Zhang and P. Schuck

256 Functional ingredients from algae for foods and nutraceuticals
Edited by H. Domínguez

257 Satiation, satiety and the control of food intake: Theory and practice
Edited by J. E. Blundell and F. Bellisle

258 Hygiene in food processing: Principles and practice Second edition
Edited by H. L. M. Lelieveld, J. Holah and D. Napper

259 Advances in microbial food safety Volume 1
Edited by J. Sofos

260 Global safety of fresh produce: A handbook of best practice, innovative commercial solutions and case studies
Edited by J. Hoorfar

261 Human milk biochemistry and infant formula manufacturing technology
Edited by M. Guo

262 High throughput screening for food safety assessment: Biosensor technologies, hyperspectral imaging and practical applications
Edited by A. K. Bhunia, M. S. Kim and C. R. Taitt

263 Foods, nutrients and food ingredients with authorised EU health claims: Volume 1
Edited by M. J. Sadler

264 Handbook of food allergen detection and control
Edited by S. Flanagan

265 Advances in fermented foods and beverages: Improving quality, technologies and health benefits
Edited by W. Holzapfel

266 Metabolomics as a tool in nutrition research
Edited by J.-L. Sébédio and L. Brennan

267 Dietary supplements: Safety, efficacy and quality
Edited by K. Berginc and S. Kreft

268 Grapevine breeding programs for the wine industry
Edited by A. G. Reynolds

269 Handbook of antimicrobials for food safety and quality
Edited by T. M. Taylor

270 Managing and preventing obesity: Behavioural factors and dietary interventions
Edited by T. P. Gill

271 Electron beam pasteurization and complementary food processing technologies
Edited by S. D. Pillai and S. Shayanfar

272 **Advances in food and beverage labelling: Information and regulations**
Edited by P. Berryman

273 **Flavour development, analysis and perception in food and beverages**
Edited by J. K. Parker, S. Elmore and L. Methven

274 **Rapid sensory profiling techniques and related methods: Applications in new product development and consumer research**
Edited by J. Delarue, J. B. Lawlor and M. Rogeaux

275 **Advances in microbial food safety: Volume 2**
Edited by J. Sofos

276 **Handbook of antioxidants for food preservation**
Edited by F. Shahidi

277 **Lockhart and Wiseman's crop husbandry including grassland: Ninth edition**
H. J. S. Finch, A. M. Samuel and G. P. F. Lane

278 **Global legislation for food contact materials**
Edited by J. S. Baughan

279 **Colour additives for food and beverages**
Edited by M. Scotter

280 **A complete course in canning and related processes 14th Edition: Volume 1**
Revised by S. Featherstone

281 **A complete course in canning and related processes 14th Edition: Volume 2**
Revised by S. Featherstone

282 **A complete course in canning and related processes 14th Edition: Volume 3**
Revised by S. Featherstone

283 **Modifying food texture: Volume 1: Novel ingredients and processing techniques**
Edited by J. Chen and A. Rosenthal

284 **Modifying food texture: Volume 2: Sensory analysis, consumer requirements and preferences**
Edited by J. Chen and A. Rosenthal

285 **Modeling food processing operations**
Edited by S. Bakalis, K. Knoerzer and P. J. Fryer

286 **Foods, nutrients and food ingredients with authorised EU health claims Volume 2**
Edited by M. J. Sadler

287 **Feed and feeding practices in aquaculture**
Edited by D. Allen Davis

288 Foodborne parasites in the food supply web: Occurrence and control
Edited by A. Gajadhar

289 Brewing microbiology: Design and technology applications for spoilage management, sensory quality and waste valorisation
Edited by A. E. Hill

290 Specialty oils and fats in food and nutrition: Properties, processing and applications
Edited by G. Talbot

291 Improving and tailoring enzymes for food quality and functionality
Edited by R. Yada

292 Emerging technologies for promoting food security: Overcoming the world food crisis
Edited by C. Madramootoo

293 Innovation and future trends in food manufacturing and supply chain technologies
Edited by C. E. Leadley

294 Functional dietary lipids: Food formulation, consumer issues and innovation for health
Edited by T. Sanders

295 Handbook on natural pigments in food and beverages: Industrial applications for improving color
Edited by R. Carle and R. M. Schweiggert

296 Integrating the packaging and product experience in food and beverages: A road-map to consumer satisfaction
Edited by P. Burgess

297 The stability and shelf life of food Second edition
Edited by Persis Subramaniam

298 Multisensory flavor perception: From fundamental neuroscience through to the marketplace
Edited by Betina Piqueras-Fiszman and Charles Spence

299 Flavor: From food to behaviors, wellbeing and health
Edited by Andrée Voilley, Christian Salles, Elisabeth Guichard and Patrick Etiévant

Introduction

Betina Piqueras-Fiszman[1] and Charles Spence[2]
[1]Marketing and Consumer Behavior Group, Wageningen University, Wageningen, The Netherlands, [2]Crossmodal Research Laboratory, Department of Experimental Psychology, University of Oxford, Oxford, United Kingdom

1 The senses and multisensory integration

Imagine that you are planning to bake a cake. Not just any cake, but your very favorite one, let's call it carrot cake. While writing down the list of ingredients that you'll need to buy while out shopping, your mind is most probably retrieving memories of those happy occasions when you had made the very same cake previously. From that moment on, your brain is already expecting the likely sensory stimulation that it will soon receive when the cake is reaching its perfect baking point in the oven; the instant when you insert the knife to cut the first slice, and most importantly when you bring that cake to your mouth. At that moment, and without even thinking about it, you just remember why it is that *this* is your favorite cake, what is in it that you like so much and that would make you take another slice, or two, melting any sorrows (not to mention pressing book deadlines) away. Note, however, that if you were accidentally to have added a little bit too much baking powder into the mix, your first thought would be: "*Hmm, it smells different than usual...*" followed by "*Yuck, it has a weird taste!*" Luckily enough, we can count upon mechanisms that let us know (or predict) in advance that something has gone wrong, alerting us to the fact that what we are about to taste might well be too bitter, disgusting even.

But what exactly happens in our mind at this point? These questions require a thorough understanding of what the term "flavor perception" really means, starting with its very definition. It turns out that what most people have in mind when they talk about "taste" is really "flavor"; that is, the result of the integration of inputs from several different senses (see Spence, 2015a,b, for recent reviews). By themselves, the gustatory receptors on the tongue only provide information about the so-called basic tastes, such as sweet, salty, bitter, sour, and umami (though see Stuckey, 2012, for a few more suggestions). In fact, according to one oft-quoted statistic, somewhere between 75% and 95% of what most people commonly describe as the "taste" of food actually comes from inputs detected at the olfactory epithelium in the nose (though see Spence, 2015d, for a critical appraisal of such scientific-sounding statistics). So, for example, the nutty, cinnamon, buttery, and citrus notes, etc. that we enjoy when eating carrot cake all derive from the detection of volatile compounds at the olfactory epithelium. That such inputs are referred to the mouth in our everyday experience of flavor is known as "oral referral" (eg, Lim & Johnson, 2011, 2012; Rozin, 1982; see chapter: Oral Referral). As Bartoshuk and Duffy (2005, p. 27) note: "'*Taste' is often used as a synonym for 'flavour'. This usage of 'taste' probably arose because the blend of true taste and retronasal olfaction is perceptually localized to the mouth via touch*" (though see Spence, 2016, for a review of the evidence showing that touch actually plays no role in oral referral).

Trigeminal inputs contribute to our perception of food and drink, giving rise to hot, cold, tingling, burning, and electric sensations (see Bryant & Mezine, 1999; Hagura, Barber, & Haggard, 2013). According to the International Standards Organization, flavor can be defined as a: "*Complex combination of the olfactory, gustatory and trigeminal sensations perceived during tasting. The flavour may be influenced by tactile, thermal, painful and/or kinaesthetic effects*" (ISO 5492, 1992, 2008). All of this implies that people perceive food as a multisensory experience (or Gestalt), rather than as a series of discrete unisensory events. This perceptual unity invokes the idea of the brain bringing together divergent sources of sensory information and combining them, that is, binding them together, into a unified multisensory experience of flavor (Rozin, 1982; Stevenson, 2014), what some want to call a "flavor object." Stevenson's chapter "Attention and Flavor Binding" examines the evidence for flavor binding by cataloging its particular features and then draws upon an explanatory framework that has arisen from studying examples of binding in the other senses (eg, Koelewijn, Bronkhorst, & Theeuwes, 2010; Navarra, Alsius, Soto-Faraco, & Spence, 2010). In this regard, the role of attentional processes (but also relevant preattentive processes) could be especially important in explaining binding in the area of flavor perception as well (Ashkenazi & Marks, 2004; Stevenson, 2014).

Moreover, our senses can be classified into two broad categories depending on when they contribute in the processing of different sensory cues: anticipatory or exteroceptive cues, and consummatory or interoceptive cues (Small, Veldhuizen, Felsted, Mak, & McGlone, 2008; see Fig. 1.1). The first of these involves those senses that are stimulated before the food or drink is placed into the mouth. These include orthonasal olfaction (ie, sniffing the food), vision (eg, color, visual texture, sheen, etc.), audition (ie, the sounds associated with cooking), and somatosensation (ie, from direct or indirect handling). The second group of sensory systems includes those senses that are stimulated once the food or drink has entered our mouth. These comprise oral-somatosensation (including touch, proprioceptive feedback, and gustation), chemesthesis, audition, and retronasal olfaction (ie, volatile chemicals perceived via the posterior nares).

To complicate matters, researchers disagree about the role of the other senses, such as oral-somatosensation, audition, and vision in flavor perception (see Spence, Smith, & Auvray, 2015, for a review). Some scientists and philosophers believe that while such sensory inputs *modulate* multisensory flavor perception, they should not be considered as *constitutive* of it. While some research suggests that tactile cues seem to play an important role in helping us to localize flavor experiences within the oral cavity (eg, Green, 2002; Hollingworth & Poffenberger, 1917; Lim & Green 2008; Murphy & Cain, 1980; Todrank & Bartoshuk, 1991; see also Lim's chapter: Oral Referral), the latest views indicate that it is the congruency between olfactory and gustatory inputs that matter, while tactile stimulation, surprisingly, does not (Lim & Johnson, 2012; see Spence, 2016, for a review). What is more, several studies have highlighted a number of complex interactions taking place between oral-somatosensory

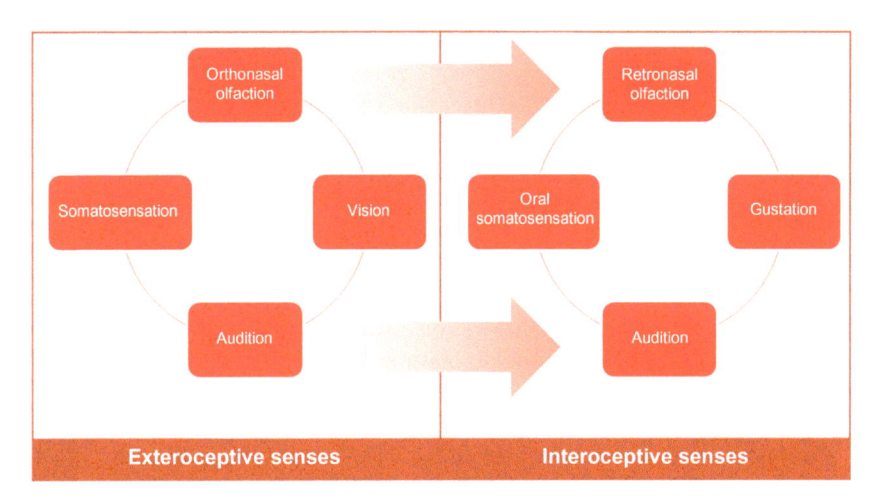

Figure 1.1 Distinguishing between the senses that contribute to (ie, are constitutive of), versus merely modulate, multisensory flavor perception. The four interoceptive sensory inputs are likely combined through a process of multisensory integration to deliver flavor percepts. Meanwhile, exteroceptive cues can also influence flavor perception, primarily by means of setting our expectations prior to consumption.

and both gustatory and olfactory perception (eg, Bult, de Wijk, & Hummel, 2007; Cerf-Ducastel, Van de Moortele, Macleod, Le Bihan, & Faurion, 2001; Francis et al., 1999; Roudnitzky et al., 2011). Stimulation of the oral-somatosensory receptors can give rise to a range of responses including everything from judgments concerning the texture of the food(s) that happen(s) to be in the mouth through to the experience of what is called "mouthfeel," a term used to describe how the consumption of certain foods changes the feeling of the oral cavity. In Spence and Piqueras-Fiszman's chapter "Oral-Somatosensory Contributions to Flavor Perception and the Appreciation of Food and Drink," we take a closer look at the various ways in which oral-somatosensory stimulation gives rise to specific food-related sensations/perceptions.

The sounds emitted by the breaking down of foods in the mouth, particularly when biting into crispy, crunchy, and crackly foods, contribute to the perception of food texture, not to mention overall enjoyment of eating and drinking (see Spence, 2015a, for a review). However, the majority of research on the topic of flavor perception tends to remain silent when it comes to audition (see Spence's chapter "Auditory Contributions to Multisensory Flavor Perception"). Indeed, when Jeannine Delwiche (2003) questioned 140 researchers working in the field about the importance of the role of the different senses on flavor perception a little over a decade ago, the majority of the experts rated "sound" as the least important aspect of flavor, coming in well behind taste, smell, temperature, texture appearance, and color (see also Schifferstein, 2006, for similar results). That said, the situation is slowly starting to change (Knight, 2012, p. 80; Spence, 2015a). Nowadays researchers are paying more attention than ever before to how consumers perceive the sounds derived from food consumption (eg, Dijksterhuis, Luyten, de Wijk, & Mojet, 2007; Varela & Fiszman, 2012) and how they impact texture perception (eg, Demattè et al., 2014; Zampini & Spence, 2004), rather than purely to characterize the texture of dry food products (see eg, Dacremont, 1995; Duizier, 2001, for a review).

2 Expectations

Taking a step back, and placing our focus on distal, anticipatory food cues, the crucial role that vision, and particularly color, plays in multisensory flavor perception, and even consumption (Piqueras-Fiszman & Spence, 2014), should be highlighted. In fact, well over 250 studies have been published on this topic since the original report by Moir, a chemist, in 1936 first documenting the fact that changing the color of a food can alter its taste and flavor (see Clydesdale, 1993; Spence, 2015c; Spence, Levitan, Shankar, & Zampini, 2010, for reviews). The research published to date clearly demonstrates that judgments of the identity of a food's taste, aroma, and flavor can all be influenced by simply changing the color (such that it is appropriate, inappropriate, or absent) of the foods that people happen to be evaluating. By contrast, those studies that have investigated the effect of varying the intensity of the color added to a food (typically colored beverages have been used) have revealed mixed results (see Spence et al., 2010, for a review; see Spence and Piqueras-Fiszman's chapter: Food Color and Its Impact on Taste/Flavor Perception). It turns out that a key part of figuring out why color sometimes does, and sometimes does not influence taste and flavor perception comes down to the question of exposure and learning in the meaning of food color (eg, see Shankar, Levitan, & Spence, 2010; Wan et al., 2014, for cross-cultural studies).

Recently, researchers have extended the effect of vision to color of product packaging (eg, Piqueras-Fiszman & Spence, 2011, 2012; Velasco et al., 2014) and even to the color of the tableware (Spence, Harrar, & Piqueras-Fiszman, 2012; Spence & Piqueras-Fiszman, 2014; Van Doorn, Wuillemin, & Spence, 2014), demonstrating the relevance of these cues in setting expectations about the product and hence shaping our subsequent food experience (Piqueras-Fiszman & Spence, 2015; Spence & Piqueras-Fiszman, 2014).

In line with the topic of expectations (both sensory and hedonic), and returning to the imagery of the cake baking in the oven, we know that the scent emanating from the vents will help us to decide when the cake is ready. The aroma perceived more clearly as we bring it out of the oven will indicate if the result is as perfect as expected or if, in contrast, we accidentally messed up with the baking soda again. But it turns out that olfactory stimuli may also influence our perception and behavior, even if we are sometimes not aware of it (see Spence, 2015e, for a review)! There is accumulating evidence focusing in particular on top-down effects on flavor perception, and on the methodologies to be able to quantify the effect of primes presented in the experimental setup. In the domain of olfactory priming, Smeets and Dijksterhuis (2014) distinguish between three different awareness circumstances under which odors can be perceived: attentively, where an odor can be clearly recognized by an observer, semiattentively, where an observer may notice, and report, something unusual in the environment or situation they are in, but cannot identify it to be olfactory, and inattentively, where observers show no evidence of being aware of something particular in the environment or situation they are in (ie, the odor is presented at a subthreshold level). The intriguing effects of the different circumstances will be presented together with a theoretical framework in Dijksterhuis' chapter "Multisensory Flavor Priming."

3 On the hedonics and memorability of flavor

Why do we, in the first place, and rather unfortunately for some of us, tend to like fatty, sweet, foods more than less calorie-dense foods? Why is it that certain combinations of sensory stimuli are, for us, no less than a match made in heaven? It would seem to be that whether we like or dislike a food is the first thing that we know, in most cases automatically, as soon as a food touches our taste buds (or more likely even beforehand when we smell it, Prescott, 2012; or see it, Spence, Okajima, Cheok, Petit, & Michel, in press). However, it turns out that flavor hedonics is based on a learning process that starts in our mother's womb (Mennella, Jagnow, & Beauchamp, 2001; Schaal & Durand, 2012; Schaal, Marlier, & Soussignan, 2000) and likely evolves throughout our lifetime (see Bremner, Lewkowicz, & Spence, 2012). Through repeated exposure to particular combinations of stimuli, we collect hedonic information about them and their consequences, which, taken together, shape our preferences and our willingness to try new foods (Pliner, Pelchat, & Grabski, 1993). The state-of-the-art as far as flavor hedonics is concerned is covered in Prescott's chapter "Flavor Liking."

Another question that often comes to mind is what is in those ingredients, in the correct proportions, that when put together take you instantly back to your childhood (or back to the first moment that you ever tried that cake)? First encounters with odors and flavors are very important, as shown by the fact that the memories evoked later in life by odors (and probably also by flavors) go further back than those evoked by visual and auditory stimuli (see Chu & Downes, 2002; Willander & Larsson, 2008). As mentioned already, our food preferences are mainly based on incidental learning from the memories of previous experiences (Laureati et al., 2008). However, the memory for particular aspects (sensations) of foods remains below the level of awareness unless, in a next encounter with the same food, these are modified or absent. It is only then that we realize that the remembered sensations and the current experience do not match (Köster, Møller, & Mojet, 2014). Thus, flavor memory is also particularly crucial to us since it helps detect the possible dangers that might be associated with different or novel foods. That said, the fact that, on occasion, we consume the exact same food some years later, in another context, and perceive that it tastes different (or at least we think we do), is an eye-opener with regard to the way in which memory can actually distort the remembered sensory features of the food. Of course, differences in the emotions and associations attached to that remembered perception and the current ones might contribute as well—this phenomenon is known as the *Provencal rose paradox* (see Spence & Piqueras-Fiszman, 2014). Mojet and Köster's chapter "Flavor Memory" will guide you down the memory lane of multisensory flavor perception.

4 Living in different taste worlds

So, up to this point, we know that our momentary experience of food, and our preferences (and also motivations), depends to a large extent on what is stored in memory. Certainly, this helps to address the question of why it is, for example, that the foods you love I hate, and vice versa (see Prescott, 2012). But, additionally, one of the most striking aspects of multisensory flavor perception is the profound individual differences that exist: it has long been known that we live in different taste worlds (eg, Bartoshuk, 1980; Spence, 2013) and some of these individual differences are genetic in origin (eg, Bartoshuk, 2000; Mauer & El-Sohemy, 2012; and see Running and Hayes' chapter: Individual Differences in Multisensory Flavor Perception). Those who are especially sensitive to certain bitter-tasting compounds are referred to as supertasters. Roughly 25% of the population fall into this category, 50% are "medium tasters," while the remainder are known as "nontasters" (Bartoshuk, 2000; Garneau et al., 2014). Supertasters are somewhat more sensitive to other tastants as well (Bajec & Pickering, 2008; Bajec, Pickering, & DeCourville, 2012). These individual differences in taster status are especially interesting in that they have been shown to be linked to various aspects related to multisensory flavor perception, such as tactile acuity or visual sensory dominance (eg, Bajec & Pickering, 2008; Bajec et al., 2012; Eldeghaidy et al., 2011; Essick, Chopra, Guest, & McGlone, 2003; Zampini, Wantling, Phillips, & Spence, 2008).

5 A closer look into our brain and bodily reactions

Neuroscientists have demonstrated some marked individual differences in the way in which the brains of different groups of individuals respond to one and the same food (see Spence et al., 2015, for a review). Eldeghaidy et al. (2011), for example, highlighted increased activity in the orbitofrontal cortex (OFC) amongst a group of supertasters in response to the delivery of a series of fat emulsions of increasing fat concentration as compared to a group of nontasters exposed to exactly the same set of stimuli. A nascent field of neuroimaging research has also started to compare the network of brain regions that are recruited by experts and regular consumers. So, for example, Pazart, Comte, Magnin, Millot, and Moulin (2014) recently demonstrated that many of the same brain areas (eg, the insula, orbitofrontal cortex, amygdala, and frontal operculum) were activated in social wine drinkers and sommeliers when tasked with evaluating the flavor of wines that had been served to them blind (though see also Castriota-Scanderbeg et al., 2005). In the chapter "Pleasure of Food in the Brain," the processing of individual sensory stimuli, the integrated multisensory processing related to the fundamental pleasure of food, and the principles of hedonic processing are described from a neuroscience perspective. Given that we rarely experience food and drink in the absence of branding, pricing, and labeling information, Spence's chapter "The Neuroscience of Flavor" extends this review of the neuroscience to cover the influence of such cues on the brain's response to flavor.

Certainly, delving inside the consumer's mind/brain and exploring which areas "light up" just by thinking about that heavenly cake helps explain what it is that makes us want to consume it, perhaps more than we should. However, it is worth noting that there are other signals in our body that also react when thinking about and consuming that cake and which can be more easily measured. de Wijk and Boesveldt's chapter "Responses of the Autonomic Nervous System to Flavors" focuses, in particular, on autonomic nervous system (ANS) responses. Researchers in the field of sensory science have recently started to investigate these in order to try to tap into unarticulated/unconscious motives and associations that may not be very well captured by traditional tests based on conscious cognitive processes. For instance, scientists have found correlations between the pleasantness of stimuli (in most cases odors) with heart rate, skin conductance (eg, Alaoui-Ismaïli, Robin, Rada, Dittmar, & Vernet-Maury, 1997; Bensafi et al., 2002; Delplanque et al., 2008; Glass, Lingg, & Heuberger, 2014; van den Bosch, van Delft, de Wijk, de Graaf, & Boesveldt, 2015). In addition, He, Boesveldt, de Graaf, and de Wijk (2014) recently documented the very rapid temporal differentiation between pleasant and unpleasant odors in terms of heart rate and skin temperature, and a somewhat slower differentiation in terms of the skin conductance response.

6 The impact of the senses in context: from the supermarket to disposal

Finally, the importance of crossmodal interactions will be explored from a broader perspective. When we drink, we nearly always come into direct contact with the receptacles—the glasses, cups, mugs, cans, and bottles—in which those drinks are contained and/or served. In Spence and Wan's chapter "Assessing the Influence of the Drinking Receptacle on the Perception of the Contents," the evidence concerning how the shape of the receptacle influences people's perception of the likely taste/flavor of a drink is reviewed (Cliff, 2001; Hummel, Delwiche, Schmidt, & Hüttenbrink, 2003; Russell, Zivanovic, Morris, Penfield, & Weiss, 2005). It turns out that the drinking vessel can affect everything from a consumer's hedonic response to a beverage through to how refreshing they find it. The color, weight, and even texture of the receptacle have also been found to influence people's perception of the contents through the notion of "sensation transference" (Cheskin, 1957; see Spence et al., 2012 and Spence & Piqueras-Fiszman, 2012, for reviews). The notion of receptacle appropriateness also plays an important role here (Schifferstein, 2009).

The research now shows that consumers ascribe a different importance to the senses at different stages of product use/experience (Fenko, Schifferstein, & Hekkert, 2010; Schifferstein, Fenko, Desmet, Labbe, & Martin, 2013). How can one capitalize on each of the senses in order to maximize the multisensory experience? Schifferstein's chapter "The Roles of the Senses in Different Stages of Consumers' Interactions With Food Products" focuses on the different roles played by the senses in the various stages of user–product interactions with food products. The focus is primarily on the multisensory aspects of the product experience, although links with aesthetic, semantic, and emotional aspects of it are described as well. Finally, have you ever wondered why it is you decided to buy those cinnamon rolls at the supermarket, and actually eat one on your way home, when actually you were not even that hungry (see Spence, 2015e)? Thomson's chapter "Sensory Branding: Using brand, pack, and product sensory characteristics to deliver a compelling brand message" describes the impact that the senses have particularly at the point of purchase (see Krishna, 2012; Spence, Puccinelli, Grewal, & Roggeveen, 2014). This chapter provides an overview of the nature of hedonic, emotional, and functional associations, and examines the importance of considering these associations in the development of new food products and brands (Thomson & MacFie, 2007). In addition, there are a variety of robust crossmodal correspondences between sounds, colors, and shapes, and the sensory attributes (specifically the taste, flavor, aroma, and oral-somatosensory attributes) of various foods and beverages. The available research now clearly suggests that marketers can enhance their consumers' product experiences by ensuring that the associations of the brand name, the labeling, and even the very shape of the packaging itself, sets up the right (ie, congruent) product-related sensory expectations in the mind of the consumer (see Spence, 2012; Velasco, Salgado-Montejo, Marmolejo-Ramos, & Spence, 2014).

This volume, then, provides state-of-the-art coverage of the latest insights from the rapidly expanding world of multisensory flavor research. Its coverage extends from the latest neuroscience insights concerning the neural underpinnings of flavor perception through to concrete applications out there in the marketplace. We hope to contribute to the growing interest and current revolution in terms of the reach of flavor research and the truly exciting applications, in areas from food product development all the way to culinary arts. We also aim to provide evidence for those interested in the emerging field of gastrophysics (Mouritsen & Risbo, 2013; Spence & Piqueras-Fiszman, 2014).

This book captures all the possible angles in which to look at our multisensory perception of foods and its impact on behavior, liking, and even emotion. Nowadays, technological advances allow researchers and practitioners to think out of the box in terms of future theoretical and empirical possibilities, and crowd-sourced internet-based testing is helping to revolutionize the way (not to mention the speed) in which research is done (Woods, Velasco, Levitan, Wan, & Spence, 2015). While this volume is theoretically motivated we hope it inspires a broadening of research practice beyond the laboratory to a more real world, dynamic, assessment of our relation/response/perception of food.

References

Alaoui-Ismaïli, O., Robin, O., Rada, H., Dittmar,, A., & Vernet-Maury, E. (1997). Basic emotions evoked by odorants: Comparison between autonomic responses and self-evaluation. *Physiology and Behavior*, *62*, 713–720.

Ashkenazi, A., & Marks, L. E. (2004). Effect of endogenous attention on detection of weak gustatory and olfactory flavors. *Perception & Psychophysics, 66*, 596–608.

Bajec, M. R. & Pickering, G. J. (2008). Thermal taste, PROP responsiveness, and perception of oral sensations. *Physiology & Behavior, 95*, 581–590.

Bajec, M. R., Pickering, G. J., & DeCourville, N. (2012). Influence of stimulus temperature on orosensory perception and variation with taste phenotype. *Chemosensory Perception, 5*, 243–265.

Bartoshuk, L. (1980). Separate worlds of taste. *Psychology Today, 14*, 48–49, 51, 54–56, 63.

Bartoshuk, L. M. (2000). Comparing sensory experiences across individuals: Recent psychophysical advances illuminate genetic variation in taste perception. *Chemical Senses, 25*, 447–460.

Bartoshuk, L. M., & Duffy, V. B. (2005). Chemical senses: Taste and smell. In C. Korsmeyer (Ed.), *The taste culture reader: Experiencing food and drink* (pp. 25–33). Oxford, UK: Berg.

Bensafi, M., Rouby, C., Farget, V., Bertrand, B., Vigouroux, M., & Holley, A. (2002). Autonomic nervous system responses to odours: The role of pleasantness and arousal. *Chemical Senses, 27*, 703–709.

Bremner, A., Lewkowicz, D., & Spence, C. (Eds.), (2012). *Multisensory development*. Oxford, UK: Oxford University Press.

Bryant, B., & Mezine, I. (1999). Alkylamides that produce tingling paresthesia activate tactile and thermal trigeminal neurons. *Brain Research, 842*, 452–460.

Bult, J. H. F., de Wijk, R. A., & Hummel, T. (2007). Investigations on multimodal sensory integration: Texture, taste, and ortho- and retronasal olfactory stimuli in concert. *Neuroscience Letters, 411*, 6–10.

Castriota-Scanderbeg, A., Hagberg, G. E., Cerasa, A., Committeri, G., Galati, G., Patria, F., et al. (2005). The appreciation of wine by sommeliers: A functional magnetic resonance study of sensory integration. *Neuroimage, 25*, 570–578.

Cerf-Ducastel, B., Van de Moortele, P.-F., Macleod, P., Le Bihan, D., & Faurion, A. (2001). Interaction of gustatory and lingual somatosensory perceptions at the cortical level in the human: A functional magnetic resonance imaging study. *Chemical Senses, 26*, 371–383.

Cheskin, L. (1957). *How to predict what people will buy*. New York, NY: Liveright.

Chu, S., & Downes, J. J. (2002). Proust nose best: Odors are better cues of autobiographical memory. *Memory and Cognition, 30*, 511–518.

Cliff, M. A. (2001). Influence of wine glass shape on perceived aroma and colour intensity in wines. *Journal of Wine Research, 12*, 39–46.

Clydesdale, F. M. (1993). Color as a factor in food choice. *Critical Reviews in Food Science and Nutrition, 33*, 83–101.

Dacremont, C. (1995). Spectral composition of eating sounds generated by crispy, crunchy and crackly foods. *Journal of Texture Studies, 26*, 27–43.

Delplanque, S., Grandjean, D., Chrea, C., Aymard, L., Cayeux, I., LeCalvé, B., et al. (2008). Emotional processing of odors: Evidence for a nonlinear relation between pleasantness and familiarity evaluations. *Chemical Senses, 33*, 469–479.

Delwiche, J. F. (2003). Attributes believed to impact flavor: An opinion survey. *Journal of Sensory Studies, 18*, 347–352.

Demattè, M. L., Pojer, N., Endrizzi, I., Corollaro, M. L., Betta, E., Aprea, E., et al. (2014). Effects of the sound of the bite on apple perceived crispness and hardness. *Food Quality and Preference, 38*, 58–64.

Dijksterhuis, G., Luyten, H., de Wijk, R., & Mojet, J. (2007). A new sensory vocabulary for crisp and crunchy dry model foods. *Food Quality and Preference, 18*, 37–50.

Duizier, L. (2001). A review of acoustic research for studying the sensory perception of crisp, crunchy and crackly textures. *Trends in Food Science and Technology, 12*, 17–24.

Eldeghaidy, S., Marciani, L., McGlone, F., Hollowood, T., Hort, J., Head, K., et al. (2011). The cortical response to the oral perception of fat emulsions and the effect of taster status. *Journal of Neurophysiology, 105*, 2572–2581.

Essick, G. K., Chopra, A., Guest, S., & McGlone, F. (2003). Lingual tactile acuity, taste perception, and the density and diameter of fungiform papillae in female subjects. *Physiology & Behavior, 80*, 289–302.

Fenko, A., Schifferstein, H. N. J., & Hekkert, P. (2010). Shifts in sensory dominance between various stages of user-product interactions. *Applied Ergonomics, 41*, 34–40.

Francis, S., Rolls, E. T., Bowtell, R., McGlone, F., O'Doherty, J., Browning, A., et al. (1999). The representation of pleasant touch in the brain and its relationship with taste and olfactory areas. *Neuroreport, 10*, 453–459.

Garneau, N. L., Nuessle, T. M., Sloan, M. M., Santorico, S. A., Coughlin, B. C., & Hayes, J. E. (2014). Crowdsourcing taste research: Genetic and phenotypic predictors of bitter taste perception as a model. *Frontiers of Integrative Neuroscience, 8*, 33.

Glass, S. T., Lingg, E., & Heuberger, E. (2014). Do ambient urban odors evoke basic emotions? *Frontiers in Psychology, 5*, 340.

Green, B. G. (2002). Studying taste as a cutaneous sense. *Food Quality and Preference, 14*, 99–109.

Hagura, N., Barber, H., & Haggard, P. (2013). Food vibrations: Asian spice sets lips trembling. *Proceedings of the Royal Society B, 280*, 20131680.

He, W., Boesveldt, S., de Graaf, C., & de Wijk, R. A. (2014). Dynamics of autonomic nervous system responses and facial expressions to odors. *Frontiers in Psychology, 5*, 110.

Hollingworth, H. L., & Poffenberger, A. T. (1917). *The sense of taste*. New York, NY: Moffat Yard.

Hummel, T., Delwiche, J. F., Schmidt, C., & Hüttenbrink, K.-B. (2003). Effects of the form of glasses on the perception of wine flavors: A study in untrained subjects. *Appetite, 41*, 197–202.

ISO, (1992). *Standard 5492: Terms relating to sensory analysis. International Organization for Standardization*. Vienna: Austrian Standards Institute.

ISO, (2008). *Standard 5492: Terms relating to sensory analysis. International Organization for Standardization*. Vienna: Austrian Standards Institute.

Knight, T. (2012). Bacon: The slice of life. In C. Vega, J. Ubbink, & E. van der Linden (Eds.), *The kitchen as laboratory: Reflections on the science of food and cooking* (pp. 73–82). New York, NY: Columbia University Press.

Koelewijn, T., Bronkhorst, A., & Theeuwes, J. (2010). Attention and the multiple stages of multisensory integration: A review of audiovisual studies. *Acta Psychologica, 134*, 372–384.

Köster, E. P., Møller, P., & Mojet, J. (2014). A "Misfit" theory of spontaneous conscious odor perception (MITSCOP): Reflections on the role and function of odor memory in everyday life. *Frontiers in Psychology, 5*, 64.

Krishna, A. (2012). An integrative review of sensory marketing: Engaging the senses to affect perception, judgment and behavior. *Journal of Consumer Psychology, 22*, 332–351.

Laureati, M., Morin-Audebrand, L., Pagliarini, E., Sulmont-Rossé, C., Köster, E. P., & Mojet, J. (2008). Influence of age and liking on food memory: An incidental learning experiment with children, young and elderly people. *Appetite, 51*, 273–282.

Lim, J., & Green, B. G. (2008). Tactile interaction with taste localization: Influence of gustatory quality and intensity. *Chemical Senses, 33*, 137–143.

Lim, J., & Johnson, M. B. (2011). Potential mechanisms of retronasal odor referral to the mouth. *Chemical Senses, 36*, 283–289.

Lim, J., & Johnson, M. (2012). The role of congruency in retronasal odor referral to the mouth. *Chemical Senses, 37,* 515–521.

Mauer, L., & El-Sohemy, A. (2012). Prevalence of cilantro (Coriandrum sativum) disliking among different ethnocultural groups. *Flavour, 1,* 8.

Mennella, J. A., Jagnow, C. P., & Beauchamp, G. K. (2001). Prenatal and postnatal flavor learning by human infants. *Pediatrics, 107,* E88.

Moir, H. C. (1936). Some observations on the appreciation of flavour in foodstuffs. *Journal of the Society of Chemical Industry: Chemistry & Industry Review, 14,* 145–148.

Mouritsen, O. G., & Risbo, J. (2013). Gastrophysics—do we need it? *Flavour, 2,* 3.

Murphy, C., & Cain, W. S. (1980). Taste and olfaction: Independence vs. interaction. *Physiology and Behavior, 24,* 601–605.

Navarra, J., Alsius, A., Soto-Faraco, S., & Spence, C. (2010). Assessing the role of attention in the audiovisual integration of speech. *Information Fusion, 11,* 4–11.

Pazart, L., Comte, A., Magnin, E., Millot, J.-L., & Moulin, T. (2014). An fMRI study on the influence of sommeliers' expertise on the integration of flavor. *Frontiers in Behavioral Neuroscience, 8,* 358.

Piqueras-Fiszman, B., & Spence, C. (2011). Crossmodal correspondences in product packaging: Assessing color-flavor correspondences for potato chips (crisps). *Appetite, 57,* 753–757.

Piqueras-Fiszman, B., & Spence, C. (2012). Sensory incongruity in the food and beverage sector: Art, science, and commercialization. *Petits Propos Culinaires, 95,* 74–118.

Piqueras-Fiszman, B., & Spence, C. (2014). Colour, pleasantness, and consumption behaviour within a meal. *Appetite, 75,* 165–172.

Piqueras-Fiszman, B., & Spence, C. (2015). Sensory expectations based on product-extrinsic food cues: An interdisciplinary review of the empirical evidence and theoretical accounts. *Food Quality and Preference, 40,* 165–179.

Pliner, P., Pelchat, M., & Grabski, M. (1993). Reduction of neophobia in humans by exposure to novel foods. *Appetite, 20,* 111–123.

Prescott, J. (2012). *Taste matters: Why we like the foods we do.* London, UK: Reaktion Books.

Roudnitzky, N., Bult, J. H. F., de Wijk, R. A., Reden, J., Schuster, B., & Hummel, T. (2011). Investigation of interactions between texture and ortho- and retronasal olfactory stimuli using psychophysical and electrophysiological approaches. *Behavioural Brain Research, 216,* 109–115.

Rozin, P. (1982). "Taste-smell confusions" and the duality of the olfactory sense. *Perception & Psychophysics, 31,* 397–401.

Russell, K., Zivanovic, S., Morris, W. C., Penfield, M., & Weiss, J. (2005). The effect of glass shape on the concentration of polyphenolic compounds and perception of Merlot wine. *Journal of Food Quality, 28,* 377–385.

Schaal, B., & Durand, K. (2012). The role of olfaction in human multisensory development. In A. J. Bremner, D. Lewkowicz, & C. Spence (Eds.), *Multisensory development* (pp. 29–62). Oxford, UK: Oxford University Press.

Schaal, B., Marlier, L., & Soussignan, R. (2000). Human foetuses learn odours from their pregnant mother's diet. *Chemical Senses, 25,* 729–737.

Schifferstein, H. N. J. (2006). The perceived importance of sensory modalities in product usage: A study of self-reports. *Acta Psychologica, 121,* 41–64.

Schifferstein, H. N. J. (2009). The drinking experience: Cup or content? *Food Quality and Preference, 20,* 268–276.

Schifferstein, H. N. J., Fenko, A., Desmet, P. M. A., Labbe, D., & Martin, N. (2013). Influence of package design on the dynamics of multisensory and emotional food experience. *Food Quality and Preference, 27,* 18–25.

Shankar, M. U., Levitan, C., & Spence, C. (2010). Grape expectations: The role of cognitive influences in color-flavor interactions. *Consciousness & Cognition, 19*, 380–390.

Small, D. M., Veldhuizen, M. G., Felsted, J., Mak, Y. E., & McGlone, F. (2008). Separable substrates for anticipatory and consummatory food chemosensation. *Neuron, 57*, 786–797.

Smeets, M. A. M., & Dijksterhuis, G. B. (2014). Smelly primes—When olfactory primes do or do not work. *Frontiers in Psychology: Cognitive Science, 5*, 96.

Spence, C. (2012). Managing sensory expectations concerning products and brands: Capitalizing on the potential of sound and shape symbolism. *Journal of Consumer Psychology, 22*, 37–54.

Spence, C. (2013). The supertaster who researches supertasters. *The BPS Research Digest* <http://www.bps-research-digest.blogspot.co.uk/2013/10/day-4-of-digest-super-week-supertaster.html>.

Spence, C. (2015a). Eating with our ears: Assessing the importance of the sounds of consumption to our perception and enjoyment of multisensory flavour experiences. *Flavour, 4*, 3.

Spence, C. (2015b). Multisensory flavour perception. *Cell, 161*, 24–35.

Spence, C. (2015c). On the psychological impact of food colour. *Flavour, 4*, 21.

Spence, C. (2015d). Just how much of what we taste derives from the sense of smell? *Flavour, 4*, 30.

Spence, C. (2015e). Leading the consumer by the nose: On the commercialization of olfactory-design for the food & beverage sector. *Flavour, 4*, 31.

Spence, C. (2016). Oral referral: Mislocalizing odours to the mouth. *Food Quality & Preference, 50*, 117–128.

Spence, C., Harrar, V., & Piqueras-Fiszman, B. (2012). Assessing the impact of the tableware and other contextual variables on multisensory flavour perception. *Flavour, 1*, 7.

Spence, C., Levitan, C., Shankar, M. U., & Zampini, M. (2010). Does food colour influence flavour identification in humans? *Chemosensory Perception, 3*, 68–84.

Spence, C., Okajima, K., Cheok, A.D., Petit, O., & Michel, C. (in press). Eating with our eyes: From visual hunger to digital satiation. *Brain & Cognition*.

Spence, C., & Piqueras-Fiszman, B. (2012). The multisensory packaging of beverages. In M. G. Kontominas (Ed.), *Food packaging: Procedures, management and trends* (pp. 187–233). Hauppauge NY: Nova Publishers.

Spence, C., & Piqueras-Fiszman, B. (2014). *The perfect meal: The multisensory science of food and dining*. Oxford, UK: Wiley-Blackwell.

Spence, C., Puccinelli, N., Grewal, D., & Roggeveen, A. L. (2014). Store atmospherics: A multisensory perspective. *Psychology & Marketing, 31*, 472–488.

Spence, C., Smith, B., & Auvray, M. (2015). Confusing tastes and flavours. In D. Stokes, M. Matthen, & S. Biggs (Eds.), *Perception and its modalities* (pp. 247–274). Oxford, UK: Oxford University Press.

Stevenson, R. J. (2014). Flavor binding: Its nature and cause. *Psychological Bulletin, 140*, 487–510.

Stuckey, B. (2012). *Taste what you're missing: The passionate eater's guide to why good food tastes good*. London, UK: Free Press.

Thomson, D., & MacFie, H. (2007). SensoEmotional optimisation of food products and brands. In H. MacFie (Ed.), *Consumer-led food product development* (pp. 281–303). Boca Raton, FL: CRC Press.

Todrank, J., & Bartoshuk, L. M. (1991). A taste illusion: Taste sensation localized by touch. *Physiology & Behavior, 50*, 1027–1031.

van den Bosch, I., van Delft, J. M., de Wijk, R. A., de Graaf, C., & Boesveldt, S. (2015). Learning to (dis) like: The effect of evaluative conditioning with tastes and faces on odor valence assessed by implicit and explicit measurements. *Physiology & Behavior, 151,* 478–484.

Van Doorn, G., Wuillemin, D., & Spence, C. (2014). Does the colour of the mug influence the taste of the coffee? *Flavour, 3,* 10.

Varela, P., & Fiszman, S. (2012). Playing with sound. In C. Vega, J. Ubbink, & E. van der Linden (Eds.), *The kitchen as laboratory: Reflections on the science of food and cooking* (pp. 155–165). New York, NY: Columbia University Press.

Velasco, C., Salgado-Montejo, A., Marmolejo-Ramos, F., & Spence, C. (2014). Predictive packaging design: Tasting shapes, typographies, names, and sounds. *Food Quality & Preference, 34,* 88–95.

Velasco, C., Wan, X., Salgado-Montejo, A., Woods, A., Andrés Oñate, G., Mu, B., et al. (2014). The context of colour-flavour associations in crisps packaging: A cross-cultural study comparing Chinese, Colombian, and British consumers. *Food Quality & Preference, 38,* 49–57.

Wan, X., Woods, A. T., van den Bosch, J., McKenzie, K. J., Velasco, C., & Spence, C. (2014). Cross-cultural differences in crossmodal correspondences between tastes and visual features. *Frontiers in Psychology: Cognition, 5,* 1365.

Willander, J., & Larsson, M. (2008). The mind's nose and autobiographical odor memory. *Chemosensory Perception, 1,* 210–215.

Woods, A. T., Velasco, C., Levitan, C. A., Wan, X., & Spence, C. (2015). Conducting perception research over the internet: A tutorial review. *PeerJ, 3,* e1058.

Zampini, M., & Spence, C. (2004). Multisensory contribution to food perception: The role of auditory cues in modulating crispness and staleness in crisps. *Journal of Sensory Science, 19,* 347–363.

Zampini, M., Wantling, E., Phillips, N., & Spence, C. (2008). Multisensory flavor perception: Assessing the influence of fruit acids and color cues on the perception of fruit-flavored beverages. *Food Quality & Preference, 19,* 335–343.

Attention and Flavor Binding

Richard J. Stevenson
Department of Psychology, Macquarie University, Sydney, NSW, Australia

1 Introduction

There can be some surprise when you tell people you study flavor. After all, what is there to know? And is not this topic the province of chefs and food scientists, not psychologists? While there are good scientific grounds to study it, most notably that it is an unusual example of crossmodal integration, a far more important reason has emerged over the last 20 years. People like foods that have fatty, sweet, and salty tastes. Habitually eating these foods leads to weight gain and obesity. One approach to this problem is to badger people into eating healthier foods. Unfortunately, however, healthier foods are often less palatable and more expensive, making this an unattractive option for many people. A better approach, then, is to try to engineer foods that are as palatable as those high in saturated fat, salt, and sucrose, but where these unhealthy ingredients have been replaced by healthier alternatives. Crucially, this approach requires an understanding of flavor perception—a central focus of this chapter.

When we eat and drink, multiple sensory systems are activated (Rozin, 1982; Stevenson, 2009). These can be considered as falling into two general classes. The first involves those that are stimulated before the food or drink enters the mouth. These are orthonasal olfaction (ie, sniffing the food), vision (eg, color, visual texture, sheen, etc.), audition (ie, associated cooking or post-cooking sounds), and somatosensation (ie, from direct or indirect handling). The second group of sensory systems involves those stimulated once the food or drink has entered the mouth. These comprise oral somatosensation (including touch, proprioceptive feedback, etc.), chemesthesis (ie, chemical irritants such as black pepper), noting that this is generally considered as part of the somatosensory system, taste (ie, tongue-based sensations of sweet, bitter, sour, salty, and umami) and retronasal olfaction (ie, volatile chemicals perceived via the internal nares).

Many authors have noted that what is perceived when we eat and drink seems to be "something more" than the sensation generated by all of those individual senses alone (eg, Auvray & Spence, 2008; McBurney, 1986; Rozin, 1982; Small & Prescott, 2005). That "something more," is routinely referred to as "taste" or "flavor"—the latter being the term used here. The use of a single label for what is a multisensory experience is suggestive. It implies that people regard eating- and drinking-related experiences as combinatorial, rather than as a series of discrete events (ie, a burst of smell, a shot of sweetness, a flash of creaminess, etc.). This suggestion of perceptual unity invokes the idea of the brain bringing together divergent sources of sensory information and combining them—binding them together that is—into a unified flavor experience.

The aim of this chapter is twofold. The first aim is to examine the evidence for flavor binding. It could simply be that upon closer examination, there is nothing more than all of the available sensory information competing for attention, with flavor being just a cascade of different sensations drawn first from one modality and then from another. A more likely alternative, however, is that some form of binding occurs, resulting in something beyond each sense operating alone. If this should occur, the question then arises as to how the senses combine to create flavor. Addressing this question forms the second aim of the chapter.

To address these two aims, the first part of the chapter examines the evidence for flavor binding by cataloging its particular features. This provides a basis for what needs to be explained in the second part of the chapter, which then draws upon an explanatory framework that has arisen from studying examples of binding in other sensory domains (eg, Koelewijn, Bronkhorst, & Theeuwes, 2010; Navarra, Alsius, Soto-Faraco, & Spence, 2010; Talsma, Senkowski, Soto-Faraco, & Woldorff, 2010). Here the primary focus is on the role of attentional processes (but also relevant preattentive processes) as these have been of major importance in explaining binding in other domains.

2 Flavor binding

The majority of authors have concluded that there is some form of flavor binding (eg, Auvray & Spence, 2008; McBurney, 1986; Rozin, 1982; Small & Prescott, 2005). That is, they assume that some integration of sensory information from the different modalities occurs, and that sensory information is not just presented in discrete cascades. Unfortunately, however, there have been relatively few examinations of the evidential basis for this claim (Rozin, 1982; Stevenson, 2014). Consequently, it is important to establish that the generally held view is, in fact, the correct one, and, more specifically, to identify the features of flavor binding—namely what is bound together and what is not. To address this, the first part of this section focuses on events in the mouth, with a smaller second part examining the influence of sensory systems that are stimulated prior to ingestion.

2.1 Retronasal olfaction, taste, and oral somatosensation

The perception of flavor has a number of established and suggestive features that imply that sensory information from retronasal olfaction, taste, and oral somatosensation is bound together. There seem to be five principal features, each of which is reviewed below.

2.1.1 The modality/content dissociation in retronasal olfaction

For over 100 years both psychologists (eg, Titchener, 1909) and those interested in culinary appreciation (eg, Brillat-Savarin, 1825/1949), have remarked that most people do not seem to recognize the role of smell in flavor perception. One immediate puzzle is that everyday experience provides many opportunities to identify the role of olfaction in flavor (Rozin, 1982). First, many children are taught by adults to pinch their nose when consuming medicines or vegetables that they dislike. Just as shutting the eyes or blocking the ears impairs vision and audition respectively, blocking the nose prevents smell (Lawless, 1995)—yet this simple act does not seem to lead to an overt recognition of smell's role in flavor. Second, there is a strong correlation between the way foods smell, and their "taste" or "flavor" in the mouth. While this correlation is imperfect, it is certainly strong enough for people to recognize many odorous substances—strawberries, coffee, cheese, etc., both inside (retronasally) and outside (orthonasally) the mouth (eg, Sun & Halpern, 2005). What is striking here is that this correlation does not lead people to conclude that smell must be involved in flavor perception.

The limited empirical work in this area also suggests people do not have much awareness of smell's role in flavor. Rozin (1982) examined whether any major language has a word used by the general population to specifically describe the two forms of olfaction recognized by science—retronasal and orthonasal perception. None of the major languages that he examined had a specific term for retronasal olfaction, while all had a word for orthonasal olfaction and a word equivalent to flavor. Of course, it could be that flavor is used to refer to retronasal olfaction, but the linguistic evidence suggests otherwise. Rozin (1982) asked his participants to complete sentences which included the name of a taste, food, or drink (eg, this has a bitter_____), and that could be completed using either the term "taste" or "flavor." If flavor refers mainly to retronasal olfactory experiences, then one would have expected this term to be used more often with foods or drinks that have a significantly greater retronasal olfactory component, yet this was not the case. In the main, the terms taste and flavor were used interchangeably, thus suggesting that the term flavor is not used especially for items with a significant retronasal component.

The importance of retronasal olfaction in flavor perception should be particularly evident in those unfortunate individuals who become anosmic. A number of studies have reported that people often present to smell and taste clinics with a combined loss (eg, Bull, 1965; Fujii et al., 2004; Gent, Goodspeed, Zagraniski, & Catalanotto, 1987). That is, the patient describes how they can no longer smell things, and that they can also no longer perceive the taste of food. These are perceived as two related but distinct deficits by the patient, with testing usually revealing a loss of smell with largely intact taste perception (Deems et al., 1991). Once again, this would suggest that people do not appreciate that smell is a major contributor to flavor perception.

While participants may not generally be aware that smell is involved in flavor perception, they do seem able to perceive smells when they are part of a food in the mouth. Three observations suggest this: (1) anosmia significantly reduces food enjoyment (eg, Hummel & Nordin, 2005); (2) the identification of foods in the mouth is impaired by disrupting retronasal olfaction (Mozell, Smith, Smith, Sullivan, & Swender, 1969); and (3) sensory panels are used extensively by industry to reliably assess olfactory-related attributes (Lawless, 1995). And while the perception of food-related odors in the mouth may be harder than perception at the nose for chemical, mechanical, and psychological reasons (about which, more later), participants can undoubtedly detect odors in the mouth under laboratory conditions (eg, Stevenson & Mahmut, 2011a). Taken together with the evidence that participants are not aware that this content is olfactory in origin, this would suggest that there is a content-modality dissociation for olfaction.

It is, of course, possible that the same distinction might also apply to taste and oral somatosensation. For tastes, setting aside label-related confusions (eg, sour-bitter), participants can readily recognize taste sensations (ie, experience taste content; eg, Lawless, 1995). They can also identify their source as being gustatory, as patients who experience gustatory loss correctly identify its locus (ie, modality awareness; Porter, Fedele, & Habbab, 2010). For somatosensation, the same broad distinction also holds true. That is, participants can readily identify and label changes in texture (eg, too soft), temperature (eg, too cold), or irritation (eg, burning), although they probably do not know that these all form part of the somatosensory system (eg, Munoz & Civille, 1987). The only case where this may break down is for fat perception. It would be especially interesting to apply Rozin's (1982) sentence completion technique to fat-related terms, as one might equally say that a food has a greasy taste or a greasy feel. Interestingly, fat perception is multisensory, having olfactory (fat-related odors), gustatory (with possibly specialized fat-taste receptors), and tactile (fatty mouth feel) components (Chale-Rush, Burgess, & Mattes, 2007; Keast & Costanzo, 2015). Setting aside fat perception, the conclusion here is that taste and somatosensory stimuli can have both their modality and content identified, while olfaction seems to contribute content but with apparently minimal modality awareness. This is one feature of flavor binding.

2.1.2 Interactions between the senses

Another feature of flavor binding is the interactions that occur between the flavor senses. Many of the large number of documented interactions have a peripheral basis (eg, see Delwiche, 2004; Stevenson, 2009, for reviews). That is, they arise from chemical, mechanical or physiological events local to the mouth. However, there are two types of interaction that have an established central basis. The first involves interactions between certain tastes and smells. Frank and Byram (1988) were among the first to show that certain combinations of odors and tastes generate interactions while others do not. They compared participants' sweetness ratings of sucrose solutions that had either peanut butter odor or strawberry odor added to them. Relative to judgments of sucrose alone, the strawberry odor enhanced ratings of sweetness, while peanut butter odor had no effect. Similar odor–taste-specific effects have been recorded now in many experiments (eg, Schifferstein & Verlegh, 1996; Stevenson, Prescott, & Boakes, 1999). While these effects have been suggested to arise from the way in which participants use rating scales (eg, Clark & Lawless, 1994), the specificity of the effect argues strongly against such an interpretation. That is, if a participant judges how sweet a strawberry-sucrose solution tastes, it will be judged as sweeter than if the participant *also* made a strawberry-flavor rating. However, the addition of an irrelevant rating (eg, peanut-butter-flavor) would have no effect. This implies that it is something special about the history of the strawberry odor and the sweet taste that is driving the taste-enhancement effect, not simply rating scale use.

A second and presumably related effect occurs when participants are asked to judge the intensity of the odorous component of a taste–odor mixture (eg, strawberry flavor). Many studies have shown that when participants are asked to evaluate specific odor–taste mixtures—where these combinations have been experienced before outside of the laboratory—participants judge the odorous component to be more intense than judgments made without the tastant being present (eg, Green, Nachtigal, Hammond, & Lim, 2012; von Sydow, Moskowitz, Jacobs, & Meiselman, 1974)—an odor-enhancement effect. This effect seems to be more potent when the tastant and odorant concentrations are lower, and there is the suspicion that this may also be the case with the taste-enhancement effect as well. More importantly, both of these odor–taste interactions have a clear psychological basis. This has been demonstrated most convincingly by those studies that have presented the odor and taste by separate routes (ie, odor via sniffing and taste via the tongue), so that peripheral effects (eg, physiochemical or peripheral physiology) could not be responsible (eg, Sakai, Kobayakawa, Gotow, Saito, & Imada, 2001).

The other type of psychological interaction that has a psychological basis involves that between oral somatosensation and retronasal olfaction. A number of studies have shown that more viscous solutions reduce the perceived intensity of concurrently presented odors (eg, a viscous fluid with an odorant added to it sampled by mouth), even when the odor is presented separately from the viscous solution, ruling out peripheral interaction effects (eg, Bult, de Wijk, & Hummel, 2007). A further phenomenon has also been identified that is more directly analogous to that between certain tastes and smells. Creamy odors seem able to increase viscosity ratings for model foods (eg, Bult et al., 2007; Weenen, Jellema, & de Wijk, 2005). Not only do these findings indicate that olfaction is a common link among these interactions, they all represent a further feature of flavor binding.

2.1.3 Perceived location

Taste and somatosensation both have their receptors located in the mouth, the same location where their associated percepts appear to arise. For somatosensation this is of little consequence, because part of the function of this sense is to spatially locate stimuli on the surface of the body. For taste, the localization of this sensation to the mouth and tongue seems to depend upon concurrent oral somatosensation (Green, 2003). A number of studies have demonstrated this effect. If half of the tongue is anesthetized, and the anesthetized half is stimulated by a tastant, participants are not able to identify that a tastant is present. However, if the tastant is sipped from a cup, thus covering the whole tongue, participants' judgments of taste perception are identical to those observed in the unanesthetized state (Lehman, Bartoshuk, Catalanotto, Kveton, & Lowlicht, 1995). A similar phenomenon is observed in patients with unilateral taste loss, who may not recognize the loss until unilateral testing is undertaken (Kveton & Bartoshuk, 1994). A similar effect can be demonstrated by asking participants to judge the intensity of a taste-saturated swab wiped just to the left or right of the tongue's most sensitive part, its tip (ie, where the highest number of taste receptors are located). At this point, participants report that taste intensity is lower than when the swab is moved over the tip of the tongue. Interestingly, when the swab is now touched to the other side of the tip (the left or right depending on which side was touched first), higher ratings are *maintained* even though receptor density is again reduced (Todrank & Bartoshuk, 1991). This again seems to result from smearing of the taste sensation across parts of the tongue that are concurrently mechanically stimulated. The upshot is that somatosensation and taste are both perceived as being located in the mouth, and the localization of taste to the mouth is a further feature of flavor binding.

Smell is rather different as its receptors are located behind the bridge of the nose and not in the mouth. Volatile chemicals arising from inside the mouth (via the internal nares) and from the external environment (via the external nares) are both processed by the same receptor sheet. Yet, when the stimulus is from the external world, it is usually perceived correctly as being located in the external environment (eg, Mainland & Sobel, 2006) and when arising from the mouth and internal nares it is judged (again correctly) as belonging to events in the mouth (eg, Rozin, 1982). Thus, during eating and drinking, olfactory sensation is localized to the mouth, and during ordinary "smelling" olfactory sensation is localized to the external world. This represents another feature of flavor binding.

2.1.4 Temporal properties

Taste and oral somatosensory stimulation are *broadly* overlapping during an eating episode. That is, tastants are available for perception during much of the period that the food is in the mouth, and the textural components are constantly available, although both may have changes in property across the course of mastication and prior to swallowing. For olfaction, the situation is different, because of the more distant location of the olfactory receptors. Volatile chemicals appear to present to the olfactory receptors in a pulsatile manner. This occurs both after swallowing, as volatile laden air is expelled through the nose (Trelea et al., 2008), and during chewing, where puffs of odorized air may pass intermittently into the internal nares (Hodgson, Linforth, & Taylor, 2003). The point here is that odorant availability is considerably more discontinuous than taste and somatosensory stimulation, and yet this is not obviously apparent when eating and drinking. With the caveat that this has not been well studied, there may either be a failure to notice lapses in smell sensation (see Sela & Sobel, 2010, for a related suggestion) or they may be "filled in," in a manner paralleling that for the visual blind spot. This represents yet another feature of flavor binding, albeit a more speculative one.

2.1.5 Capacity to perceive parts and wholes

A further aspect of flavor binding is that it may result in limited access to particular types of sensory information. Two examples of this have been documented. The first concerns the ability of participants to detect odors and tastes in mixtures sampled by mouth (Laing, Link, Jinks, & Hutchinson, 2002; Marshall, Laing, Jinks, & Hutchinson, 2006). A key finding to emerge from these studies concerns the effect of adding extra odors or tastes to a mixture containing one odor and one taste. The addition of tastes or smells has only a limited effect on taste identification, but a significantly deleterious effect on detection of the odor component(s). Somehow, the recognition of odor components is harder as mixture complexity increases, relative to taste recognition (Laing et al., 2002; Marshall et al., 2006).

A second effect concerns the influence of perceptual expertise. Three studies have examined—indirectly—whether experts (sensory panellists and chefs) are less susceptible to odor–taste interaction effects than naïve participants. That is, can experts better tell apart the odor and taste components of a mixture, thus recognizing their individual contribution to the flavor? While an early study found some evidence favorable to this hypothesis, the results are, unfortunately, hard to interpret because different tests of odor-induced taste enhancement were used on the naïve and expert participants (Bingham, Birch, de Graaf, Behan, & Perring, 1990). Two further studies have reported that experts (chefs in one study and trained sensory panellists in another) are as prone as naïve participants to odor–taste interaction effects (Boakes & Hemberger, 2012; Davidson, Linforth, Hollowood, & Taylor, 1999). This would suggest it might not be possible to attend to odors in the mouth without also attending to the accompanying taste (and vice versa).

2.2 The broader perspective

The experience of eating and drinking clearly involves more than just the sensations arising in the mouth (Spence & Piqueras-Fiszman, 2014). Food is generally seen, smelled, and felt before it is eaten, and on some occasions it is heard as well. These experiences, which precede each mouthful of food, can clearly influence the flavor experience in the mouth. The color of a food, and many other aspects of its visual presentation, can affect how the food is perceived, and these types of effect have been well documented in the literature and reviewed in a number of articles (eg, Shermer & Levitan, 2014; Spence, 2015; Spence, Levitan, Shankar, & Zampini, 2010). The more important question here is whether these various effects can be considered as being bound in some way to the broader flavor experience. Functionally, it could be argued that the main virtue of pre-consumption smelling and viewing is to identify safe, palatable, and nutrient-rich sources of energy. While this would therefore necessitate the formation of associations between the immediate and delayed consequences of consuming a particular food (or drink) and its smell and visual appearance, does this constitute binding?

This is not an easy question to answer for two reasons. First, there is clear temporal and spatial separation between viewing, smelling, and touching food and the flavor experience, which would suggest that the two experiences are discrete. Second, a number of studies suggest that flavor events, such as repeated experience of a particular taste and smell together in the mouth, come to affect how that odor is experienced when it is later sniffed in the external environment. In this case, events in the mouth are linked with later perception of food odors preceding ingestion, suggesting rather an intimate link, but one that is historical rather than current. For vision (and touch), these may influence flavor perception *in the now* (and *earlier* flavor perception may also influence our later reaction to a visual or tactile food cue) but this may be by more explicit means, such that we expect something that looks like strawberry ice-cream to *taste* like strawberry ice-cream (Yeomans, Chambers, Blumenthal, & Blake, 2008). The olfactory and visual experiences prior to ingestion may then be linked to flavor by learning, and the way that this manifests *may* be more implicit for smell and more explicit for vision. However, the presence of clear spatial and temporal separation between food on the plate and food in the mouth seems to suggest that these experiences are discrete and are not perceptually bound in the way that flavor appears to be.

2.3 Discussion

Flavor has a number of characteristics that reflect binding. These are: (1) the modality/content dissociation for odors (ie, content awareness but not modality awareness); (2) the capacity of tastes and somatosensory stimuli to influence odor perception and vice versa; (3) the localization of smell and taste to a common spatial location, the mouth; (4) the apparent failure to notice that smell experiences may only be intermittently present during eating and drinking; and (5) the limited capacity to perceive certain parts of the flavor experience, notably smells, even amongst experts. These would seem to be the principal features of flavor binding and the following section examines how they may be explained.

3 How might flavor binding occur?

The main part of this section focuses on events in the mouth. This is divided into two parts, the first dealing with preattentive processes that contribute to flavor binding and the second with attentional accounts.

3.1 Preattentive accounts of flavor binding

The localization of taste experience to the tongue and mouth is probably a result of a preattentive binding process. The main reason for making this claim is that taste and somatosensory information are co-processed from the receptor level upwards (Green, 2003). In animals, certain first-order taste neurons are multisensory, being as responsive to somatosensory-related inputs as to taste (Oakley, 1985; Whitehead & Kachele, 1994). This neural co-processing extends all the way to the primary taste cortex, where there are roughly as many neurons devoted to taste as there are to somatosensory stimulation (Ogawa, Ito, Murayama, & Hasegawa, 1990; Plata-Salaman, Scott, & Smith-Swintosky, 1993). This close physiological intertwining is naturally complemented by the fact that the onset and offset of taste stimulation broadly co-occurs with the onset and offset of concurrent somatosensory information. Together these characteristics would seem to favor a preattentive and largely automatic form of binding (Green, 2003).

When considering preattentive effects, these could arise in two different ways. The first, as described above for taste and somatosensation, is largely hard-wired. The second can occur as a consequence of repeated experience, thus becoming automatic by repetition. While the consequence may be preattentive, it is only so by virtue of learning, a process that *may* initially be attention-demanding and explicit. As noted earlier, certain taste and smell stimuli interact, generating taste- and odor-enhancement effects. It has been suggested that the origin of these effects may lie in the co-experience of tastes and smells in the mouth, which can additionally lead to the odor component coming to acquire certain taste-like qualities—odor–taste learning (Prescott, 1999).

Prior experience of tastes and smells in the mouth may then result in preattentive processing of the previously co-experienced pair. One way this idea has been tested is to see whether detection of a threshold-level smell can be facilitated by the presentation of a congruent (ie, a taste with which the odor has previously been experienced) threshold-level taste. Evidence suggests that it can facilitate detection (Dalton, Doolittle, Nagata, & Breslin, 2000), but the early attempts to test this did not control for probability summation effects (ie, better odds of detection when two modalities are stimulated concurrently). Later attempts that controlled for this found evidence of preattentive processing, but interestingly they have only used previously co-experienced tastes and smells (Veldhuizen, Shephard, Wang, & Marks, 2010). A key unanswered question is whether some form of preattentive processing occurs when tastes and smells are used that have no prior history of co-occurrence. This would point to an additional form of odor–taste binding discrete from any effects of learning.

A further and related issue concerns the status of odor–taste learning. Under laboratory conditions, the pairing of an unfamiliar odor with a sweet taste by mouth as a flavor results in that odor coming to smell sweeter, while pairing it with citric acid results in it smelling sourer (relative to pairing with water which has no effect; Stevenson, Boakes, & Prescott, 1998; Stevenson, Prescott, & Boakes, 1995). It has been suggested that this may all occur with minimal conscious awareness (eg, Brunstrom, 2004), in other words that it may also be a preattentive process. Evidence for this is circumstantial. Odors are hard to name and so forming any verbal representation of the contingency (eg, "lemon went with sour") is also going to be hard (eg, Davis, 1977). Participants do not appear to have contingency-related knowledge when this is investigated in laboratory-based odor-learning tasks—noting that such tests of contingency awareness can be weak (eg, Stevenson, Boakes, & Wilson, 2000). Finally, odor–taste learning is resistant to extinction, in that presenting an odor that was previously paired with a taste, with water, has no effect on how "tasty" that odor smells (Stevenson et al., 2000). Usually, the observation of extinction under these types of condition is an indication that the participant has consciously noted that the cue is no longer predictive of a particular consequence. The failure to obtain extinction also implies a lack of awareness of the changing contingencies. While this might all be favorable to the view that odor–taste learning is preattentive, the problem is that this is contrary to current thinking about human associative learning. The current view, which is well supported empirically, is that people come to notice the contingency between two events and that noticing—thus attending to—this set of events results in the formation of an association (eg, Shanks, 2010).

One further line of evidence that bears on this issue comes from examining the impact of directing attention either to the "to-be-learned" odor and taste, or to the overall configuration, the flavor. It has been suggested that elemental processing, which may be necessary for supporting the standard view of human associative learning (ie, where each of the to-be-learned events can be adequately discriminated from the other) may actively disrupt odor–taste learning. While there has been evidence both favoring this proposal (eg, Prescott & Murphy, 2009) and evidence against it (eg, Stevenson & Case, 2003), only one study has directly determined whether participants can spot the to-be-learned odors and tastes (Stevenson & Mahmut, 2011a). Under these conditions, where participants can discriminate the odor and the taste, learning (as indexed by the degree to which the odor comes to smell of the paired taste) was as good in the discrimination condition as it was in the control condition (where participants did not have to tell the taste and smell apart). This would suggest that the attentional stance adopted by participants has little impact, implying that odor–taste learning is probably not dependent on volitional conscious processing.

3.2 Attentional accounts of flavor binding

Attention-based accounts of binding currently dominate the human experimental literature for vision and audition (eg, Koelewijn et al., 2010; Talsma et al., 2010; Treisman & Gelade, 1980). For flavor, there are now two broad theoretical approaches, both of which involve a central role for attention (Stevenson, 2014). The first is more empirically derived and concerns attentional capture, both spatially by stimulation in the mouth (Ashkenazi & Marks, 2004) and by modality, specifically by taste (Stevenson, 2012). The second derives primarily from a theoretical stance that perception is better understood by focusing on function rather than individual senses (Gibson, 1966). Several authors have suggested that this is a useful way of thinking about flavor (Auvray & Spence, 2008; Gibson, 1966; Prescott, 1999). One way in which this type of approach can be instantiated is by theorizing the existence of a common attentional channel shared by taste and smell in the mouth. This account can be envisaged in several ways—as a hard-wired variant, where all tastes and smells in the mouth are processed together, or as a learnt variant, where only tastes and smells that have previously been experienced together come to be processed via the common attentional channel. Importantly, various intermediate positions are also possible, such as all tastes and smells in the mouth being weakly processed through a common attentional channel, with learning strengthening this process for frequently co-experienced tastes and smells.

3.2.1 Attentional capture

There are two basic premises here, which are not mutually exclusive. The first is that somatosensory/gustatory events in the mouth move spatial attention to this location (Ashkenazi & Marks, 2004). The second is that the occurrence of biologically salient taste events focus modality-specific attention on gustation (Stevenson, 2012). The implications of this are relevant for explaining several effects outlined in the first section of this chapter. If spatial attention is located in the mouth during flavor perception, while smells are actually sensed retronasally via the nose, then it requires a shift of spatial attention to the nose to focus on olfaction. If modality-specific attention is focused on gustation, it again requires a shift of attention to the sense of smell to achieve a focus on this sense. One outcome of this may be that it is routinely difficult to attend to the olfactory channel when experiencing a flavor, and thus smell may receive less attention than taste (noting this is curious given olfaction's role in defining flavor quality—a point returned to further below). This could then account for the finding of a common location for flavor in the mouth, because smell is largely unattended. It could also account for the difficulty encountered in recognizing odors in the mouth, relative to tastes, as modality-specific attention would also need to be switched from taste to smell.

Attending to smell in the mouth may be further hampered by the absence of a strategy to focus attention on it. For orthonasal olfaction, we sniff when we wish to attend to smell (Sobel et al., 1998). However, for retronasal smell most individuals lack a strategy to focus on it, which may exaggerate difficulties in attending to this modality in the mouth. Although there may be some strategies that do aid this process, they do not appear to be widely known or used (see Burdach & Doty, 1987).

That spatial attention may be drawn to the mouth by the presence of oral-somatosensation, and that this may be important in binding smell to that location, seems to be a reasonable suggestion (eg, Green, 2003; Murphy, Cain, & Bartoshuk, 1977). The key to testing this is to see whether concurrent oral somatosensation alone or oral somatosensation *and* taste, is necessary to bind concurrent olfactory stimulation to the mouth. One way that has been developed to explore this has been to deliver odorants orthonasally, while inducing oral somatosensation with and without concurrent taste (eg, Lim & Johnson, 2011; Stevenson, Oaten, & Mahmut, 2011). Using two different but convergent techniques, it has become apparent that a somatosensory stimulus alone (ie, water; water vigorously moved around the mouth; fluids of varying viscosity) is not sufficient to alter where participants perceive a sniffed smell to be—that is, at the nose (eg, Stevenson, Oaten et al., 2011). However, when a taste is added participants then report that the odor feels as if it is located in their mouth, even though it is still being sniffed through the nose (eg, Lim & Johnson, 2011).

That this is fundamentally an attentional phenomenon is suggested by several findings. First, more intense tastes are better able to produce the illusion that the odor is located in the mouth rather than at the nose (Stevenson, Oaten et al., 2011). Second, adding an irritant to the odor effectively removes the illusion of localization to the mouth, presumably by shifting attention to the trigeminal (somatosensory) modality (and nose; Stevenson, Mahmut, & Oaten, 2011). Third, tasks that require attention (odor naming and discrimination) are performed more poorly when participants perceive the odor to be located in the mouth (Stevenson & Mahmut, 2011b). Not only do these findings suggest that localization of smell to the mouth is affected by attention-related manipulations, they also suggest that modality-specific attention (ie, focusing on taste) may be more important in dictating this effect than spatial attention (ie, focusing on the mouth). This is because it is the presence of taste—not oral somatosensation alone (the sense that presumably drives spatial attention)—that is required for a shift in perceived location to occur.

The idea that gustation may be able to capture modal attention at the expense of smell seems theoretically reasonable (ie, exogenous capture; see Spence, Nicholls, &

Driver, 2001). First, attentional capture by one modality (or submodality) over another is generally accomplished when the capturing modality is biologically more significant (Knudsen, 2007), which is why pain may be so effective at capturing attention (eg, Eccleston & Crombez, 1999; and relatedly, see Van Damme, Crombez, & Spence, 2009). There are good grounds to think that taste is often biologically significant. Functionally, it can detect both nutrient sources and poisons in the foods that are placed in the mouth (Scott & Mark, 1987). This functional importance is reflected in the strong affective reaction that tastes engender (eg, sweet vs bitter) and their innate basis (Steiner, Glaser, Hawilo, & Berridge, 2001). In addition to being good at commanding attention, biologically salient events are also good at holding attention—again like pain (Van Damme, Crombez, Eccleston, & Goubert, 2004). This may make it especially hard to shift attention away from the taste modality, making detection of concurrent olfactory events more difficult.

Gustation's suspected capacity to command attention at the expense of smell offers one type of explanation as to how the smell component of flavor is localized to the mouth, and why it might be hard to focus attention on it (relative to taste). The focus of attention on taste might also serve to mask fluctuations in olfactory sensation generated by the intermittent nature of olfactory delivery to the receptors, and contribute to the more general phenomenon of lack of olfactory awareness. Lack of olfactory awareness in the mouth is also consistent with claims that odor–taste learning can occur independently of awareness. Thus, it would seem that this type of account could potentially accommodate a number of the phenomena identified in the first half of this chapter.

This account does, however, have its problems. The first concerns an important conceptual issue. At its boldest, one could claim that taste captures attention at the *complete* expense of smell. No attention to smell would be regarded by many theorists to imply no conscious perception of the olfactory part of flavor, yet clearly this is not the case (ie, equating attention to the modality, to perception of its content; O'Regan & Noe, 2001). There could, of course, be special grounds for assuming a dissociation between attention and conscious perception in olfaction. This possibility is suggested by the absence of a mandatory thalamic relay for olfaction (see Tham, Stevenson, & Miller, 2009), which implies that attentional processing may be different in this sense (see Mahmut & Stevenson, 2015), perhaps allowing content to be experienced in the absence of modality-specific attention. A further possibility here is that attentional capture by taste is not absolute, but only relative, leaving some limited attentional capacity for smell. One reason to suspect there may be such a limited attentional capacity derives from people who experience concurrent visual experiences when they smell things—odor–visual synesthetes (Russell, Stevenson, & Rich, 2015). Interestingly, such individuals *only* spontaneously report experiencing their visual experiences with orthonasal smells, but testing reveals that similar experiences can occur with retronasally presented odors. As attention seems to be necessary for the occurrence of synesthetic experiences (eg, Mattingley & Rich, 2004), this would suggest at least some attentional capacity to smells in the mouth—but probably less than at the nose (hence the greater capacity of sniffing to generate visual concurrents).

A second set of concerns are empirical. Ashkenazi and Marks (2004) examined participants' capacity to detect tastes and smells at threshold both with and without attention to a target modality. When a taste was placed in the mouth, selectively attending to the taste enhanced detection ability, but this did not occur when an odor was placed in the mouth and participants were asked to selectively attend this stimulus. However, when a taste and smell were placed in the mouth, attempts to selectively attend either to the taste or to the smell had no beneficial effect on detection performance, nor did it matter whether the taste and smell had been experienced together before or not. Troublingly for the taste capture model, attention to taste should not have been adversely affected by the presence of a smell. Moreover, the absence of a taste should have been (presumably) beneficial to detecting a smell in the mouth because there would be no attentional capture by the taste modality. Finally, while prior history of odor–taste co-occurrence could have explained the failure to detect a beneficial effect of focusing on taste, when a taste–smell mixture was present (with learning acting as the binding mechanism), whether the odor and taste had occurred together did not affect performance. While this is just one experiment, it is a particularly well-executed test of the capacity to attend to taste and smell in a flavor mixture and so its findings are worryingly at odds with predictions derived from the taste capture model (Stevenson, 2014).

3.2.2 A common attentional channel

A common attentional channel account owes its genesis to the ideas of Gibson (1966), who championed the idea of considering sensory systems from a functional perspective. On this basis, it may be useful to think of taste and smell in the mouth as a functional system, which is psychologically instantiated through a common attentional channel. Put more bluntly, in the mouth taste and smell function as a single sense. As noted earlier, this account can vary depending upon how hard-wired one considers this common channel to be. Arguably, a key source of evidence in determining whether or not it is hard-wired is to examine the impact of learning. One place this has been done is where a smell sniffed at the nose appears to be located in or nearer to the mouth when a taste is present—the localization illusion. If the common channel is hard-wired, any taste or smell combination should generate this effect to the same extent—that is, produce an illusion of localization to the mouth. Research findings here are divergent, possibly due to differing methods to assess localization. One group found no evidence that prior history of co-occurrence affected the localization illusion (Stevenson, Mahmut et al., 2011), while another group found that it did (Lim & Johnson, 2011). That prior co-occurrence is needed to generate a localization effect would suggest that one may have to learn each taste–odor pair to generate a unitary attentional channel. That no prior co-occurrence is required is more consistent with the idea that the channel is hard-wired and will operate whenever a taste and a smell are jointly present.

A further line of evidence comes from the study of Ashkenazi and Marks (2004). Recall they found that participants were not able to benefit from selectively attending to the taste or the smell, when an odor–taste mixture was present and that this effect occurred irrespective of whether the taste and smell had a history of prior co-occurrence. This would suggest that the capacity to attend to either element of the common channel is reduced irrespective of any prior learning. Currently, then, it is not possible to differentiate the learned from the hard-wired common channel account.

Irrespective of how the common attentional channel comes about, it requires some additional characteristic in order to explain the binding features documented in the first half of the chapter. First, it would, presumably, have to be initiated by taste. That is, a taste in the mouth would be needed to activate the channel, while a smell alone in the mouth would not. Empirically, a number of studies suggest this: (1) Ashkenazi and Marks (2004) found that while a taste alone in the mouth could be selectively attended, a smell alone in the mouth could not; (2) somatosensory stimuli alone do not generate the localization illusion—it requires a taste for this to occur; and (3) odors presented retronasally appear to markedly increase in perceived intensity when a small amount of tastant is added to them (Green et al., 2012).

As taste and smell share a common channel, it should be more difficult to attend to either of these modalities when both are present, as it is typically more difficult to attend to two events in one channel than to two events from different channels. The common channel may also be asymmetric, with more bandwidth (see Gallace, Ngo, Sulaitis, & Spence, 2012, for discussion of this concept) devoted to taste than to smell. This may be out of biological necessity, as many foods have multiple tastes, and detecting bitterness amongst them may be especially important in avoiding poisoning, just as detecting sweetness may signal nutrients. In contrast, olfactory perception is not an analytical sense (Stevenson & Wilson, 2007), rather what is learned is the pattern of neural activation associated with an external event (eg, the myriad volatiles that together make coffee). Thus it may make more functional sense to devote greater bandwidth to taste than to smell.

The common channel account can encompass many of the same phenomena as the taste capture account, whilst providing an arguably more satisfying explanation of the modality/content dissociation for smell. Here, taste and smell in the mouth *are* taste—a common sense—and so the modality here is functionally different from smell alone. Yet, the content is still available, and here at least the capacity to attend to it should be somewhat inferior to that of the nose alone, but better than if an odor is placed into the mouth in the absence of a taste. The latter reflects the need for the taste channel to be activated to allow perception of odor in the mouth and the former reflects the greater bandwidth available to taste over smell in the mouth. These broadly accord with the finding that olfactory ability in the mouth tends to be more constrained than at the nose (noting that various physiological and physiochemical factors may also contribute to this; eg, see Stevenson, 2009).

Localization of olfaction to the mouth can be accounted for through the preattentive mechanism for taste discussed earlier. This is because smell and taste in the mouth are *taste*, and taste is bound preattentively to the tongue by concurrent somatosensory stimulation. Similarly, the putative failure to notice shifts in olfactory intensity due to variations in delivery of odorant to the receptors might be masked by both the continuous presence of taste activity in the channel and by reduced olfactory bandwidth relative to taste. Finally, interactions between taste and smell can be seen as an integral part of this model under the "learned" variant or as an additional feature that *possibly* complements binding via the hard-wired channel. For the latter, odor–taste learning could act *solely* to benefit decisions to ingest, based upon smelling odors in the environment (ie, making food odors smell of their taste, without needing to sample them). Not only does this account for the general failure to find a strong link between the consequences of odor–taste learning and taste/odor enhancement effects (eg, Stevenson et al., 1995, 1998), it also suggests a rather attractive symmetry. When food odors are smelled at the nose they generate taste-like sensations that are perceived as being smells, while odors in the mouth are perceived as being tastes.

A further potential line of evidence for the common channel account comes from contemplating its neural underpinning. The anterior insula appears to be involved in the perception of odor (eg, Veldhuizen & Small, 2011). If the brain has dedicated processing space devoted to the common channel, loss of one of the input senses might be expected to gradually reduce its processing efficiency through partial disuse. This would then predict that loss of the sense of smell should start to affect gustatory performance. Of the four studies that have examined this possibility, three have found evidence that anosmia is associated with a somewhat poorer sense of taste (Gudziol, Rahneberg, & Burkert, 2007; Hummel et al., 2001; Landis et al., 2010; the exception being Stinton, Atif, Barkat, & Doty, 2010). This type of finding has also been observed in animal models (see Fortis-Santiago, Rodwin, Neseliler, Piette, & Katz, 2010). Perhaps, then, anosmic participants who present as losing both their sense of smell and taste are really reporting something more than described in the first part of this chapter.

3.3 The other senses

As noted earlier, the spatial and temporal separation of events prior to ingestion, with events in the mouth during a flavor experience, suggests that some form of strong binding (ie, unitization) is unlikely. Nonetheless, there are clearly strong bidirectional influences that need to be explained (eg, the effects of color on flavor perception; and of flavor perception on orthonasal smell). One obvious starting point is to assume that events prior to ingestion are a major factor in constructing flavor expectancies, which are then used to compare the flavor experience. This may be a much more powerful influence on perception than one might at first imagine, as suggested for example by the impact of coloring white wine red (Morrot, Brochet, & Dubourdieu, 2001). However, expectancy effects have their limits. Yeomans et al. (2008) found that a salmon-based mousse that *looked* like strawberry ice-cream evoked a markedly negative hedonic reaction when participants wrongly assumed it was strawberry ice-cream. There is an envelope in which the visual and other external attributes of a food or drink can affect flavor—assimilation—that might be considered a weak form of binding.

Finally, and as described earlier, learning in the mouth can affect both the sensory and affective perceptions of odors later smelled outside of the mouth, again providing information about the likely flavor/hedonic impact of the food (Stevenson &

Tomiczek, 2007). Learning might be considered as a means of bridging the spatial and temporal gap between events prior to a meal (sniffing) and flavor. This could be argued to be a form of binding, in that taste and smell are unitized.

3.4 Discussion

As with binding in other modalities, attention is a potentially important explanatory tool for flavor. The two attentional models discussed in this section remain poorly explored, even though certain lines of enquiry could arguably clarify which has the greater explanatory power.

4 Conclusions

This chapter started by describing the putative features of flavor binding, and then examined how these features might be best accounted for. Five different binding features were specified, and only one of these—odor–taste interactions—has been studied in any depth. Of the remainder, two—the suggested dissociation between modality and content for olfaction, and variation in olfactory content during the course of a mouthful of food—have received almost no attention, while there has been some limited work on the perception of parts and wholes and the localization of flavor to the mouth. A similar lack of work is reflected in both theory development and testing of accounts of flavor binding—with some notable exceptions (eg, Auvray & Spence, 2008; Rozin, 1982; Small & Prescott, 2005).

A key difference between flavor binding and binding observed in the other senses relates to function. For most examples outside of flavor, binding acts to facilitate identification and detection of external events (Calvert, Spence, & Stein, 2004). For flavor the functional benefits seem to be rather different. Foods are identified by vision and smell, well before they are placed in the mouth. The key importance of flavor is in *learning* associations between: (1) the immediate and delayed consequences of that food or drink and (2) its external attributes. Thus, seeing or smelling a particular food again can provide a good idea of the likely nutritional benefits and safety of consuming that food, without the risk of placing any of it in the mouth. This emphasizes the functional importance of learning in linking the flavor percept with its external attributes. There seem to be few functional benefits from identifying or recognizing food in the mouth when one can see and smell it beforehand, although a retained capacity to recognize gross deviations from expectation or to detect particular tastes is clearly important. Thus, just as Gibson (1966) suggested, function may dictate the properties of this perceptual system.

It should be clear from this chapter that much remains to be done. It is likely that the attentional theories examined here are deficient in several respects. They have not been subjected to detailed study, and indeed, there are no studies that have as yet set out expressly to test them. More broadly, little is known about many of the features of flavor binding. It is possible that there are additional features. It is also possible that some of the features, notably fluctuations in the delivery of the odorant to the receptors, may, when better measurement techniques are available, not be a "problem" after all. The great pleasure of working on flavor, apart of course from combining its study with the enjoyment of eating and drinking, is that so little is known about it. With overconsumption of palatable food and drink a significant public health issue, there is now a compelling practical reason to understand it too.

References

Ashkenazi, A., & Marks, L. E. (2004). Effect of endogenous attention on detection of weak gustatory and olfactory flavors. *Perception and Psychophysics, 66*, 596–608.
Auvray, M., & Spence, C. (2008). The multisensory perception of flavor. *Consciousness and Cognition, 17*, 1016–1031.
Bingham, A. F., Birch, G. G., de Graaf, C., Behan, J. M., & Perring, K. D. (1990). Sensory studies with sucrose-maltol mixtures. *Chemical Senses, 15*, 447–456.
Boakes, R. A., & Hemberger, H. (2012). Odour-modulation of taste ratings by chefs. *Food Quality and Preference, 25*, 81–86.
Brillat-Savarin, J. A. (1825/1949). The physiology of taste, or meditations on transcendental gastronomy. In M. Fisher (Ed.),. New York, NY: George Macy and Company.
Brunstrom, J. M. (2004). Does dietary learning occur outside awareness? *Consciousness and Cognition, 13*, 453–470.
Bull, T. (1965). Taste and the chorda tympani. *Journal of Laryngology and Otology, 79*, 479–493.
Bult, J. H. F., de Wijk, R. A., & Hummel, T. (2007). Investigations on multimodal sensory integration: Texture, taste, and ortho- and retronasal olfactory stimuli in concert. *Neuroscience Letters, 411*, 6–10.
Burdach, K., & Doty, R. L. (1987). The effects of mouth movements, swallowing and spitting on retronasal odor perception. *Physiology and Behavior, 41*, 353–356.
Calvert, G. A., Spence, C., & Stein, B. E. (Eds.). (2004). *The handbook of multisensory processes*. Cambridge, MA: The MIT Press.
Chale-Rush, A., Burgess, J. R., & Mattes, R. D. (2007). Multiple routes of chemosensitivity to free fatty acids in humans. *American Journal of Physiology—Gastrointestinal and Liver Physiology, 292*, G1206–G1212.
Clark, C. C., & Lawless, H. T. (1994). Limiting response alternatives in time-intensity scaling: An examination of the halo-dumping effect. *Chemical Senses, 19*, 583–594.
Dalton, P., Doolittle, N., Nagata, H., & Breslin, P. A. S. (2000). The merging of the senses: Integration of subthreshold taste and smell. *Nature Neuroscience, 3*, 431–432.
Davidson, J. M., Linforth, R. S. T., Hollowood, T. A., & Taylor, A. J. (1999). Effect of sucrose on the perceived flavor intensity of chewing gum. *Journal of Agricultural and Food Chemistry, 47*, 4336–4340.
Davis, R. G. (1977). Acquisition and retention of verbal associations to olfactory stimuli of varying familiarity and to abstract stimuli of varying similarity. *Journal of Experimental Psychology: Human Learning and Memory, 3*, 37–51.
Deems, D. A., Doty, R. L., Settle, R. G., Moore-Gillon, V., Shaman, P., Mester, A., et al. (1991). Smell and taste disorders, a study of 750 patients from the university of Pennsylvania smell and taste center. *Archives of Otolaryngology Head and Neck Surgery, 117*, 519–528.

Delwiche, J. (2004). The impact of perceptual interactions on perceived flavor. *Food Quality and Preference*, *15*, 137–146.

Eccleston, C., & Crombez, G. (1999). Pain demands attention: A cognitive-affective model of the interruptive function of pain. *Psychological Bulletin*, *125*, 356–366.

Fortis-Santiago, Y., Rodwin, B., Neseliler, S., Piette, C., & Katz, D. B. (2010). State dependence of olfactory perception as a function of taste cortical activation. *Nature Neuroscience*, *13*, 158–159.

Frank, R. A., & Byram, J. (1988). Taste-smell interactions are tastant and odorant dependent. *Chemical Senses*, *13*, 445–455.

Fujii, M., Fukazawa, K., Hashimoto, Y., Takayasu, S., Umemoto, M., Negoro, A., et al. (2004). Clinical studies of flavor disturbance. *Acta Otolaryngologica*, *533*(Suppl.), 109–112.

Gallace, A., Ngo, M. K., Sulaitis, J., & Spence, C. (2012). Multisensory presence in virtual reality: Possibilities & limitations. In G. Ghinea, F. Andres, & S. Gulliver (Eds.), *Multiple sensorial media advances and applications: New developments in MulSeMedia* (pp. 1–40). Hershey, PA: IGI Global.

Gent, J. F., Goodspeed, R. B., Zagraniski, R. T., & Catalanotto, F. A. (1987). Taste and smell problems: Validation of questions for the clinical history. *Yale Journal of Biology and Medicine*, *60*, 27–35.

Gibson, J. J. (1966). *The senses considered as perceptual systems*. London: Allen & Unwin.

Green, B. G. (2003). Studying taste as a cutaneous sense. *Food Quality and Preference*, *14*, 99–109.

Green, B. G., Nachtigal, D., Hammond, S., & Lim, J. (2012). Enhancement of retronasal odors by taste. *Chemical Senses*, *37*, 77–86.

Gudziol, H., Rahneberg, K., & Burkert, S. (2007). Anosmiker schmecken schlechter als gesunde [Anosmics are more poorly able to taste than normal persons]. *Laryngo-Rhino-Otologie*, *86*, 640–643.

Hodgson, M., Linforth, R. S. T., & Taylor, A. (2003). Simultaneous real-time measurements of mastication, swallowing, nasal airflow and aroma release. *Journal of Agricultural Food Chemistry*, *51*, 5052–5057.

Hummel, T., Nesztler, C., Kallert, S., Kobal, G., Bende, M., & Nordin, S. (2001). Gustatory sensitivity in patients with anosmia. *Chemical Senses*, *26*, 1118.

Hummel, T., & Nordin, S. (2005). Olfactory disorders and their consequences for quality of life. *Acta Oto-Laryngologica*, *125*, 116–121.

Keast, R., & Costanzo, A. (2015). Is fat the sixth taste primary? Evidence and implications. *Flavour*, *4*, 5.

Knudsen, E. (2007). Fundamental components of attention. *Annual Review of Neuroscience*, *30*, 57–78.

Koelewijn, T., Bronkhorst, A., & Theeuwes, J. (2010). Attention and the multiple stages of multisensory integration: A review of audiovisual studies. *Acta Psychologica*, *134*, 372–384.

Kveton, J., & Bartoshuk, L. (1994). The effect of unilateral chorda tympani damage on taste. *Laryngoscope*, *104*, 25–29.

Laing, D., Link, C., Jinks, A. L., & Hutchinson, I. (2002). The limited capacity of humans to identify the components of taste mixtures and taste-odour mixtures. *Perception*, *31*, 617–635.

Landis, B., Scheibe, M., Weber, C., Berger, R., Bramerson, A., Bende, M., et al. (2010). Chemosensory interaction: Acquired olfactory impairment is associated with decreased taste function. *Journal of Neurology*, *257*, 1303–1308.

Lawless, H. T. (1995). Flavor. In M. Friedman & E. Carterette (Eds.), *Handbook of perception and cognition. Volume 16, Cognitive ecology* (pp. 325–380). San Diego, CA: Academic Press.

Lehman, C., Bartoshuk, L., Catalanotto, F., Kveton, J., & Lowlicht, R. (1995). Effect of anesthesia of the chorda tympani nerve on taste perception in humans. *Physiology and Behavior, 57*, 943–951.

Lim, J., & Johnson, M. B. (2011). Potential mechanisms of retronasal odor referral to the mouth. *Chemical Senses, 36*, 283–289.

Mahmut, M., & Stevenson, R. J. (2015). Failure to obtain reinstatement of an olfactory representation. *Cognitive Science, 30*, 1940–1949.

Mainland, J., & Sobel, N. (2006). The sniff is part of the olfactory percept. *Chemical Senses, 31*, 181–196.

Marshall, K., Laing, D. G., Jinks, A. L., & Hutchinson, I. (2006). The capacity of humans to identify components in complex odor-taste mixtures. *Chemical Senses, 31*, 539–545.

Mattingley, J. B., & Rich, A. N. (2004). Behavioural and brain correlates of multisensory experience in synesthesia. In G. A. Calvert, C. Spence, & B. E. Stein (Eds.), *Handbook of multisensory processing* (pp. 851–866). Cambridge, MA: MIT Press.

McBurney, D. H. (1986). Taste, smell, and flavor terminology: Taking the confusion out of fusion. In H. Meiselman & R. Rivlin (Eds.), *Clinical measurement of taste and smell* (pp. 117–125). New York, NY: Macmillan.

Morrot, G., Brochet, F., & Dubourdieu, D. (2001). The color of odors. *Brain and Language, 79*, 309–320.

Mozell, M., Smith, B., Smith, P., Sullivan, R., & Swender, P. (1969). Nasal chemoreception in flavor identification. *Archives of Otolarynogology, 90*, 367–373.

Munoz, A. M., & Civille, G. V. (1987). Factors affecting perception and acceptance of food texture by American consumers. *Food Reviews International, 3*, 285–322.

Murphy, C., Cain, W. S., & Bartoshuk, L. M. (1977). Mutual action of taste and olfaction. *Sensory Processes, 1*, 204–211.

Navarra, J., Alsius, A., Soto-Faraco, S., & Spence, C. (2010). Assessing the role of attention in the audio-visual integration of speech. *Information Fusion, 11*, 4–11.

Oakley, B. (1985). Taste responses of human chorda tympani nerve. *Chemical Senses, 10*, 469–481.

Ogawa, H., Ito, S., Murayama, N., & Hasegawa, K. (1990). Taste area in granular and dysgranular insular cortices in the rat identified by stimulation of the entire oral cavity. *Neuroscience Research, 9*, 196–201.

O'Regan, J. K., & Noe, A. (2001). A sensorimotor account of vision and visual consciousness. *Behavioral and Brain Sciences, 24*, 939–1031.

Plata Salaman, C. R., Scott, T. R., & Smith-Swintosky, V. L. (1993). Gustatory neural coding in the monkey cortex: The quality of sweetness. *Journal of Neurophysiology, 69*, 482–493.

Porter, S. R., Fedele, S., & Habbab, K. M. (2010). Taste dysfunction in head and neck malignancy. *Oral Oncology, 46*, 457–459.

Prescott, J. (1999). Flavour as a psychological construct: Implications for perceiving and measuring the sensory qualities of foods. *Food Quality and Preference, 10*, 349–356.

Prescott, J., & Murphy, S. (2009). Inhibition of evaluative and perceptual odour-taste learning by attention to the stimulus elements. *Quarterly Journal of Experimental Psychology, 62*, 2133–2140.

Rozin, P. (1982). "Taste-smell confusions" and the duality of the olfactory sense. *Perception and Psychophysics, 31*, 397–401.

Russell, A., Stevenson, R. J., & Rich, A. N. (2015). Chocolate smells pink and stripy: Exploring olfactory-visual synesthesia. *Cognitive Neuroscience, 6*, 77–88.

Sakai, N., Kobayakawa, T., Gotow, N., Saito, S., & Imada, S. (2001). Enhancement of sweetness ratings of aspartame by a vanilla odor presented either by orthonasal or retronasal routes. *Perception and Motor Skills, 92*, 1002–1008.

Schifferstein, H., & Verlegh, P. (1996). The role of congruency and pleasantness in odor-induced taste enhancement. *Acta Psychologica, 94*, 87–105.

Scott, T. R., & Mark, G. P. (1987). The taste system encodes stimulus toxicity. *Brain Research, 414*, 197–203.

Sela, L., & Sobel, N. (2010). Human olfaction: A constant state of change-blindness. *Experimental Brain Research, 205*, 13–29.

Shanks, D. R. (2010). Learning: From association to cognition. *Annual Review of Psychology, 61*, 273–301.

Shermer, D., & Levitan, C. (2014). Red hot: The crossmodal effect of color intensity on piquancy. *Multisensory Research, 27*, 207–223.

Small, D. M., & Prescott, J. (2005). Odor/taste integration and the perception of flavor. *Experimental Brain Research, 166*, 345–357.

Sobel, N., Prabhakaran, V., Desmond, J., Glover, G., Goode, R., Sullivan, E., et al. (1998). Sniffing and smelling: Separate subsystems in the human olfactory cortex. *Nature, 392*, 282–286.

Spence, C. (2015). On the psychological impact of food colour. *Flavour, 4*, 21.

Spence, C., Levitan, C., Shankar, M., & Zampini, M. (2010). Does food color influence taste and flavor perception in humans? *Chemosensory Perception, 3*, 68–84.

Spence, C., Nicholls, M. E. R., & Driver, J. (2001). The cost of expecting events in the wrong sensory modality. *Perception & Psychophysics, 63*, 330–336.

Spence, C., & Piqueras-Fiszman, B. (2014). *The perfect meal: The multisensory science of food and dining*. Oxford: Wiley-Blackwell.

Steiner, J. E., Glaser, D., Hawilo, M. E., & Berridge, K. C. (2001). Comparative expression of hedonic impact: Affective reactions to taste by human infants and other primates. *Neuroscience and Biobehavioral Reviews, 25*, 53–74.

Stevenson, R. J. (2009). *The psychology of flavour*. Oxford: OUP.

Stevenson, R. J. (2012). The role of attention in flavour perception. *Flavour, 1*, 2.

Stevenson, R. J. (2014). Flavor binding: Its nature and cause. *Psychological Bulletin, 140*, 487–510.

Stevenson, R. J., Boakes, R. A., & Prescott, J. (1998). Changes in odor sweetness resulting from implicit learning of a simultaneous odor-sweetness association: An example of learned synesthesia. *Learning and Motivation, 29*, 113–132.

Stevenson, R. J., Boakes, R. A., & Wilson, J. P. (2000). The persistence of conditioned odor perceptions: Evaluative conditioning is not unique. *Journal of Experimental Psychology: Learning Memory and Cognition, 26*, 423–440.

Stevenson, R. J., & Case, T. I. (2003). Preexposure to the stimulus elements, but not training to detect them, retards human odour-taste learning. *Behavioural Processes, 61*, 13–25.

Stevenson, R. J., & Mahmut, M. (2011a). Discriminating the stimulus elements during human odor-taste learning: A successful analytic stance does not eliminate learning. *Journal of Experimental Psychology: Animal Behavior Processes, 37*, 477–482.

Stevenson, R. J., & Mahmut, M. (2011b). Olfactory test performance and its relationship with the perceived location of odors. *Attention, Perception and Psychophysics, 73*, 1966–1973.

Stevenson, R. J., Mahmut, M., & Oaten, M. (2011). The role of attention in the localisation of odors to the mouth. *Attention, Perception and Psychophysics, 73*, 247–258.

Stevenson, R. J., Oaten, M., & Mahmut, M. (2011). The role of taste and oral somatosensation in olfactory localisation. *Quarterly Journal of Experimental Psychology, 64*, 224–240.

Stevenson, R. J., Prescott, J., & Boakes, R. A. (1995). The acquisition of taste properties by odors. *Learning and Motivation, 26*, 433–455.

Stevenson, R. J., Prescott, J., & Boakes, R. A. (1999). Confusing tastes and smells: How odours can influence the perception of sweet and sour tastes. *Chemical Senses, 24*, 627–635.
Stevenson, R. J., & Tomiczek, C. (2007). Olfactory-induced synesthesias: A review and model. *Psychological Bulletin, 133*, 294–309.
Stevenson, R. J., & Wilson, D. (2007). Olfactory perception: An object recognition approach. *Perception, 36*, 1821–1833.
Stinton, N., Atif, M., Barkat, N., & Doty, R. L. (2010). Influence of smell loss on taste function. *Behavioral Neuroscience, 124*, 256–264.
Sun, B. C., & Halpern, B. P. (2005). Identification of air phase retronasal and orthonasal odorant pairs. *Chemical Senses, 30*, 693–706.
Talsma, D., Senkowski, D., Soto-Faraco, S., & Woldorff, M. G. (2010). The multifaceted interplay between attention and multisensory integration. *Trends in Cognitive Sciences, 14*, 400–410.
Tham, W., Stevenson, R. J., & Miller, L. A. (2009). The functional role of the medio dorsal thalamic nucleus in olfaction. *Brain Research Reviews, 62*, 109–126.
Titchener, E. B. (1909). *A text-book of psychology. Part 1*. New York, NY: Macmillan.
Todrank, J., & Bartoshuk, L. M. (1991). A taste illusion: Taste sensation localized by touch. *Physiology and Behavior, 50*, 1027–1031.
Treisman, A., & Gelade, G. (1980). A feature-integration theory of attention. *Cognitive Psychology, 12*, 97–136.
Trelea, I. C., Atlan, S., Deleris, I., Saint-Eve, A., Marin, M., & Souchon, I. (2008). Mechanistic mathematical model for in vivo aroma release during eating of semiliquid foods. *Chemical Senses, 33*, 181–192.
Van Damme, S., Crombez, G., Eccleston, C., & Goubert, L. (2004). Impaired disengagement from threatening cues of impending pain in a crossmodal cueing paradigm. *European Journal of Pain, 8*, 227–236.
Van Damme, S., Crombez, G., & Spence, C. (2009). Is the visual dominance effect modulated by the threat value of visual and auditory stimuli? *Experimental Brain Research, 193*, 197–204.
Veldhuizen, M., Shephard, T. G., Wang, M.-F., & Marks, L. E. (2010). Coactivation of gustatory and olfactory signals in flavor perception. *Chemical Senses, 35*, 121–133.
Veldhuizen, M., & Small, D. M. (2011). Modality specific neural effects of selective attention to taste and odor. *Chemical Senses, 36*, 747–760.
von Sydow, E., Moskowitz, H., Jacobs, H., & Meiselman, H. (1974). Odor-taste interactions in fruit juices. *Lebensmittel-Wissenschaft & Technologie, 7*, 18–24.
Weenen, H., Jellema, R. H., & de Wijk, R. A. (2005). Sensory sub-attributes of creamy mouthfeel in commercial mayonnaises, custard desserts and sauces. *Food Quality and Preference, 16*, 163–170.
Whitehead, M. C., & Kachele, D. L. (1994). Development of fungiform papillae, taste-buds, and their innervation in the hamster. *Journal of Comparative Psychology, 340*, 515–530.
Yeomans, M. R., Chambers, L., Blumenthal, H., & Blake, A. (2008). The role of expectancy in sensory and hedonic evaluation: The case of smoked salmon ice-cream. *Food Quality and Preference, 19*, 565–573.

Oral Referral

Juyun Lim
Department of Food Science and Technology, Oregon State University, Corvallis, OR, United States

3

1 Introduction

Several years ago, I was sitting in a local Thai restaurant in New Haven, Connecticut, with a close colleague and mentor. We were talking about the science (of chemical senses, mainly gustation back then), when our yellow curry dish arrived. The curry had a bright yellow color with clearly visible fresh red and green vegetables on top. It smelled appetizing for sure. When I took the first bite, I noticed the warmth and spiciness as well as the creamy and crunchy textures right away. My colleague asked what the curry tasted like to me. I told him about those sensations. But then, "Is that it?" he said. "I don't know; it tastes like a yellow curry." Of course, my answer did not satisfy him (nor myself). So we had a long conversation about the flavor of foods and our sensory systems, which only helped us realize that we didn't know much about how a flavor percept is formed and perceived.

Yellow curry, just like any other dish, consists of many ingredients: cumin, coriander, turmeric, bay leaf, lemongrass, fish sauce, sugar, garlic, ginger, cayenne pepper, and coconut milk. The list is long, even without including meats or vegetables that are commonly added. Anyone who has eaten yellow curry knows how it tastes, but describing its flavor to someone else who has not tasted it is not an easy task. Why this should be the case is an interesting question. What is even more interesting is the fact that we always talk about its taste, even after knowing that the key ingredients of yellow curry are mostly aromatic herbs and spices. Why we do is definitely worth looking into.

In order to answer these questions, I start this chapter by describing the basics of the sensory modalities and systems that are directly related to flavor perception, and then discuss how the sensory information is processed and merged in the central nervous system along with perceptual evidence for such integration including the oral referral of flavor components. I then present the illustrations of oral referral followed by their underlying conditions and mechanisms supported by perceptual and neural evidence.

2 Conceptual background

A good starting point is to consider what constitutes a flavor or a flavor percept. As discussed in the other chapters in this volume (see chapters: Food Color and Its Impact on Taste/Flavor Perception and Color and Flavor, in particular), there is no doubt that the appearance, sound, and smell of a food, even before placing it in the mouth (see Spence, 2013b), affect its flavor perception. Accordingly, some scientists regard all sensory inputs related to a food as flavor components (eg, Auvray & Spence, 2008; Verhagen & Engelen, 2006). In contrast, in this chapter, only those modalities that are evoked from food in the mouth are considered flavor components (see also Small, 2012); more specifically, those include taste, retronasal odor, and oral-somatosensation. This definition is based on whether the absence of a component could change the perception of a flavor and thereby disrupt its recognition and/or identification.

2.1 Flavor modalities

2.1.1 Gustation

When food is placed in the mouth and masticated, the matrix of food is broken down, and tastants, odorants, and chemical irritants are released and mixed with saliva (Noble, 1996). Among these, the tastants are penetrated into the taste buds, and taste transduction is mediated by collections of specialized epithelial cells (see Liman, Zhang, & Montell, 2014, for details) that are arrayed in taste buds. The resulting taste qualities include the sensations of sweet, salty, umami, sour, and bitter, and potentially others. The notion that each taste quality signals the presence of potential nutrients or poisons is well accepted. In particular, sweetness signals the presence of sugars, saltiness signals electrolytes, umami signals amino acids, sourness signals organic acids or low pH, and bitterness signals potential toxins (Scott & Plata-Salaman, 1991). Recent evidence also suggests that humans can detect hydrolysis products of fat (see Gaillard et al., 2008; Gilbertson, 1998; Running, Craig, & Mattes, 2015) and complex carbohydrates (see Lapis, Penner, & Lim, 2014; Lim, Balto, Lapis, & Penner, 2015; Sclafani, 2004).

2.1.2 Olfaction

When the breakdown of food is mixed with saliva, volatiles are released and distributed to the headspace in the mouth (Noble, 1996). During exhalation, the volatiles are transported by the current of air to the receptors on the olfactory epithelium via the nasopharynx (see Fig. 3.1). This type of smelling is referred to as retronasal olfaction, in contrast to orthonasal olfaction, which occurs during inhalation through the nose. While taste plays an important role in providing information regarding the nutritive values of food, olfactory inputs carry information regarding the identity of a flavor. For example, blends of volatiles, such as particular esters, terpenes, and furans, together make up the characteristics of strawberry (Schwieterman et al., 2014).

2.1.3 Somesthesis

Oral somesthesis provides information about the physicochemical makeup of food and substances taken into the mouth. While tastants in food substances are sensed by the gustatory system, tactile properties of foods are detected by the somatosensory system. Somatosensory nerves contain three different classes of sensory receptors

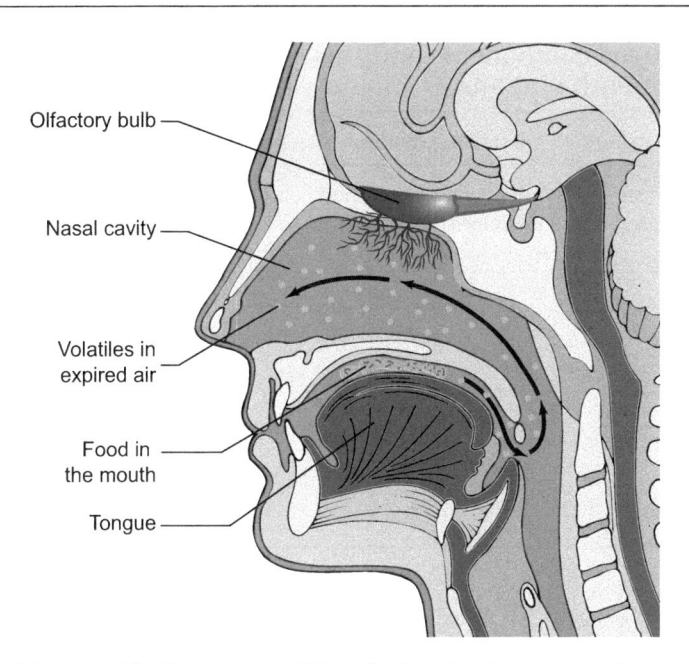

Figure 3.1 Diagram of the flavor system. When food is taken into the mouth, taste and tactile sensations are perceived in the mouth while retronasal smell and tactile sensations produced by volatile chemicals are perceived in the nose.

(see Table 3.1): mechanoreceptors, thermoreceptors, and nociceptors, which are responsible for the perception of touch, temperature, and pain, respectively (Bryant & Silver, 2000). Notably, some of these receptors also respond to chemical irritants such as capsaicin and menthol, which can elicit some of the same sensations as temperature and painfully intense stimuli (eg, cold, heat, burning, pricking; Green, 2002). When such sensations are evoked by chemicals, they are referred to as chemesthetic sensations (Green, Mason, & Kare, 1990). This means that chemesthesis is not a separate sense like touch, temperature, and pain, but instead derived from the chemical sensitivity of all three senses. Most chemesthetic sensations are mediated by chemically sensitive thermoreceptors and nociceptors (Green, 1996; Viana, 2011), but certain chemicals (eg, hydroxyl-α-sanshool in Szechuan peppers) can stimulate mechanoreceptors that serve the sense of touch, resulting in unusual sensations such as tingling or numbing (Bryant & Mezine, 1999; Hagura, Barber, & Haggard, 2013).

Sensory receptors of the trigeminal nerve innervate the skin and mucosal membranes of not only the mouth, but also the nose (as well as the eye and the respiratory tract). Thus, somesthetic sensations can be perceived in the nose as well. In particular, volatile chemical irritants from substances in the mouth can cause chemesthetic sensations in the nose area (eg, think only of the pungency; see Table 3.1). However, physical temperature and touch in the nose are not considered constitutes of flavor perception (ie, the texture and temperature of foods are perceived only in the mouth).

Table 3.1 **Flavor components and their relevant receptors in the organs of mouth and nose**

	Gustation	Olfaction	Somesthesis/chemesthesis		
Receptors **Organs**	Taste receptors	Olfactory receptors	Mechanoreceptors	Thermoreceptors	Nociceptors
Mouth	Tastants	–	Texture, pressure, and vibration	Temperature and chemical irritants	Intense pressure, temperature, and chemical irritants
Nose	–	Odorants	–	Volatile chemical irritants	Volatile chemical irritants

2.2 Pathways and convergence of sensory inputs

For much of the 20th century, research on chemosensory perception considered taste, olfaction, and somesthesis as independent senses due to their distinct receptor mechanisms (see Table 3.1). However, more recent evidence supports the idea that those senses interact within a larger, oro-nasal perceptual system (Gibson, 1966), that is, the flavor system. More specifically, multisensory interaction and integration seems to occur when taste, retronasal smell, and somatosensory inputs convey related information regarding the same object in the mouth. Note that it has also been shown that integration can occur between taste and orthonasal smell when they convey information about a known percept (see Section 2.3.2). One obvious question is then at what level do these three inputs converge? Note the brief summary below is based on human and nonhuman primate data whenever available.

As much as taste and tactile sensations are mediated by two physiologically independent sensory systems, taste and tactile sensations in the mouth are rather intimately related to one another. Most notably, taste stimulation almost always accompanies tactile stimulation under normal circumstances. In addition, evidence suggests that taste and oral somatosensation, in particular touch and temperature, are integrated at virtually every level of sensory processing (see Green, 2002; Small, 2012, for details). The taste signals produced in taste receptor cells are carried by branches of the facial (VII), glossopharyngeal (IX), or vagus (X) nerves. In the meantime, the somatosensory inputs in the oral and nasal cavities are carried by the trigeminal nerve (V). Importantly, cranial nerves IX and X have not only taste-specific but also general sensory neurons, which contain mechano-, thermo-, and nociceptors mentioned above. Moreover, somatosensory fibers of the trigeminal (V) and glossopharyngeal nerves (IX) also innervate taste buds in humans (Gairns, 1952). This may explain, at least partially, why some tastants can evoke chemesthetic sensations (Dessirier, O'Mahony, Iodi-Carstens,

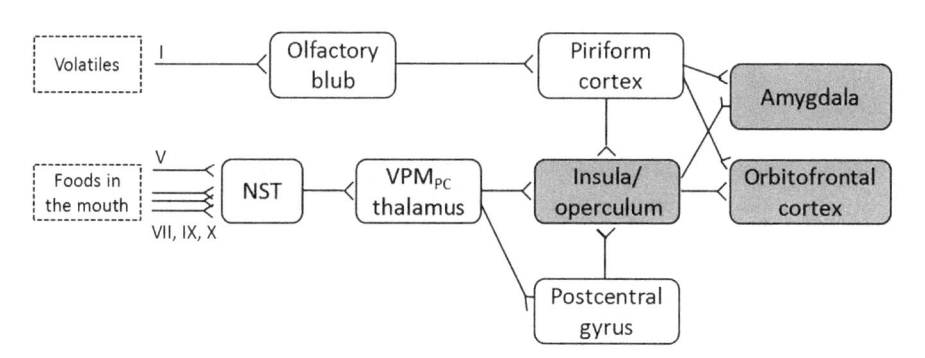

```
┌──────────┐    ┌──────────┐        ┌──────────┐
│Volatiles │─I─<│Olfactory │────────│ Piriform │
└──────────┘    │   blub   │        │  cortex  │>──┐────┐
                └──────────┘        └──────────┘   │ ┌──────────┐
                                                   ├─│ Amygdala │
           V                        ┌──────────┐   │ └──────────┘
┌──────────┐  ┌───┐  ┌──────────┐   │ Insula/  │   │ ┌──────────┐
│ Foods in │══│NST│─<│  VPM_PC  │──<│operculum │>──┴─│Orbitofr. │
│the mouth │  └───┘  │ thalamus │   └──────────┘     │  cortex  │
└──────────┘         └──────────┘        │           └──────────┘
         VII, IX, X                      ▼
                                    ┌──────────┐
                                    │Postcentral│
                                    │  gyrus   │
                                    └──────────┘
```

Figure 3.2 Diagram of oronasal sensory pathways. Information on taste properties of food in the mouth is conveyed via facial (VII), glossopharyngeal (IX), and vagus (X) nerves to the nucleus of the solitary tract (NST) in the brainstem, which projects to the parvicellular part of the ventroposterior medial nucleus of the thalamus (VPM$_{PC}$). Thalamic afferents then project to the anterior insular and operculum areas (ie, the primary gustatory cortex). Information regarding the physicochemical properties of food in the mouth as well as the chemical irritants perceived in the nose is carried by the trigeminal nerve (V) and general sensory neurons of cranial nerves IX and X to the NST, which then also projects to VPM$_{PC}$. Somatosensory information is then relayed to the opercular region of the postcentral gyrus (ie, the somatosensory cortex), which sends a projection to the primary gustatory cortex. The primary gustatory cortex then projects to the central nucleus of the amygdala and also to the secondary taste cortex (ie, orbitofrontal cortex). Olfactory information is conveyed via cranial nerve I to the olfactory bulb, which projects to the primary olfactory cortex, which includes the piriform cortex. The piriform cortex makes the second synapses to a number of cortical areas including the insular and operculum complex, amygdala, and orbitofrontal cortex. Thus, taste and somatosensory information converge at almost every level of sensory processing (light and dark gray boxes). In contrast, olfactory signals converge on to taste and somatosensory inputs in higher-order cortical regions (dark gray boxes).

Yao, & Carstens, 2001; Gilmore & Green, 1993; Green & Gelhard, 1989) and some chemical irritants, such as capsaicin and menthol, can evoke a bitter taste (Green & Schullery, 2003; Lim & Green, 2007). Evidence also suggests that heating or cooling small areas of the tongue can cause taste sensations (Cruz & Green, 2000).

The taste signals are first sent to the nucleus of the solitary tract (NST) in the brainstem, while the somatosensory inputs are carried to the spinal trigeminal nucleus, which projects to the NST (see Fig. 3.2; Verhagen, & Engelen, 2006). While the precise locations of the trigeminal projections seem to vary across species (Small, 2012), there is evidence that such projections overlap with gustatory areas of the NST (King, Travers, Rowland, Garcea, & Spector, 1999; Scott, Yaxley, Sienkiewicz, & Rolls, 1986; South & Ritter, 1986; Whitehead, 1990). In primates, afferents conveying taste and oral somatosensory information travel from the NST to the parvicellular part of the ventroposterior medial nucleus of the thalamus (VPM$_{PC}$) (Pritchard, Hamilton, & Norgren, 1989). Thus, taste and somatosensory inputs seem to be integrated in the thalamus as well. From the VPM$_{PC}$, taste information projects to the primary gustatory cortex (ie, the anterior insular and operculum areas), which, in turn, project to the orbitofrontal cortex (Verhagen & Engelen, 2006), the secondary gustatory cortex. Somatosensory information is then sent to the primary somatosensory cortex (ie, the opercular region of the postcentral gyrus), which projects to all regions of the primary gustatory cortex (eg, Cerf-Ducastel, Van De Moortele, Macleod, Le Bihan, & Faurion, 2001; de Araujo & Rolls, 2004; Guest et al., 2007; Kadohisa, Rolls, & Verhagen, 2004; Pritchard, Hamilton, Morse, & Norgren, 1986; Rudenga, Green, Nachtigal, & Small, 2010). Interestingly, the primary gustatory cortex contains unimodal neurons that are either taste- or somatosensory-specific, and also bimodal neurons responding to both somatosensory and taste stimulation (Cerf-Ducastel et al., 2001; Smith-Swintosky, Plata-Salaman, & Scott, 1991).

Unlike taste and oral somatosensory inputs that are integrated at such early stages of neural coding, olfactory signals do not converge on to taste and oral somatosensory inputs until higher-order cortical regions (Small, 2012). Olfactory signals are initially carried through cranial nerve I (ie, the olfactory nerve) to the olfactory bulb, which projects to higher olfactory centers where they make the second synapse in the pathway (see Fig. 3.2). These centers include a number of cortical areas, such as the piriform cortex, insula, and orbitofrontal cortex (Christensen & White, 2000; Verhagen & Engelen, 2006). Specifically, the evidence suggests that responses to taste and retronasally perceived olfactory stimulation not only overlap (Cerf-Ducastel & Murphy, 2001; de Araujo, Rolls, Kringelbach, Mcglone, & Phillips, 2003; Veldhuizen, Nachtigal, Teulings, Gitelman, & Small, 2010) but also show supra-additivity (Small et al., 2004) in the anterior ventral insula. Overlap was also observed in the amygdala, caudal orbitofrontal cortex, and far posterior medial orbitofrontal cortex (de Araujo et al., 2003; Rolls, 1997; Rolls & Baylis, 1994). Combined, it is clear that as much as the three sensory systems are autonomous in terms of their transduction mechanisms, postsynaptic currents evoked by different modalities converge at some points of the neural coding. Considering neural evidence of multimodal convergence, it would also be of interest to consider perceptual evidence of multimodal interactions before discussing how flavor percept is created in the brain (Small, 2012) and relayed to the mouth.

2.3 Perceptual evidence of multisensory interaction and integration

There have been growing indications that sensations of taste, smell, and oral somesthesis interact with one another and also are integrated at the perceptual level. Accordingly, there are excellent systematic reviews on psychophysical evidence of multisensory interaction and integration (see Delwiche, 2004; Stevenson, 2009; Verhagen & Engelen, 2006).

2.3.1 Multisensory interactions

When two or more stimuli above threshold levels are mixed together, the intensity can be less than or more than the sum of the individual components. These kinds of phenomena, called mixture suppression and enhancement, have been studied extensively within the taste modality (see Keast & Breslin, 2002, for a review). While there are exceptions and complications, when two compounds eliciting a similar taste quality are mixed together, enhancement seems to occur especially at low intensities. When two compounds with different qualities are mixed, the outcome seems to be harder to generalize. However, interactions between bitter and sweet (Bonnans & Noble, 1993) and also sour and sweet (Bonnans & Noble, 1993; McBurney & Bartoshuk, 1973) are known to be mutually suppressive, especially at medium and high intensities. While some of these effects can be peripheral in origin (eg, umami tastes of MSG and disodium 5′-inosinate; Yamaguchi, 1967), some others are a central cognitive effect. Kroeze and Bartoshuk (1985) showed that sweet and bitter tastes can suppress each other, whether the stimuli were applied independently to either side of the tongue, or together as a mixture.

Studies have shown that enhancement and suppression effects occur between tastes and retronasally perceived odors as well. While there are some contradictions and disagreements in the conditions under which taste and odor enhancement occur as well as the nature of such phenomena (reviewed in Green, Nachtigal, Hammond, & Lim, 2012; Linscott & Lim, 2016), it is generally agreed that taste and odor enhancement occur when taste and odor stimuli are congruent (eg, vanilla and sweetness; Frank & Byram, 1988; Kuo, Pangborn, & Noble, 1993; Lim, Fujimaru, & Linscott, 2014; Linscott & Lim, 2016; Schifferstein & Verlegh, 1996). Suppression between taste and retronasal odor has received less attention. However, they have been seen as well (Stevenson, Prescott, & Boakes, 1999); for example, the intensity of vanilla odor can be suppressed by sour and salty taste (Kuo et al., 1993), while the intensity of bitter taste can be suppressed by chicken or soy sauce odor (Linscott & Lim, 2016). It has also been shown that the burning sensation of capsaicin can reduce the perceived sweetness of sucrose (Prescott, Allen, & Stephens, 1984; Prescott & Stevenson, 1995), but not saltiness of sodium chloride nor sourness of citric acid (Cowart, 1987; Prescott et al., 1984; Prescott & Stevenson, 1995). More work is needed to determine the generality of enhancement and suppression between those modalities contributing to flavor perception. However, the findings of the studies that have been reported thus far suggest that flavor modalities can be functionally similar in that perceptually congruent qualities seem to enhance (especially when the component intensities are weak), while the opposing (eg, beneficial vs dangerous) qualities seem to suppress each other.

Further evidence of multisensory interactions at the perceptual level concerns quality association. It is commonly observed that tasteless odor stimuli elicit taste-like sensations. Or rather, odor sensations can possess taste-like qualities (eg, isoamyl acetate, the primary aroma compound of banana, being "sweet" smelling; Burdach, Kroeze, & Koster, 1984; Frank & Byram, 1988; Stevenson et al., 1999). This phenomenon of quality association (or quality illusion) has been explained as the outcome of associative learning; when relatively novel odors were repeatedly paired with either sweet or sour tastes, sucrose-paired odors smelled sweeter and citric-acid-paired odors smelled more sour (Stevenson, Boakes, & Prescott, 1998; Stevenson, Prescott, & Boakes, 1995). The attribution of taste quality to odor by associative learning has been demonstrated at the neural level as well (see Rolls, 2011); for example, some of the odor-responsive single neurons in the orbitofrontal cortex and surrounding areas are responsive to associated taste stimuli after taste–odor associative learning (eg, Critchley & Rolls, 1996; Rolls, Critchley, Mason, & Wakeman, 1996). Rolls (2011) explained that bimodal taste/odor neurons may be developed from olfactory unimodal neurons through repeated exposures to particular tastes with odors. In turn, it is possible that an odor, even in the absence of the related taste component, could weakly simulate a neuron that responds optimally to the taste–odor combination (Rolls & Baylis, 1994).

2.3.2 Multisensory integration

One of the most important functions of the flavor system is to detect and identify known, safe food substances. The flavor modalities seem to work together to perform this critical task. When a subthreshold concentration of a sweet-tasting compound (sodium saccharin) was presented together with a subthreshold concentration of a cherry/almond-like odorant (benzaldehyde), participants were able to detect the combination (Dalton, Doolittle, Nagata, & Breslin, 2000). Such cross-modal summation of subthreshold stimuli, however, did not occur when benzaldehyde was presented with monosodium glutamate, an umami compound (Breslin, Doolittle, & Dalton, 2001). These two studies clearly demonstrate that two extremely weak signals from different modalities can be combined when they usually contribute together to a known percept (ie, sweet-tasting cherry). Unfortunately, no study so far has attempted to demonstrate such cross-modal summation between gustation/olfaction and somesthesis.

Further evidence of multisensory integration is a more obvious one, at least in daily life. When we place a food substance in the mouth, we do not necessarily perceive an array of sensations (until someone starts asking about its elements; see Chapter 4 in Stevenson, 2009 for details). Rather, we perceive an emergent percept (like yellow curry!). This, so-called "quality fusion," occurs when two or more sensory inputs are "bound" (see chapter: Attention and Flavor Binding for details) to form a perceptual unit of Gestalt. Rozin is one of the pioneers who discussed quality fusion. In his argument that olfaction is the only dual-sensory modality, Rozin (1982, p. 397) pointed out several relevant lines of evidence supporting the existence of taste–odor illusions. First, retronasal olfaction is frequently confused with taste: "The always surprising loss of ability to 'taste' or discriminate common foods when olfactory receptors are blocked by a head cold is everyday evidence for this." Second, the retronasal olfactory sensations which are associated with substances in the mouth are referred to the mouth. In addition, (as a consequence) people are largely unaware of the distinction between the terms taste and flavor, and commonly use the term "taste" to describe their experience of flavor. Such perceptual and behavioral evidence suggests that multisensory inputs are integrated and that the resulting percept is localized in the mouth. These phenomena of quality/spatial fusion will be discussed in the next section.

3 Oral referral

When sensory inputs of different spatial origin present related information simultaneously, quality/spatial fusion occurs. The ventriloquist effect, which we experience regularly when watching television or movie, is a well-known example; the voices seem to emanate from the actors' lips rather than the actual source of sound, that is, speakers (Alais & Burr, 2004). Such "visual capture" has been explained in terms of visual information being more spatially precise and thus dominating the perceived location of the multisensory event (Bonath et al., 2007). Similar phenomena also occur in flavor perception.

3.1 Illustrations of oral referral

3.1.1 Taste referral

Taste receptors are located in discrete regions of the tongue: fungiform papillae on the front and edges, foliate papillae on the sides, and circumvallate papillae on the back of the tongue. The middle surface of the tongue is covered with filliform papillae, which do not contain taste buds. Yet, we perceive taste as if it arises throughout the mouth. In order to understand the sensory mechanism underlying this taste "illusion," Todrank and Bartoshuk (1991) performed one of the first formal experiments. They had their participants rate the perceived intensity of taste as a taste solution was applied on one side of the tongue (an area of low receptor density) and then moved it past the tip (an area of high receptor density) to the opposite side of the tongue. The taste sensation reported was weak on the first side, became stronger at the tip and maintained its intensity on the other side. This finding was interpreted as indicating that the strong taste from the tip followed the tactile path of the stimulus sweep, referring the intensity from the tip to the less sensitive area of the second side. The ability of touch to "capture" taste sensation has also been demonstrated in a study by Delwiche, Lera, and Breslin (2000). They provided four gelatin cubes of the same size ($1 \, cm^3$) in a cup: one sweetened cube and three tasteless cubes, or three sweetened cubes and one tasteless cube. The participants were asked to place all four cubes in the mouth, taste them simultaneously, and identify a target cube by selectively spitting it out. The task proved significantly more difficult when the participants had to search for a tasteless cube from amongst three sweetened cubes. The authors speculated that tactile capture of taste might have occurred for the tasteless target cube, which then interfered with the localization of taste.

Green (2002) followed up on these studies by stimulating the tongue with three cotton swabs in two conditions: in one condition, the middle swab contained a taste stimulus ("veridical" condition), and in another condition, the outer two swabs contained the taste stimulus ("referral" condition). When the perceived intensities of the middle swab were compared, the taste intensities of the tasteless tactile stimuli were quantitatively indistinguishable from the ratings for the actual taste stimuli (ie, sucrose, sodium chloride, citric acid, and quinine). This effect of taste referral was seen in all test stimuli with the only exception being citric acid. Green suggested that the principal function of taste referral may be to bind tastes to foods and beverages, which also elicit tactile sensation, in return, the tastes are perceived as intrinsic qualities of the foods and beverages.

3.1.2 Retronasal odor referral

Another example of oral referral is retronasal odor referral to the mouth (see Stevenson, Mahmut, & Oaten, 2011; Stevenson, Oaten, & Mahmut, 2011 for conditions in which the origin of orthonasally perceived odors can be referred to the mouth). It has long been recognized that confusion between taste and retronasal odor occurs and that the uncertainly about the stimulation locus is always resolved in favor of taste (Hollingworth & Poffenberger, 1917; Murphy & Cain, 1980; Murphy, Cain, & Bartoshuk, 1977; Rozin, 1982). Hollingworth and Poffenberger (1917, pp. 13–14) raised a question: "why should it be the rule that, since the taste and (retronasal) smell qualities are to be confused, (retronasal) smell should so commonly sacrifice its claim, so that odors are called tastes rather than vice versa?" They suspected that retronasal odors are referred to as tastes because of "the customary presence of sensations of pressure, temperature, movement, and resistance which are localized in the mouth and in the organ of taste." Murphy and Cain (1980) and Murphy et al. (1977) formally investigated the nature of taste–odor interactions by asking subjects to estimate the perceived intensities of taste and odor of various tastants, odorants, and their mixtures. Interestingly, subjects attributed considerable amount of taste magnitude to odor solutions (eg, ethyl butyrate), but such misattribution was minimal for taste solutions. In recognition that misattribution of taste and retronasal odor occurs only in one direction, Murphy and Cain (1980) suggested that such confusion may be mediated through the tactile stimulation, which always accompanies taste stimulation.

Shortly after, Rozin (1982) argued that the quality of the same olfactory stimulation can be perceived differently depending on whether it is referred to the mouth (through the retronasal olfaction) or to the external world (through the orthonasal olfaction). In an effort to explain possible mechanisms underlying the duality of the olfactory sense, he suggested that retronasal odor input may combine with taste inputs, or alternatively cutaneous stimulation, into an emergent percept, in which the olfactory information loses its own identity. Such speculation, however, had not been tested until recently. To understand the underlying conditions and sensory mechanisms, we systematically investigated the sensory inputs that drive retronasal odor referral to the mouth.

In our first study (Lim & Johnson, 2011), participants were asked to inhale two food odors through a straw while the experimenter delivered air, water (tactile control), or a taste to the mouth using a pipette (see Fig. 3.3). The participants were then asked to report the location of either vanilla or soy sauce odor after consulting a diagram of the cross-section of the head with regions labeled nose, oral cavity, tongue, and throat. When the odors were perceived alone, they were localized remarkably often to the oral cavity and the tongue, accounting for about 40–45% of total localizations (see Fig. 3.4, "air" condition). What was even more striking was that, contrary to longstanding speculation, the presence of water in the mouth did not increase the degree of retronasal odor referral to the mouth (see Fig. 3.4, "air" vs "water" conditions), suggesting that somatosensory stimulation itself does not cause retronasal odor referral. Instead, when a congruent taste (eg, sweetness for vanilla and saltiness for soy sauce) was presented simultaneously, the degree of retronasal odor referral to the mouth, primarily to the tongue, increased significantly (Fig. 3.4). Note that the odors were rarely perceived in the throat area.

Given the role of tactile sensation in taste referral (described in Section 3.1.1) and also the longstanding speculation about its role in odor referral, the effect of tactile stimulation in retronasal odor referral was further investigated in another study.

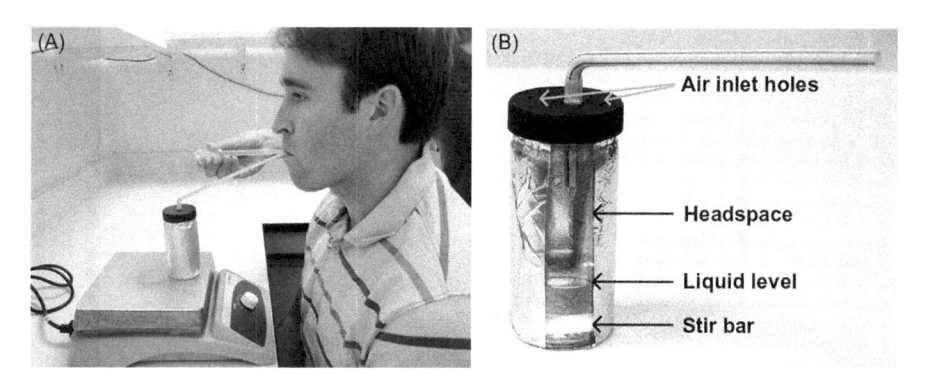

Figure 3.3 (A) Participant performing the task using the odor delivery device. The device was placed on a stir plate rotating 600 rpm during testing sessions. The disposable pipette and straw were positioned in the participant's mouth simultaneously. The participant then inhaled through the straw and exhaled through the nose. The taste stimulus was then deposited from the pipette onto the subject's tongue while the subject continued to breathe. This allowed for the taste and odor sensations to be delivered separately but experienced simultaneously. (B) The retronasal odor delivery device used in the experiments. A portion of aluminum foil has been cut-away to show an interior view.
Source: Figures taken from Lim et al. (2014) and Lim and Johnson (2011).

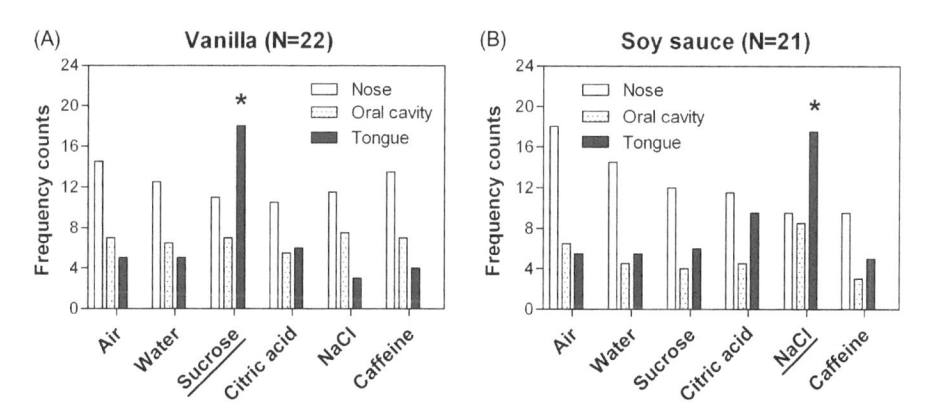

Figure 3.4 Averaged frequency responses for the odor localization tasks for each vanilla–air, vanilla–water, or vanilla–taste pair and soy sauce–air, soy sauce–water, or soy sauce–taste pair. Note that the participants were allowed to report none, one location, or multiple locations for each test pair and thus the total frequency counts across the test pairs are not exactly the same. In one condition (air control), neither taste nor tactile stimulation was presented. In another condition, water was presented simultaneously in the mouth to provide tactile stimulation without any taste input. The frequency responses for each vanilla–taste pair were compared to those for the vanilla–water pair by a two-tailed Chi-squared test. The asterisk indicates a significant difference at $P < 0.0001$.
Source: Figure taken from Lim and Johnson, 2011.

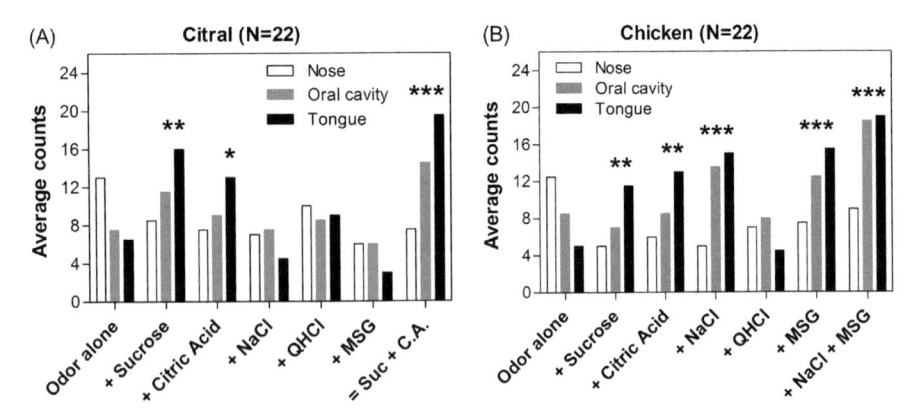

Figure 3.5 Averaged frequency responses for the odor localization tasks for each citral–blank or citral–taste pair and chicken–blank or chicken–taste pair. Note that the participants were allowed to report none, one location, or multiple locations for each test pair and thus the total frequency counts across the test pairs are not exactly the same. The frequency responses for each citral–taste pair were compared to those for the citral–blank pair by a two-tailed Chi-squared test. The asterisks indicate a significant difference (*$P < 0.05$, **$P < 0.01$, ***$P < 0.0001$).
Source: Lim and Johnson, 2012.

We reasoned that we saw such a negative effect in the earlier study because the psychophysical paradigm used did not represent a normal eating/drinking circumstance (ie, the flavor components were teased apart and delivered separately instead of as a single unit). Accordingly, we hypothesized that the presence of a congruent tactile stimulation (ie, food-like texture) would increase odor referral to the mouth. In the follow-up study (Lim & Johnson, 2012), participants performed the same odor localization tasks after sampling gelatin stimuli that contained either odor alone or various congruent and incongruent taste–odor combinations. The results revealed that when tasteless tactile stimulation was presented simultaneously, the retronasal odors were localized in the nose at a similar rate (about 50% of total localization), whether it was presented as a separate entity (water, see Fig. 3.4) or as a food substance (a tasteless gelatin disk, see Fig. 3.5). At the same time, odor referral to the mouth was significantly augmented when a congruent taste was added to the gelatin disk. Interestingly, the localization of odors in the oral cavity was even more enhanced when the binary taste mixtures closely mimic a familiar food source (ie, citral and sucrose + citric acid, a lemony gelatin dessert; chicken odor and NaCl + MSG, a piece of chicken). Combined together, the findings of this study suggest that tactile stimulation itself is not a primary factor in retronasal odor referral to the mouth, but rather that retronasal odor referral to the mouth depends strongly on the presence of a congruent taste(s). The mechanism responsible for the retronasal odor referral in the absence of taste or tactile stimulations (Fig. 3.4, "air" and "water" conditions and Fig. 3.5, "odor alone") is currently unclear. Yet, one possible explanation can be found in the fact that volatiles were exhaled through the mouth, instead of inhaled from the nose. Mozell (1964) proposed that the sorption of odors to the olfactory epithelium in relation to the direction of airflow changes the pattern of mucosal activation and results in perceptual differences between ortho- and retronasal odors. One of the perceptual differences between ortho- and retronasal olfaction could be localization of the odor in the mouth rather than in the nose.

3.2 Underlying conditions for oral referral

There are important conditions to consider for the occurrence of oral referral. These include congruency and temporal/spatial synchrony.

3.2.1 Congruency

Based on the study findings described earlier, congruency between taste and odor quality is critical for the occurrence of taste–odor interactions (eg, Frank & Byram, 1988; Lim et al., 2014; Linscott & Lim, 2016; Schifferstein & Verlegh, 1996) and integration (Dalton et al., 2000; Lim & Johnson, 2011, 2012; Lim et al., 2014; White & Prescott, 2001). Congruency between taste and odor quality seems to be a crucial factor for retronasal odor referral to the mouth as well; the test food odors were localized to the oral cavity and tongue significantly more often when a congruent taste(s) was present in the mouth. In order to test for its role in retronasal odor referral, we measured the degree of taste–odor congruency and compared the ratings directly to the degree of odor referral (Lim et al., 2014). The results indicated that the addition of a highly congruent taste or a taste mixture significantly augmented the degree of retronasal odor referral to the mouth and more specifically to the tongue. For example, the degree of odor referral was enhanced by sucrose and sucrose + caffeine mixture for "sweet" coffee odor and by caffeine and caffeine + sucrose mixture for "bitter" coffee odor. In contrast, caffeine and sucrose did not augment the degree of odor referral to the mouth for "sweet" and "bitter" coffee odors, respectively. When compared more directly, the degree of congruency between taste and odor was positively correlated with the degree of odor referral to the mouth ($r=0.88$–0.98). Thus, the data suggest that the degree of congruency between tastes and odors further modulates the degree of retronasal odor referral to the mouth. Unfortunately, and once again, the role of congruency between taste/odor and tactile sensation in oral referral has not been directly investigated. It is currently hypothesized that tactile and thermal dimensions of foods may also contribute to the concept of flavor object.

An interesting question to consider here concerns the nature of congruence. One may consider that some qualities are naturally consonant (ie, that go well together) and thus congruence is more of an inherent condition. However, the evidence so far suggests that congruency between taste and odor arises out of associative learning (Frank & Byram, 1988; Prescott, 1999; Prescott, Johnstone, & Francis, 2004; Stevenson et al., 1995, 1998). For example, when relatively novel odors, initially low in smelled sweetness and sourness, are repeatedly paired with either sweet or sour tastes in solution, the odors were perceived to be significantly more sweet or sour, depending on the taste with which they were paired (Stevenson et al., 1995, 1998). Accordingly, in our studies we have defined the term congruency as qualities that commonly appear (or are experienced) together and thus are highly associated with one another. However, the extent to which associative learning plays in the development of taste–odor congruency remains to be further explored.

3.2.2 Temporal and spatial synchrony

Perceptual experiences in real life often offer sensory information across different senses and such information is almost always synchronized temporarily (although the temporal synchrony may never be perfect; see, Spence & Squire, 2003). Flavor perception is no exception. In fact, multisensory experience in flavor perception might be a special case since the event occurs at the same place within the body. This temporal synchrony and spatial coincidence between sensory attributes is highly advantageous for information processing because it increases perceptual reliability and saliency (see Keetels & Vroomen, 2012 for more detail). In other words, multisensory interaction and integration are more likely to occur if information from different modalities originating from the same (or a similar) place arrives in the central nervous system at around the same time and thereby is processed concurrently (Stein, Huneycutt, & Meredith, 1988). When this temporal and/or spatial proximity is violated, then the information we gather may become confusing, the information processing (and thereby recognition of the information) is delayed, and/or even the information is perceived as originating from two separate events (although, see Spence, 2013a). While investigating the effect of time on the direction of an olfactory source, Von Békésy (1964) showed that the perceived location of an orthonasal odor can shift from the nose to the mouth, depending on the time between taste and odor stimulation. When an odor and a taste were presented simultaneously, the taste–odor mixture was perceived as a combined sensation and the sensation was localized on the back of the tongue and throat. An important question to ask is, then, "What is considered to be temporal synchrony?" (ie, the "temporal window"; Sarko et al., 2012). Recently, there has been a substantial amount of research (primarily on vision and audition) on how the brain handles temporal lags between senses (Keetels & Vroomen, 2012; Vatakis & Spence, 2010). The general conclusion seems to be that small lags (ie, several hundred milliseconds) go unnoticed (Colonius & Diederich, 2004; Spence & Squire, 2003; van Wassenhove, Grant, & Poeppel, 2007), because intermodal timing is rather flexible and adaptive. Thus, while there might be small lags in the registration of taste and smell information at the central level (due to the differences in transmission times among taste, smell, and tactile stimulation), the synchronous presentation of pairs seems to be sufficient for oral referral to occur.

3.3 The sensory and neural mechanisms

Two fundamental questions were raised at the start of this chapter. The first is why describing the flavor of foods (such as curry) is so hard, although we usually have a good idea of what it should taste like. The findings from the above-mentioned studies suggest that sensory attributes from different sensory modalities are integrated and thereby an emergent percept is created centrally. Unfortunately, the exact neural mechanisms for the integration process are currently unknown, although this area of research is burgeoning. Nevertheless, the study findings provide compelling evidence that the perceptual congruency between taste and odors plays a critical role in the process of "flavor binding" (Stevenson, 2009; see also chapter: Attention and Flavor Binding). It has been proposed that when flavor components are congruent, the qualities of taste and smell fuse (Auvray & Spence, 2008; McBurney, 1986; Small & Green, 2012) into a coherent, unitary percept. This higher-order, cortical binding mechanism might be based on the proximity of the encoded neural pattern to the pattern of a known flavor object (eg, yellow curry; Auvray & Spence, 2008; Gibson, 1966; Small & Prescott, 2005). In other words, if the encoding of neural activities of a food taken into the mouth match with those of a known flavor object, perceptual binding may occur. In that sense, oral referral is probably a manifestation of flavor convergence (Lim & Johnson, 2012).

The second question is why is it that the retronasal odor is perceived as if it is sensed in the mouth instead of the nose? It was speculated that retronasally perceived odors are referred to the mouth because tactile sensations dominate and they "capture" odor sensations (Hollingworth & Poffenberger, 1917; Murphy & Cain, 1980; Rozin, 1982). As discussed earlier, however, our data suggest that tactile stimulation does not "capture" retronasal odors. In the study of the ventriloquist effect, Alais and Burr (2004) explained that when two sensory stimuli are presented simultaneously at different locations, the more spatially precise information dominates the perceived location of the multimodal event. For flavor perception, taste is perhaps the dominating information since the mouth is where we give the most attention while eating. In terms of neural explanation, there is indirect evidence that flavor binding may occur in the somatomotor mouth area of the cortex. Small and Prescott (2005) measured brain responses using fMRI after delivering vaporized odorants via orthonasal and retronasal routes. Results showed that there was preferential activity during retronasal delivery at the base of the central sulcus, a brain region that is responsive to oral cavity somatosensory stimulation in humans (Boling, Reutens, & Olivier, 2002; Pardo, Wood, Costello, Pardo, & Lee, 1997; Yamashita et al., 1999). It appears thus that when perceptual components approximate a known food, they are recognized as a unitary percept, which is then projected to the mouth area.

4 Conclusions

In this chapter, the fundamentals of sensory modalities and systems that are directly related to flavor perception have been described and how those modalities interact with one another at the central and perceptual levels were discussed. In addition, oral referral that commonly occurs in flavor perception was illustrated while the underlying conditions and sensory mechanisms were discussed. One related topic that has not been covered so far is the ecological benefits of multisensory integration. The primary function of our sensory systems is to recognize and identify objects and events that are meaningful in our daily life. In reality, our senses are constantly stimulated and sources of stimulation may or may not even be related. For example, when we are about to try the first bite of a visually appetizing dish at a restaurant, we might hear well-coordinated background music, conversation from the next table, or even a loud ambulance passing by. Of course we may also perceive all sorts of other sensations at the same time: the smell of different foods in the restaurant or even a nice cool breeze (or hot and humid air) on our face. Taking all this information in is an experience for sure, but the perceptual systems still have to perform their tasks. For that matter, multisensory integration within a related object or event is highly beneficial for a comprehensive awareness of the environment. Multisensory integration also has an adaptive purpose of enhancing each system's performance in terms of its speed and accuracy of recognizing (identifying) objects. Flavor perception would appear to be no exception in this regard. When we recognize a known, safe food that is placed in the mouth, quality and spatial fusion occur immediately at the central level. The emergent percept is then projected to the somatomotor mouth area. In contrast, our attention is more likely to be placed on specific attributes, which might originate from different modalities, so that the information is processed appropriately.

In recent years, we have made great strides in understanding perceptual processes and their neural correlates underlying the multisensory flavor perception. Yet, we still have a lot more to discover. One area of great interest is to further investigate the role of the somatosensory system in flavor perception. As described earlier, the somatosensory system is a complex sensory system that is made up of three different senses (ie, touch, temperature, and pain). Much of how these modalities interact with taste and retronasal odor has yet to be investigated.

References

Alais, D., & Burr, D. (2004). The ventriloquist effect results from near-optimal bimodal integration. *Current Biology, 14*, 257–262.
Auvray, M., & Spence, C. (2008). The multisensory perception of flavor. *Consciousness and Cognition, 17*, 1016–1031.
Boling, W., Reutens, D. C., & Olivier, A. (2002). Functional topography of the low postcentral area. *Journal of Neurosurgery, 97*, 388–395.
Bonath, B., Noesselt, T., Martinez, A., Mishra, J., Schwiecker, K., Heinze, H. J., et al. (2007). Neural basis of the ventriloquist illusion. *Current Biology, 17*, 1697–1703.
Bonnans, S., & Noble, A. C. (1993). Effect of sweetener type and of sweetener and acid levels on temporal perception of sweetness, sourness and fruitiness. *Chemical Senses, 18*, 273–283.
Breslin, P. A., Doolittle, N., & Dalton, P. (2001). Subthreshold integration of taste and smell: The role of experience in flavor integration. *Chemical Senses, 26*, 1035.
Bryant, B., & Silver, W. L. (2000). Chemesthesis: The common chemical senses. In T. E. Finger, W. L. Silver, & D. Restrepo (Eds.), *The neurobiology of taste and smell*. New York, NY: Wiley-Liss.
Bryant, B. P., & Mezine, I. (1999). Alkylamides that produce tingling paresthesia activate tactile and thermal trigeminal neurons. *Brain Research, 842*, 452–460.
Burdach, K. J., Kroeze, J. H. A., & Koster, E. P. (1984). Nasal, retronasal, and gustatory perception: An experimental comparison. *Perception & Psychophysics, 36*, 205–208.

Cerf-Ducastel, B., & Murphy, C. (2001). fMRI activation in response to odorants orally delivered in aqueous solutions. *Chemical Senses, 26*, 625–637.

Cerf-Ducastel, B., Van De Moortele, P. F., Macleod, P., Le Bihan, D., & Faurion, A. (2001). Interaction of gustatory and lingual somatosensory perceptions at the cortical level in the human: A functional magnetic resonance imaging study. *Chemical Senses, 26*, 371–383.

Christensen, T. A., & White, J. (2000). Representation of olfactory information in the brain. In T. E. Finger, W. L. Silver, & D. Restrepo (Eds.), *The neurobiology of taste and smell*. New York, NY: Willey-Liss.

Colonius, H., & Diederich, A. (2004). Multisensory interaction in saccadic reaction time: A time-window-of-integration model. *Journal of Cognitive Neuroscience, 16*, 1000–1009.

Cowart, B. J. (1987). Oral chemical irritation: Does it reduce perceived taste intensity? *Chemical Senses, 12*, 467–479.

Critchley, H. D., & Rolls, E. T. (1996). Olfactory neuronal responses in the primate orbitofrontal cortex: Analysis in an olfactory discrimination task. *Journal of Neurophysiology, 75*, 1659–1672.

Cruz, A., & Green, B. G. (2000). Thermal stimulation of taste. *Nature, 403*, 889–892.

Dalton, P., Doolittle, N., Nagata, H., & Breslin, P. A. (2000). The merging of the senses: Integration of subthreshold taste and smell. *Nature Neuroscience, 3*, 431–432.

de Araujo, I. E., & Rolls, E. T. (2004). Representation in the human brain of food texture and oral fat. *The Journal of Neuroscience, 24*, 3086–3093.

de Araujo, I. E., Rolls, E. T., Kringelbach, M. L., Mcglone, F., & Phillips, N. (2003). Taste-olfactory convergence, and the representation of the pleasantness of flavour, in the human brain. *European Journal of Neuroscience, 18*, 2059–2068.

Delwiche, J. F. (2004). The impact of perceptual interactions on perceived flavor. *Food Quality and Preference, 15*, 137–146.

Delwiche, J. F., Lera, M. F., & Breslin, P. A. S. (2000). Selective removal of a target stimulus localized by taste in humans. *Chemical Senses, 25*, 181–187.

Dessirier, J. M., O'Mahony, M., Iodi-Carstens, M., Yao, E., & Carstens, E. (2001). Oral irritation by sodium chloride: Sensitization, self-desensitization, and cross-sensitization to capsaicin. *Physiology & Behavior, 72*, 317–324.

Frank, R. A., & Byram, J. (1988). Taste-smell interactions are tastant and odorant dependent. *Chemical Senses, 13*, 445–455.

Gaillard, D., Laugerette, F., Darcel, N., El-Yassimi, A., Passilly-Degrace, P., Hichami, A., et al. (2008). The gustatory pathway is involved in CD36-mediated orosensory perception of long-chain fatty acids in the mouse. *FASEB Journal, 22*, 1458–1468.

Gairns, F. W. (1952). Sensory nerve endings other than taste buds in the human tongue. *Journal of Physiology, 121*, 33–34.

Gibson, J. J. (1966). *The senses considered as perceptual systems*. Boston, MA: Houghton Mifflin.

Gilbertson, T. A. (1998). Gustatory mechanisms for the detection of fat. *Current Opinion in Neurobiology, 8*, 447–452.

Gilmore, M. M., & Green, B. G. (1993). Sensory irritation and taste produced by NaCl and citric acid: Effects of capsaicin desensitization. *Chemical Senses, 18*, 257–272.

Green, B. G. (1996). Chemesthesis: Pungency as a component of flavor. *Trends in Food Science & Technology, 7*, 415–420.

Green, B. G. (2002). Studying taste as a cutaneous sense. *Food Quality and Preference, 14*, 99–109.

Green, B. G., & Gelhard, B. (1989). Salt as an oral irritant. *Chemical Senses, 14*, 259–271.

Green, B. G., Mason, J. R., & Kare, M. R. (1990). *Chemical senses, vol. 2: Irritation*. New York, NY: Marcel Dekker, Inc.

Green, B. G., Nachtigal, D., Hammond, S., & Lim, J. (2012). Enhancement of retronasal odors by taste. *Chemical Senses, 37*, 77–86.

Green, B. G., & Schullery, M. T. (2003). Stimulation of bitterness by capsaicin and menthol: Differences between lingual areas innervated by the glossopharyngeal and chorda tympani nerves. *Chemical Senses, 28*, 45–55.

Guest, S., Grabenhorst, F., Essick, G., Chen, Y. S., Young, M., Mcglone, F., et al. (2007). Human cortical representation of oral temperature. *Physiology & Behavior, 92*, 975–984.

Hagura, N., Barber, H., & Haggard, P. (2013). Food vibrations: Asian spice sets lips trembling. *Proceedings of the Royal Society. Biological Sciences, 280*, 20131680.

Hollingworth, H. L., & Poffenberger, A. T. (1917). *The sense of taste*. New York, NY: Moffat, Yard and Company.

Kadohisa, M., Rolls, E. T., & Verhagen, J. V. (2004). Orbitofrontal cortex: Neuronal representation of oral temperature and capsaicin in addition to taste and texture. *Neuroscience, 127*, 207–221.

Keast, R. S. J., & Breslin, P. A. S. (2002). An overview of binary taste-taste interactions. *Food Quality and Preference, 14*, 111–124.

Keetels, M., & Vroomen, J. (2012). Perception of synchrony between the senses. In M. M. Murray & M. T. Wallace (Eds.), *The neural bases of multisensory processes* (pp. 147–178). Boca Raton, FL: CRC Press.

King, C. T., Travers, S. P., Rowland, N. E., Garcea, M., & Spector, A. C. (1999). Glossopharyngeal nerve transection eliminates quinine-stimulated fos-like immunoreactivity in the nucleus of the solitary tract: Implications for a functional topography of gustatory nerve input in rats. *The Journal of Neuroscience, 19*, 3107–3121.

Kroeze, J. H. A., & Bartoshuk, L. M. (1985). Bitterness suppression as revealed by split-tongue taste stimulation in humans. *Physiology & Behavior, 35*, 779–783.

Kuo, Y. -L., Pangborn, R. M., & Noble, A. C. (1993). Temporal patterns of nasal, oral, and retronasal perception of citral and vanillin and interaction of these odourants with selected tastants. *International Journal of Food Science & Technology, 28*, 127–137.

Lapis, T. J., Penner, M. H., & Lim, J. (2014). Evidence that humans can taste glucose polymers. *Chemical Senses, 39*, 737–747.

Lim, J., Balto, A. S., Lapis, T. J., & Penner, M. H. (2015). Influence of saccharide length on detection of glucose polymers in humans. *Chemical Senses, 40*, 616.

Lim, J., Fujimaru, T., & Linscott, T. D. (2014). The role of congruency in taste-odor interactions. *Food Quality and Preference, 34*, 5–13.

Lim, J., & Green, B. G. (2007). The psychophysical relationship between bitter taste and burning sensation: Evidence of qualitative similarity. *Chemical Senses, 32*, 31–39.

Lim, J., & Johnson, M. B. (2011). Potential mechanisms of retronasal odor referral to the mouth. *Chemical Senses, 36*, 283–289.

Lim, J., & Johnson, M. B. (2012). The role of congruency in retronasal odor referral to the mouth. *Chemical Senses, 37*, 515–522.

Liman, E. R., Zhang, Y. V., & Montell, C. (2014). Peripheral coding of taste. *Neuron, 81*, 984–1000.

Linscott, T. D., & Lim, J. (2016). Retronasal odor enhancement by salty and umami tastes. *Food Quality and Preference, 48*, 1–10.

McBurney, D. H. (1986). Taste, smell and flavor terminology: Taking the confusion out of confusion. In H. L. Meiselman & R. S. Rivkin (Eds.), *Clinical measurement of taste and smell*. New York, NY: Macmillan.

McBurney, D. H., & Bartoshuk, L. M. (1973). Interactions between stimuli with different taste qualities. *Physiology & Behavior, 10*, 1101–1106.

Mozell, M. M. (1964). Evidence for sorption as a mechanism of the olfactory analysis of vapours. *Nature, 203*, 1181–1182.
Murphy, C., & Cain, W. S. (1980). Taste and olfaction: Independence vs interaction. *Physiology & Behavior, 24*, 601–605.
Murphy, C., Cain, W. S., & Bartoshuk, L. M. (1977). Mutual action of taste and olfaction. *Sensory Processes, 1*, 204–211.
Noble, A. C. (1996). Taste-aroma interactions. *Trends in Food Science & Technology, 7*, 439–444.
Pardo, J. V., Wood, T. D., Costello, P. A., Pardo, P. J., & Lee, J. T. (1997). PET study of the localization and laterality of lingual somatosensory processing in humans. *Neuroscience Letters, 234*, 23–26.
Prescott, J. (1999). Flavour as a psychological construct: Implications for perceiving and measuring the sensory qualities of foods. *Food Quality and Preference, 10*, 349–356.
Prescott, J., Allen, S., & Stephens, L. (1984). Interactions between oral chemical irritation, taste and temperature. *Chemical Senses, 18*, 389–404.
Prescott, J., Johnstone, V., & Francis, J. (2004). Odor-taste interactions: Effects of attentional strategies during exposure. *Chemical Senses, 29*, 331–340.
Prescott, J., & Stevenson, R. J. (1995). Effects of oral chemical irritation on tastes and flavors in frequent and infrequent users of chili. *Physiology & Behavior, 58*, 1117–1127.
Pritchard, T. C., Hamilton, R. B., Morse, J. R., & Norgren, R. (1986). Projections of thalamic gustatory and lingual areas in the monkey, *Macaca fascicularis*. *Journal of Comparative Neurology, 244*, 213–228.
Pritchard, T. C., Hamilton, R. B., & Norgren, R. (1989). Neural coding of gustatory information in the thalamus of *Macaca mulatta*. *Journal of Neurophysiology, 61*, 1–14.
Rolls, E. T. (1997). Taste and olfactory processing in the brain and its relation to the control of eating. *Critical Reviews in Neurobiology, 11*, 263–287.
Rolls, E. T. (2011). Chemosensory learning in the cortex. *Frontiers in Systems Neuroscience, 5*, 78.
Rolls, E. T., & Baylis, L. L. (1994). Gustatory, olfactory, and visual convergence within the primate orbitofrontal cortex. *The Journal of Neuroscience, 14*, 5437–5452.
Rolls, E. T., Critchley, H. D., Mason, R., & Wakeman, E. A. (1996). Orbitofrontal cortex neurons: Role in olfactory and visual association learning. *Journal of Neurophysiology, 75*, 1970–1981.
Rozin, P. (1982). "Taste-smell confusions" and the duality of the olfactory sense. *Perception & Psychophysics, 31*, 397–401.
Rudenga, K., Green, B., Nachtigal, D., & Small, D. M. (2010). Evidence for an integrated oral sensory module in the human anterior ventral insula. *Chemical Senses, 35*, 693–703.
Running, C. A., Craig, B. A., & Mattes, R. D. (2015). Oleogustus: The unique taste of fat. *Chemical Senses, 40*, 507–516.
Sarko, D. K., Nidiffer, A. R., Powers, A. R., III, et al., Ghose, D., Hillock-Dunn, A., Fister, M. C., et al. (2012). Spatial and temporal features of multisensory processes: Bridging animal and human studies. In M. M. Murray & M. T. Wallace (Eds.), *The neural basese of multisensory processes* (pp. 191–216). Boca Raton, FL: CRC Press.
Schifferstein, H. N. J., & Verlegh, P. W. J. (1996). The role of congruency and pleasantness in odor-induced taste-enhancement. *Acta Psychologica, 94*, 87–105.
Schwieterman, M. L., Colquhoun, T. A., Jaworski, E. A., Bartoshuk, L. M., Gilbert, J. L., Tieman, D. M., et al. (2014). Strawberry flavor: Diverse chemical compositions, a seasonal influence, and effects on sensory perception. *PLoS One, 9*, e88446.
Sclafani, A. (2004). The sixth taste? *Appetite, 43*, 1–3.

Scott, T. R., & Plata-Salaman, C. R. (1991). Coding of taste quality. In T. V. Getchell, L. M. Bartoshuk, R. L. Doty, & J. B. J. Snow (Eds.), *Smell and taste in health and disease* (pp. 345–368). New York, NY: Raven Press.

Scott, T. R., Yaxley, S., Sienkiewicz, Z. J., & Rolls, E. T. (1986). Gustatory responses in the nucleus tractus solitarius of the alert cynomolgus monkey. *Journal of Neurophysiology, 55*, 182–200.

Small, D. M. (2012). Flavor is in the brain. *Physiology & Behavior, 107*, 540–552.

Small, D. M., & Green, B. G. (2012). A proposed model of a flavor modality. In M. M. Murray & M. T. Wallace (Eds.), *The neural bases of multisensory processes* (pp. 717–738). Boca Raton, FL: CRC Press.

Small, D. M., & Prescott, J. (2005). Odor/taste integration and the perception of flavor. *Experimental Brain Research, 166*, 345–357.

Small, D. M., Voss, J., Mak, Y. E., Simmons, K. B., Parrish, T., & Gitelman, D. (2004). Experience-dependent neural integration of taste and smell in the human brain. *Journal of Europhysiology, 92*, 1892–1903.

Smith-Swintosky, V. L., Plata-Salaman, C. R., & Scott, T. R. (1991). Gustatory neural coding in the monkey cortex: Stimulus quality. *Journal of Neurophysiology, 66*, 1156–1165.

South, E. H., & Ritter, R. C. (1986). Substance P-containing trigeminal sensory neurons project to the nucleus of the solitary tract. *Brain Research, 372*, 283–289.

Spence, C. (2013a). Just how important is spatial coincidence to multisensory integration? Evaluation the spatial rule. *Annals of the New York Academy of Sciences, 1296*, 31–49.

Spence, C. (2013b). Multisensory flavour perception. *Current Biology, 23*, R365–R369.

Spence, C., & Squire, S. (2003). Multisensory integration: Maintaining the perception of synchrony. *Current Biology, 13*, R519–R521.

Stein, B. E., Huneycutt, W. S., & Meredith, M. A. (1988). Neurons and behavior: The same rules of multisensory integration apply. *Brain Research, 448*, 355–358.

Stevenson, R. J. (2009). *The psychology of flavour*. New York, NY: Oxford University Press.

Stevenson, R. J., Boakes, R. A., & Prescott, J. (1998). Changes in odor sweetness resulting from implicit learning of a simultaneous odor-sweetness association: An example of learned synesthesia. *Learning and Motivation, 29*, 113–132.

Stevenson, R. J., Mahmut, M. K., & Oaten, M. J. (2011). The role of attention in the localization of odors to the mouth. *Attention, Perception, & Psychophysics, 73*, 247–258.

Stevenson, R. J., Oaten, M. J., & Mahmut, M. K. (2011). The role of taste and oral somatosensation in olfactory localization. *Quarterly Journal Of Experimental Psychology, 64*, 224–240.

Stevenson, R. J., Prescott, J., & Boakes, R. A. (1995). The acquisition of taste properties by odors. *Learning and Motivation, 26*, 433–455.

Stevenson, R. J., Prescott, J., & Boakes, R. A. (1999). Confusing tastes and smells: How odours can influence the perception of sweet and sour tastes. *Chemical Senses, 24*, 627–635.

Todrank, J., & Bartoshuk, L. M. (1991). A taste illusion: Taste sensation localized by touch. *Physiology & Behavior, 50*, 1027–1031.

van Wassenhove, V., Grant, K. W., & Poeppel, D. (2007). Temporal window of integration in auditory-visual speech perception. *Neuropsychologia, 45*, 598–607.

Vatakis, A., & Spence, C. (2010). Audiovisual temporal integration for complex speech, object-action, animal call, and musical stimuli. In M. J. Naumer & J. Kaiser (Eds.), *Multisensory object perception in the primate brain* (pp. 65–98). New York, NY: Springer.

Veldhuizen, M. G., Nachtigal, D., Teulings, L., Gitelman, D. R., & Small, D. M. (2010). The insular taste cortex contributes to odor quality coding. *Frontiers in Human Neuroscience, 4*, 1–11.

Verhagen, J. V., & Engelen, L. (2006). The neurocognitive bases of human multimodal food perception: Sensory integration. *Neuroscience and Biobehavioral Reviews, 30*, 613–650.
Viana, F. (2011). Chemosensory properties of the trigeminal system. *ACS Chemical Neuroscience, 2*, 38–50.
Von Bekesy, G. (1964). Olfactory analogue to directional hearing. *Journal of Applied Physiology, 19*, 369–373.
White, T., & Prescott, J. (2001). Odors influence speed of taste naming. *Chemical Senses, 26*, 1119.
Whitehead, M. C. (1990). Subdivisions and neuron types of the nucleus of the solitary tract that project to the parabrachial nucleus in the hamster. *Journal of Comparative Neurology, 301*, 554–574.
Yamaguchi, S. (1967). Synergistic taste effect of monosodium glutamate and disodium 5′-inosinate. *Journal of Food Science, 32*, 473–478.
Yamashita, H., Kumamoto, Y., Nakashima, T., Yamamoto, T., Inokuchi, A., & Komiyama, S. (1999). Magnetic sensory cortical responses evoked by tactile stimulations of the human face, oral cavity and flap reconstructions of the tongue. *European Archives of Oto-Rhino-Laryngology, 256*(Suppl. 1), S42–S46.

Oral-Somatosensory Contributions to Flavor Perception and the Appreciation of Food and Drink

Charles Spence[1] and Betina Piqueras-Fiszman[2]
[1]Crossmodal Research Laboratory, Department of Experimental Psychology, University of Oxford, Oxford, United Kingdom, [2]Marketing and Consumer Behavior Group, Wageningen University, Wageningen, The Netherlands

1 Introduction

While oral-somatosensation undoubtedly plays a central role in multisensory flavor perception, it is one of the senses that has not (yet) made it into the International Standards Organization's (see ISO 5492, 1992, 2008) definition of flavor.[1] That said, in this chapter, we argue that the oral-somatosensory attributes of food and drink really do play an important role in determining our perception, and hence our enjoyment, of many of our most preferred foods and drinks (as well as our dislike of certain others).

It has to be said, though, at the outset that the majority of researchers have simply not given oral-somatosensation and food texture the attention that it so surely deserves (see Lyman, 1989, p. 85, for a similar criticism). On the one hand, it is striking how those researchers who are interested more generally in the topic of tactile perception rarely delve into the oral cavity (see Gallace & Spence, 2014, on this point). On the other, it has to be admitted that it is simply much more difficult to change the texture of a food or drink, say, than it is to change its color or taste. Generating carefully calibrated sets of stimuli that vary in no way other than their oral-somatosensory properties (ie, texture or viscosity) has proved to be something of a challenge for researchers. Furthermore, it has always been difficult to demonstrate convincingly that the effects of, for instance, a change in the viscosity of a solution, resulted from a change in the multisensory integration of the component unisensory signals rather than, say, from just a (more mundane) physicochemical change in the release of volatiles in the mouth.

---

[1] According to ISO 5492 (1992, 2008), flavor can be defined as a: *"Complex combination of the olfactory, gustatory and trigeminal sensations perceived during tasting. The flavour may be influenced by tactile, thermal, painful and/or kinaesthetic effects."*

2 Multisensory flavor perception

While the consensus view has always been that flavor is mostly about the taste perceived via the gustatory receptors that can be found on the tongue (and elsewhere in the oral cavity; see also Trivedi, 2012), it turns out that *all* of the senses actually contribute to the perception of what we eat and drink, and how much we enjoy the experience (see Spence, 2012, 2015b; Stevenson & Mahmut, 2011). Olfactory cues play a crucial (constitutive; see Spence, Smith, & Auvray, 2015) role in our perception or flavor; they give rise to attributes such as meaty, fruity, floral, burnt, etc. By contrast, the activation of the gustatory receptors only gives rise to basic taste sensations, including sweet, sour, bitter, salty, umami, and possibly also a few others (see Stuckey, 2012).

Oral-somatosensation informs us about everything from the temperature of the food or drink through to its texture and viscosity (eg, Kramer, 1973; Mony et al., 2013; Szczesniak, 1990; Szczesniak & Kahn, 1971, 1984; Szczesniak & Kleyn, 1963; see Green, 2002; Green & Lawless, 1991; Stevenson, 2009, for reviews). It is now more than three decades since Bourne (1982, p. 259) defined food texture as: "*the response of the tactile senses to physical stimuli that result from contact between some part of the body and the food.*" In the years that have followed, some researchers have been tempted to include a contribution from the other senses, such as vision, hearing, olfaction, and kinesthesia in their definitions (Szczesniak, 1990), since several studies have reported robust effects of these other cues in the perception of texture attributes (see Chylinski, Northey, & Ngo, 2015, for recent findings).

As we hope to make clear in this chapter, the oral-somatosensory experiences that result from the consumption of food and drink play an important role in determining our perception of both the sensory discriminative attributes of whatever it is that we happen to be eating or drinking and our hedonic response to them. That said, oral-somatosensation does not typically rate all that highly in people's rankings of the relative importance of each of the senses to multisensory flavor perception. Indeed, according to the results of a survey completed by 140 people working in various capacities in the area of food/chemical senses, temperature and texture ranked below color, appearance, and sound in people's assessment of the importance of various cues to flavor (see Delwiche, 2003; see Table 4.1).

That said, it is important to remember that it is not always so easy to figure out exactly which sense is really doing the work in terms of being responsible for various aspects of our multisensory experience of food and drink (see Spence et al., 2015, for a review). For example, just take attributes such as carbonation, fattiness, and astringency. While the intuitive view might well be to suggest that the experience of carbonation in the mouth is attributable to the *feel* of the bubbles bursting on the tongue, this most pleasing of oral sensations actually results from the stimulation of the sour *taste* receptors (Chandrashekar et al., 2009). Most people also believe that the perception of fattiness in a food results from the stimulation of the oral-somatosensory receptors in the oral cavity. However, it turns out that our experience of this hedonically pleasing food property is modulated by stimulation of the olfactory

Table 4.1 Summary of findings from a study by Delwiche (2004) in which the participants had to rate which of the seven attributes were (1) important to flavor, (2) essential, and (3) changeable without impacting flavor. The bottom row represents the results of a rank task of the attributes in terms of their importance to flavor perception

	Taste	Smell	Chemesthesis	Temperature	Texture	Color	Appearance	Sound
% Important	97	94	–	78	64	40	37	21
% Essential	96	90	–	37	34	12	16	6
% Changeable	0	2	–	19	41	68	68	82
Mean Ranked	1.5	1.7		4	4.4	5	4.8	6.6

and/or gustatory receptors (eg, Bult, de Wijk & Hummel, 2007; see also Roudnitzky et al., 2011; Tournier, Sulmont-Rossé, & Guichard, 2007). Indeed, a growing number of researchers now believe that there is also a fatty acid taste (alongside sweet, sour, salty, bitter, and umami; see Keast & Costanzo, 2015; Mattes, 2009; see also chapter: Individual Differences in Multisensory Flavor Perception). Conversely, it has been suggested that astringency is a component of the taste/flavor of a beverage. However, astringency—just think of an unripe banana, a tannic young red wine that has been fermented in new oak barrels, or worse still, an overstewed cup of tea—is also a tactile sensation (again one's mind is brought back to the inadequacy of the ISO's definition of flavor perception; see Breslin, Gilmore, Beauchamp, & Green, 1993).

As we will see below, stimulation of the oral-somatosensory receptors can give rise to a range of responses including everything from a judgment about the texture of the food(s) that happen to be in the mouth through to the experience of what is called "mouthfeel", a term used to describe how the consumption of certain foods changes the way the oral cavity feels. In the sections that follow, we will take a closer look at the various ways in which oral-somatosensory stimulation gives rise to specific sensations/perceptions.

3 Oral-somatosensation

3.1 Mouthfeel

This term is used to describe the feeling in the mouth that results from eating certain foods or drinks (eg, Christensen, 1984; Gawel, Oberholster, & Francis, 2000; Kappes, Schmidt, & Lee, 2007; Langstaff, Guinard, & Lewis, 1991; Marsilli, 1993; Szczesniak, 1979). Olive oil, for example, gives rise to an oily mouth-coating or mouthfeel, whereas foods containing menthol give rise to a cool mouthfeel (Nagata, Dalton, Doolittle, & Breslin, 2005). Jowitt (1974, p. 356) defined mouthfeel as *"those textural attributes of a food or beverage responsible for producing characteristic tactile sensations on the surfaces of the oral cavity."* Typical mouthfeel characteristics include sticky, astringent, stinging, oily, etc.[2]

Mouthfeel is also an important aspect of our appreciation of many alcoholic beverages, such as, for example, beer and wine (eg, Gawel et al., 2000; Langstaff et al., 1991). Take the following quote from Richard Gawel, developer of the Mouth-feel Wheel: *"Just listen to red wine consumers when they explain why they like, or don't like a particular red wine. Wines that they perceive as 'soft' and 'smooth' in the mouth are frequently at the top of their shopping lists. This convinced me of the merits of compiling an extensive list of defined terms that could be used by wine-tasters to describe red wine texture"* (Gawel, 2006).

---

[2] As one commentator writing for the *New Scientist* put it: *"Think of the creaminess of mayonnaise, the viscosity of toffee and the greasiness of cold lard. Or the brittleness of butterscotch, the elasticity of jelly and the crunchiness of fresh carrots"* (Anon, 2005, p. 47).

3.2 Oral stereognosis

This is the name given to describe the determination of the shape of objects on the basis of oral stimulation (see Jacobs, Serhal, & van Steenberghe, 1998, for a review). In order to assess an individual's stereognostic ability, a series of small objects are placed in the mouth and, in most experimental set-ups, the free manipulation of the objects is encouraged. This will obviously give rise to the activation of a large number of receptors (periodontal, mucosal, muscular, and articular). Since the tip of the tongue is one of the most densely innervated areas of the human body, it should come as little surprise to find that it plays an important role in oral stereognosis. An individual's oral stereognotic abilities provide an indicator of functional sensibility, including the synthesis of numerous sensory inputs in higher brain centers; perhaps unsurprisingly, however, no clear relationship has been demonstrated between manual and oral stereognosis.

Moreover, research by Topolinski and Türk Pereira (2012) suggests that our perception of the size of an item of food in the mouth might, at least to a certain degree, depend on how hungry we are. These researchers investigated the effect of hunger versus satiation on their participants' oral and manual perception of the length of objects (measured by means of straw segments). According to Topolinski and Türk Pereira's findings, objects were perceived to be longer when their participants were hungry as compared to when they were satiated. Interestingly, while this effect occurred for orally perceived stimuli, no such effect was documented for those stimuli that were experienced manually. These results highlight the important relationship that exists between oral stimulation of the mucosa, feelings of hunger, and the perception of the characteristics of food (such as, eg, its size).

3.3 Viscosity and its interaction with taste/odor perception

The viscosity of a food or drink exerts a significant influence over the multisensory perception of flavor (eg, Bult et al., 2007; Frost & Janhoj, 2007; Weel et al., 2002). Here, it is perhaps worth bearing in mind that while the results of a number of early studies suggested that increased viscosity impaired the perception of taste and aroma in a foodstuff (eg, Christensen, 1980), it has always been difficult to disentangle whether such effects had a physicochemical, as opposed to a neurophysiological, origin (since increased viscosity is likely to reduce volatility at the food–air interface; eg, see Delwiche, 2004). However, that said, the technological advances that have been seen over the last decade or so now mean that it has been possible for food science researchers to isolate, and thus to convincingly demonstrate, the genuinely psychological nature of at least a part of this particular crossmodal effect (eg, Bult et al., 2007; Kutter, Hanesch, Rauh, & Delgado, 2011).

So, for example, in one laboratory study, the participants were presented with a creamy odor using a computer-controlled olfactometer (Bult et al., 2007). At the same time, milk-like foods with different viscosities were delivered to the participant's mouth. The participants rated the creaminess and thickness of the resulting experience as well as the intensity of the overall flavor experience. The results revealed that as the viscosity of the liquid increased, the participants' ratings of flavor intensity decreased. Interestingly, this was true no matter whether the odor was delivered orthonasally or retronasally.[3] Given the independent delivery of texture and odor, these results clearly highlight the importance of texture (mouthfeel) to the multisensory perception of flavor. These results also suggest that the presence of a retronasal odor can alter the perceived thickness of a foodstuff in the mouth (cf. Kilcast, 2008; Sundqvist, Stevenson, & Bishop, 2006; Tournier et al., 2009).

The hope among some researchers is that the systematic study of the impact of texture on taste/odor perception may one day help contribute to the development of more effective strategies for sugar and salt reduction (Stieger & van de Velde, 2013). For instance, Mosca, van de Velde, Bult, van Boekel, and Stieger (2012) advanced this idea by investigating the effects of texture and the spatial distribution of sucrose on the perception of sweetness. Those gels with a nonuniform distribution of sucrose were perceived to taste sweeter than those gels in which the sucrose had been distributed homogeneously. What is more, this did not depend on the texture of the gel matrix itself (though see also Slocombe, Carmichael, & Simner, in press).

Tactile stimulation of the oral cavity is important for another reason too. Everyone has heard of the ventriloquist—the illusionist who seemingly projects his/her voice to the articulated lips of the dummy. This illusion provides a particularly well-known example of the visual capture, or bias, of perceived auditory location (eg, Alais & Burr, 2004). A very similar effect may well occur in the mouth when a person eats/drinks (Auvray & Spence, 2008). It turns out that our perception of the origin of a taste sensation tends to follow the location of a tactile stimulus drawn across the tongue and not the point at which the taste stimulus itself happens to be transduced on the receptor surface (Green, 2002; Lim & Green, 2008, 2011; Todrank & Bartoshuk, 1991; though see also Stevenson, 2012, chapter: Attention and Flavor Binding; and Michel, Velasco, & Spence, 2014). However, while researchers traditionally believed that the same may also be true for olfactory stimuli—that is, that retronasal olfactory inputs were also localized to the origin of oral-somatosensory inputs (see Hollingworth & Pofferberger, 1917; Murphy & Cain, 1980; Rozin, 1982)—subsequent research has revealed this not to be the case (see Lim & Johnson, 2011; Stevenson, Oaten & Mahmut, 2011; see Spence, 2016, for a review).[4]

Intriguingly, the ventriloquism illusion just might play a role in one of the classic dishes that appeared on the tasting menu at Heston Blumenthal's, The Fat Duck

---

[3] Remember that there are two relatively distinct senses of smell: One, the orthonasal system, associated with the inhalation of external odors, and the other, the retronasal system, associated with the detection of the olfactory stimuli emanating from the food we eat, as odors are periodically forced out of the nasal cavity when ever we chew or swallow. A growing body of empirical research now highlights a number of salient differences between these two kinds of olfaction, at both the subjective/perceptual (eg, Diaz, 2004; Rozin, 1982) and neural levels (eg, Small, Gerber, Mak, & Hummel, 2005).

[4] Given the pronounced differences in transduction latencies between the senses (see Spence & Squire, 2003), the tactile sensations that are associated with eating and drinking will normally arrive centrally in the brain before either the associated gustatory or retronasal olfactory stimuli, and hence this "prior entry" of tactile inputs may also play some role in helping to localize the combined multisensory flavor experience to the mouth (Spence, 2012).

restaurant (in Bray, UK). When bacon-and-egg-flavored ice cream was first developed, it was only rated as being moderately pleasant; the subjective report of many diners was that the flavors did not appear to stand out one from the other. Part of the breakthrough here came when a piece of crispy fried brioche was added to the plate. While the bread does not, in-and-of-itself, impart much of a flavor to the dish, it somehow seems to help bring the dish alive by giving rise to a perceptual separation of the bacon and eggy flavors. What may be going on in this case, although the proper psychophysical experiment has yet to be done, is that the bacon flavor is ventriloquized toward, and hence becomes perceptually localized within, the crispy brioche (because it is congruent with the texture of crispy bacon). By contrast, the eggy flavor tends to "stay behind" in the more texturally appropriate soft ice-cream instead (see Blumenthal, 2008).

3.4 Oral-somatosensation: Temperature

The temperature of a foodstuff in the mouth influences the perception of taste (eg, Anon, 2013; McBurney, Collins, & Glanz, 1973; Powers, Howell, Lillard, & Vacinek, 1971; Sato, 1967; Snyder, Prescott, & Bartoshuk, 2006; Talavera, Ninomiya, Winkel, Voets, & Nilius, 2007). Most of us have likely experienced the cloying sweetness of melted ice-cream or a warm cola drink, or the noticeably bitter taste of a beer when tasted at room temperature. Although the story here is somewhat complicated (see Bajec, Pickering, & de Courville, 2012), generally speaking the available research shows that the threshold for the detection of bitter, sweet, salty, and sour stimuli shows a U-shaped response as a function of temperature. The lowest thresholds are documented in the 20–30°C temperature range. As yet, though, less research has been conducted on the effect of temperature on the perception of umami. The perception of the temperature of foods in the mouth also appears to be influenced by the temperature of the mouth itself (see Engelen, De Wijk, & Prinz, 2002).

Another kind of crossmodal interaction between temperature and taste that is worth mentioning here is the "*thermal-taste*" illusion (Cruz & Green, 2000; Green & George, 2004). This term refers to an effect whereby simply by raising or lowering the temperature at various points on the tongue, different taste sensations can be elicited.[5] It has been estimated that around a third to a half of the population experience this particular crossmodal illusion.

3.5 The sound of texture

The sounds that we hear when eating contribute to our perception of crispness and freshness in foods such as crisps (potato chips), biscuits, breakfast cereals, and vegetables/fruit (Dematté et al., 2014; Masuda & Okajima, 2011; Masuda, Yamaguchi,

---

[5] One can try this at home by taking an ice cube and placing it against different parts of one's tongue. On so doing, some people experience sweetness and/or saltiness when the cube is rubbed across the tip and front sides of the tongue, respectively. Note that this taste experience is illusory (given that the water that goes into making the ice cubes has no taste of its own).

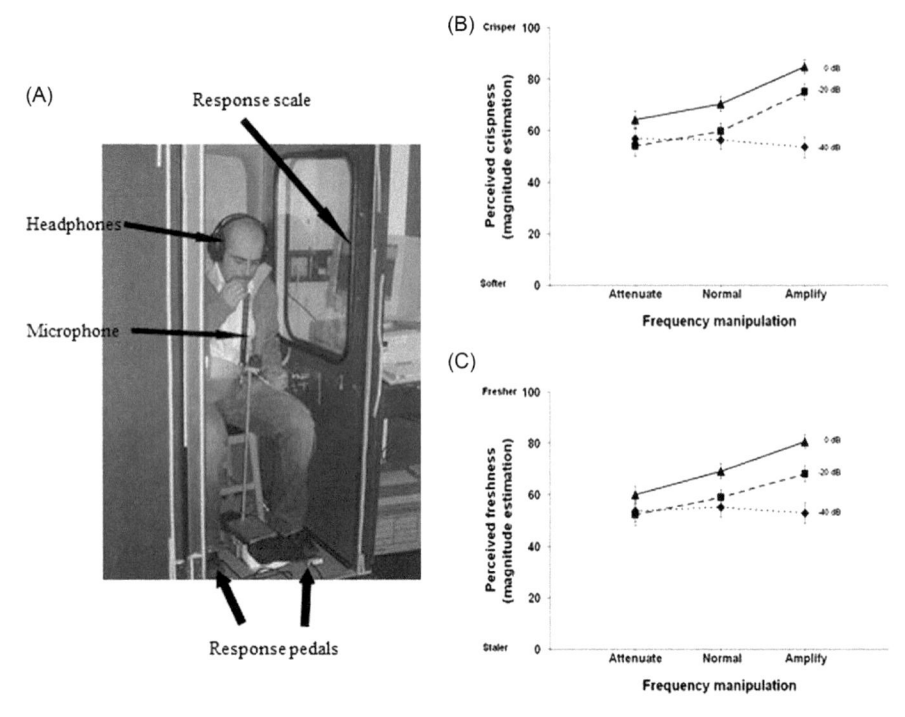

Figure 4.1 (A) Schematic view of apparatus and participant in Zampini and Spence's (2004) study. Note that the door of the experimental booth was closed during the experiment and the response scale was viewed through the window shown in the side wall of booth (on the participant's left). Mean responses for soft–crisp (B) and fresh–stale (C) response scales for three overall attenuation levels (0, −20, or −40 dB) against three frequency manipulations (high frequencies attenuated, veridical auditory feedback, or high frequencies amplified) are reported. The error bars represent between-participants standard errors of the means.
Source: Reprinted from Zampini and Spence (2004), with permission.

Arai, & Okajima, 2008; Szczesniak, 1988; Zampini & Spence, 2004; see Spence, 2015a, for a review). The participants in a classic study by Zampini and Spence had to bite into a large number of potato chips (around 180 in total) and rate each one in terms of its perceived crispness and freshness (see Fig. 4.1).[6] The sounds of the participants biting into the crisps were picked up by microphone, modified, and then played back in real-time over headphones. Importantly, the crisps were rated as tasting significantly crisper and fresher if the overall sound level was increased, or if just the high-frequency components of the crisp-biting sound were boosted. Intriguingly, no such crossmodal effect was obtained if the crunching sounds were temporally desynchronized from the act of biting into the potato chips, a common indicator that

---

[6] This research garnered the authors the 2008 IG Nobel prize for nutrition.

the effect one is looking at relies on multisensory integration (see Calvert, Spence, & Stein, 2004). Intriguingly, subsequent research has revealed that people's perception of the crispness of potato chips can also be modified, albeit more subtly, by changing the sound of the packaging that people hold (Spence, Shankar & Blumenthal, 2011) as well.

Elsewhere researchers have noted that the perception of carbonation in a beverage in a cup can also be modulated by what a person happens to hear (Zampini & Spence, 2005). Such results demonstrate that much of our perception of the texture of foods in the mouth can sometimes depend on the sounds that we hear while eating and drinking (both the sound of the food or drink itself, and, on occasion, the sound of the packaging in which that food or drink comes). Note here that even the sounds that one hears when a drink is poured into a glass may provide useful information about the drink's likely temperature (hot vs cold) when grasped or brought to the lips (see Velasco, Jones, King, & Spence, 2013a,b). There have also been reports that the sounds of carbonation can be used to distinguish between different kinds of beer (Stummerer & Hablesreiter, 2010), or between soda water, prosecco, and champagne (Spence & Wang, 2015). To the extent that our expectations (both sensory and hedonic) anchor our subsequent experiences when it comes to tasting (see Piqueras-Fiszman & Spence, 2015), these informative cues may well be expected to influence a person's subsequent rating of the experienced temperature of what they are drinking. That said, though, it should be noted that the effects of modulating the sounds of carbonation in Zampini and Spence's (2005) study were only documented when the beverage was held in the hand, not once the liquid had been transferred to the participant's mouth. However, it is worth bearing in mind that one of the limitations with Zampini and Spence's study is that their manipulation of the carbonation sounds had little ecological validity. Hence, a different result might well have been obtained had the sonic manipulation been more realistic.

4 Oral-somatosensation: Neural substrates

Oral-somatosensory information regarding the food or liquid is transferred from the mouth to the brain by means of the trigeminal nerve, which projects directly to the primary somatosensory cortex (see Simon, de Araujo, Gutierrez, & Nicolelis, 2006; Wang, Volkow, Felder, Fowler, Levy, Pappas, et al., 2002). Thereafter, the projections extend throughout the primary gustatory cortex (Small, 2012). These projections carry information concerning touch, texture (mouthfeel), temperature, and proprioception (as well as nociception or oral pain, and chemical irritation) from the relevant receptors within the oral cavity. Oral texture appears to be represented in the orbitofrontal cortex (OFC) as well as in other parts of the brain (Cerf-Ducastel, van de Moortele, Macleod, Le Bihan, & Faurion, 2001; Eldeghaidy Marciani, McGlone, Hollowood, Hort, Head, et al., 2011; Shepherd, 2012; Small, 2012; Verhagen & Engelen, 2006). For example, De Araujo and Rolls (2004) demonstrated that the texture of fatty foods "lights-up" the cingulate cortex—a part of the brain known to be involved in the formation and processing of emotions as well as memories. Indeed, evolutionarily speaking, it may have been particularly important for our ancestors to detect the textures of fatty foods, since that would normally have been a good signal that the food constituted a worthwhile source of energy (cf. Allen, 2012).

The neuroimaging data also reveal that an individual's taster status impacts on their experience of oral-somatosensory food qualities (Eldeghaidy et al., 2011). In particular, those individuals who are referred to as "supertasters" (Bartoshuk, 2000; though see Stuckey, 2012) tend to be a little more sensitive to fat in foods such as salad dressing and ice-cream (Kirkmeyer & Tepper, 2005; Tepper & Nurse, 1997; though see Yackinous & Guinard, 2001) and to creaminess (and sweetness) in sweetened water or milk-based products (Hayes & Duffy, 2007, 2008; see also chapter: Individual Differences in Multisensory Flavor Perception). Intriguingly, Wang et al. (2002) have even reported that obese individuals exhibit increased resting activity in the oral-somatosensory cortex as compared to those individuals of normal weight.

5 The role of oral-somatosensation in determining our food likes/dislikes

It is worth pausing here to note how many of our food likes and dislikes are contingent on the oral-somatosensory texture of particular foodstuffs (see Prescott, 2012; see also Munoz & Civille, 1987). So, for example, many Asian consumers find the texture of rice pudding to be more off-putting than its taste (or flavor). By contrast, for the Westerner breakfasting in Japan, the fermented black natto has a texture and consistency that isn't quickly forgotten. As pointed out by Prescott (2012, pp. 25–26): *"Other less obvious tactile sensations are also important in food acceptability. In particular, problems with texture are a common reason for rejecting foods."* Just take, for example, the oyster—it is this shellfish's slippery, slimy texture, not its taste or flavor, that people typically find so objectionable (Lyman, 1989). As Prescott puts it: *"The oyster is a pre-eminent example of the role that texture often plays as a reason for rejection of a food."*[7]

It should, though, be noted that a food's textural properties can also constitute a key part of what we find so pleasing about the foods that we love. Indeed, a number of researchers have argued that is part of the appeal of chocolate, one of the few foods to melt at mouth temperature (Stuckey, 2012).[8] Texture, then, plays a crucial role in determining our perception of a food's quality, its acceptability, and ultimately our food and beverage preferences (Guinard & Mazzuchelli, 1996; Szczesniak, 2002). In fact, it is worth noting that comfort foods typically have a soft texture (just think, for

---

[7] What A.A. Gill, the British food critic, memorably described as *"Sea-snot on a half-shell"* (Gill, 2009; see Prescott, 2012, p. 26).
[8] Just try eating a very cold versus a warm piece of chocolate to experience this difference.

instance, of mashed potatoes, apple sauce, and many puddings). Foods having this texture tend to be thought of as comforting and nurturing (Dornenburg & Page, 1996, p. 31). By contrast, many snack foods are crispy, like chips and pretzels. Texture contrast is something that many chefs and food developers work with (Stuckey, 2012, pp. 93–95). More generally, it is known to be something that consumers value in food (Szczesniak & Kahn, 1984). As Stuckey (2012, p. 93) puts it: *"Good chefs go to great lengths to add texture contrast to their plates, utilizing four different approaches: within a meal, on the plate, within a complex food, and within a simple food."*

6 The role of oral-somatosensation in expected and experienced satiety

Expectations about the capacity of foods to give rise to satiation clearly differ across a broad range of food products. Such expectations also depend on the energy content of the selected food, a consumer's familiarity with the food, and its appropriateness for the specific eating occasion, among other factors (Brunstrom, Shakeshaft, & Scott-Samuel, 2008). Expectations may also differ between foods within a given product category where the familiarity and appropriateness of the foods may be similar. In this situation, food texture has been demonstrated to be a key dimension modulating people's expectations about how filling a food is, were it to be consumed in its entirety. Drinks, such as juice, tend to be less satiating than iso-caloric semisolid and solid products (such as, eg, the apples that are used to make that juice; eg, de Wijk, Zijlstra, Mars, de Graaf, & Prinz, 2008; see Fig. 4.2). It has been suggested that this difference in satiating capacity may, among other things, be attributable to the reduced sensory stimulation that occurs during the consumption of liquid foods as compared to their semisolid counterparts (see also de Wijk, Polet, & Engelen, 2004).

Hogenkamp, Stafleu, Mars, Brunstrom, and de Graaf (2011) investigated the role of sensory attributes and the means of consumption (straw or spoon) in the expected satiation associated with a variety of dairy-based products. A strong effect of texture was obtained, but not of flavor, nor of the means of consumption, on expected satiation. Increased thickness was positively correlated with expected satiation. In one study, McCrickerd, Chambers, Brunstrom, and Yeomans (2012) demonstrated that adding subtle thick and creamy sensory cues to a beverage (without changing its nutrient content) increased the expectation that the beverage would be filling and would suppress hunger to a greater extent than the same drink without these added characteristics. So why, exactly, does this happen? It would seem likely that creamy textural cues have previously been associated with nutrient-rich foods (Bertenshaw, Lluch, & Yeomans, 2008, 2013), and so we tend to rely on satiety-relevant sensory cues when it comes to estimating the satiating power of other foods that have similar sensory characteristics. Such a view would certainly support the notion that oral-somatosensory cues act as a nutrient sensor (Woods, 2009) which eventually helps us determine when we have eaten enough.

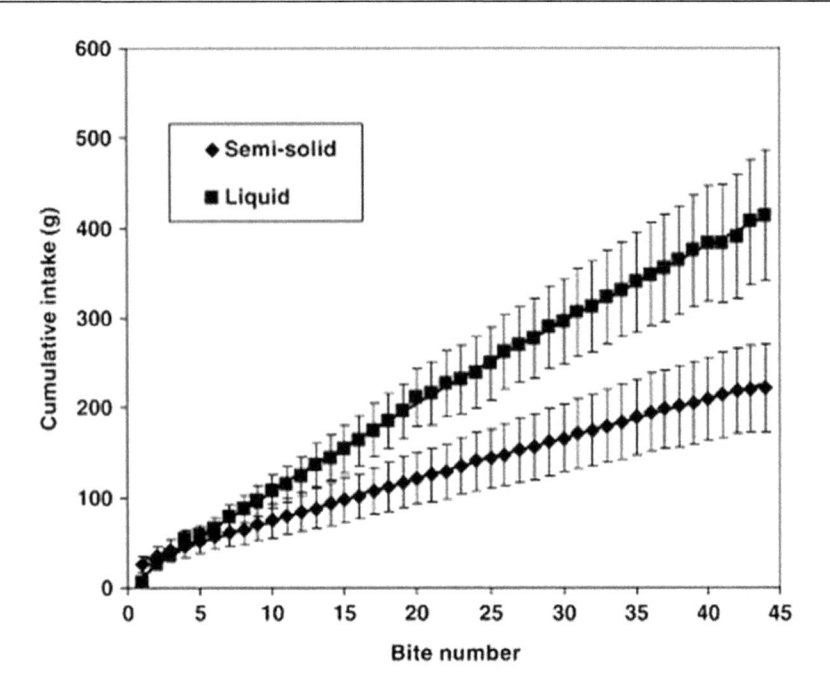

Figure 4.2 Cumulative amount (± SEM) of liquid and semi-solid milk-based products ingested over 15 min (± SEM) in the first study, where subjects sipped the liquid or semi-solid by means of a thick straw.
Source: Adapted from de Wijk et al. (2008).

7 On the contribution of somatosensory stimulation outside the mouth

While the focus of this chapter has been on the specific contribution of oral-somatosensation to flavor perception and our enjoyment of food and drink, it is important to note that what we feel in the hand, not to mention on other skin surfaces, also influences our perception of food texture, and expected satiation (see Gallace & Spence, 2014; Spence, Hobkinson, Gallace, & Piqueras-Fiszman, 2013; Spence & Piqueras-Fiszman, 2014, for reviews). To give just a few examples here, Barnett-Cowan (2010) demonstrated that the felt texture of a pretzel in the hand biased the participants' judgments of the pretzel they were evaluating in the mouth. In this study, the two halves of one and the same pretzel were given different textures (somewhat soggy, or kept fresh and crunchy).[9] Elsewhere, Piqueras-Fiszman and her colleagues have conducted

---

[9] Given a single substrate, the "unity assumption" may have been evoked (see Woods, Poliakoff, Lloyd, Dijksterhuis, & Thomas, 2010). Potentially also relevant here when trying to interpret the likely neural substrates underlying such effects, Francis, Rolls et al. (1999) have shown previously that the pleasant feeling associated with touching affectively positive surfaces (eg, velvet) with the hand gives rise to activation of the OFC.

Figure 4.3 The heavy textured cutlery served as part of the last course at the recently reopened The Fat Duck restaurant in Bray.
Source: Figure taken by Charles Spence.

a number of studies to show how the weight and texture of the food packaging and plateware we eat from also influence our judgments of the sensory attributes of the food, as well as influencing expected satiation (Piqueras-Fiszman, Harrar, Roura, & Spence, 2011; Piqueras-Fiszman & Spence, 2012; Spence & Piqueras-Fiszman, 2011). Intriguingly, even the weight of the cutlery in the hand can influence our expectations concerning, what we eat and hence our experience of it (Michel, Velasco, & Spence, 2015; Piqueras-Fiszman, & Spence, 2011). Thus, just as suspected by the Italian Futurists early in the last century,[10] what we experience in-mouth often depends, at least in part, on what we experience elsewhere on the skin surface (see also Biggs, Juravle, & Spence, 2016). These insights were brought together in the heavy furry spoon that diners are presented with when consuming their final course on The Fat Duck restaurant tasting menu (see Fig. 4.3).

One intriguing recent trend that has emerged at a number of the world's top three Michelin starred restaurants (think Noma in Copenhagen, The Fat Duck in Bray, and Mugaritz in San Sebastian) is that the diners have to eat the first three, four, or even five courses on the tasting menu with their hands/fingers (see Fig. 4.4). Eating with one's hands would have been unthinkable in a top restaurant last century (see http://en.wikipedia.org/wiki/Debretts; Furness, 2012). It can be argued that the introduction of "finger food" in these restaurants hints at a growing interest in the role of touch, especially the sensations conveyed by the fingers and hands, in setting our expectations concerning food.

---

[10] F.T. Marinetti's suggestion was that in order to maximally stimulate the senses, diners should wear pyjamas made of (or covered by) differently textured materials such as cork, sponge, sandpaper, and/or felt and eat without the aid of knives and forks to enhance the tactile sensations (see David, 1987, p. 61; Harrison, 2001, pp. 1–2).

Figure 4.4 (A) Beef tartare à la Viking with wood sorrel, bread crumbs, and tarragon cream. One of the early dishes served at Noma, and eaten with the hands. (B) Nasturtium blossoms. Another of the innovative new forms of cutlery/ways of eating with the hands from Noma. *Source*: Figure taken from http://megzimbeck.com/2011/10/getting-over-noma/.

8 Conclusions

The evidence that has been reviewed in this chapter clearly demonstrates just how important oral-somatosensation is to our experience and enjoyment of food and drink. The difficulties associated with working in this area, not to mention the intuitive idea that this is one of the senses that does not really contribute much to the pleasure that we derive from food (when compared to taste and odor), have most likely limited the development of our understanding in this area. Nevertheless, whether we realize it or not (and mostly the evidence suggests that we do not), oral-somatosensory inputs are involved in localizing flavors to the mouth in the first place, and thereafter play a key role in determining why we like (or dislike) what we do. Mouthfeel is clearly an important attribute for many foods/drinks, just think, for example, of a young oaked wine, or the sensation associated with running your tongue against your palate after taking a mouthful of coffee with cream or tasting an olive oil (see van Aken, 2013a, 2013b).

Most importantly, and beyond hedonics, understanding the relationships that exist between food microstructure, texture, and sensory perception may one day allow us to design foods with specific functional properties, such as foods that deliver an enhanced perception of sweetness or saltiness. Tracking tongue movements in the oral cavity, although technically challenging, is of great interest when it comes to quantifying oral processing in relation to texture and taste perception (see de Wijk, Engelen, & Prinz, 2003; Prinz, 2008). Currently, experimental methodologies to monitor tongue movements under normal eating conditions are limited. Techniques such as electromagnetic articulography will need to be further developed in order to quantify the spatial movement of the tongue during the oral processing of foods. Eventually, this might allow us to gain a better understanding of the oral processes and perceptual mechanisms involved in the sensory perception of foods which could one day be applied in the development of salt- and sugar-reduced foods, as well as in terms of foods/meals with enhanced textural interest (Shepherd, 2012). A large proportion of the population, including the elderly and children, for whom the texture of foods is a key element for their enjoyment of a meal, will then hopefully benefit from advances in this area of research. Finally, although beyond the remit of this chapter, one probably also needs to consider the role of salivation in changing the texture of (especially dry) food during oral mastication (Lyman, 1989; Spence, 2011).

References

Alais, D., & Burr, D. (2004). The ventriloquist effect results from near-optimal bimodal integration. *Current Biology, 14*, 257–262.
Allen, J. S. (2012). *The omnivorous mind: Our evolving relationship with food*. London, UK: Harvard University Press.
Anon. (2005). Mouth feel. *New Scientist, July 23*, 47.
Anon. (2013). Hot or not? How serving temperature affects the way food tastes. The Guardian (Word of Mouth Blog), September 17th. Downloaded from <http://www.theguardian.com/lifeandstyle/wordofmouth/2013/sep/17/serving-temperature-affects-taste-food> on 18.11.15.
Auvray, M., & Spence, C. (2008). The multisensory perception of flavor. *Consciousness and Cognition, 17*, 1016–1031.
Bajec, M. R., Pickering, G. J., & DeCourville, N. (2012). Influence of stimulus temperature on orosensory perception and variation with taste phenotype. *Chemosensory Perception, 5*, 243–265.
Barnett-Cowan, M. (2010). An illusion you can sink your teeth into: Haptic cues modulate the perceived freshness and crispness of pretzels. *Perception, 39*, 1684–1686.
Bartoshuk, L. M. (2000). Comparing sensory experiences across individuals: Recent psychophysical advances illuminate genetic variation in taste perception. *Chemical Senses, 25*, 447–460.
Bertenshaw, E. J., Lluch, A., & Yeomans, M. R. (2009). Dose-dependent effects of beverage protein content upon short-term intake. *Appetite, 52*, 580–587.
Bertenshaw, E. J., Lluch, A., & Yeomans, M. R. (2013). Perceived thickness and creaminess modulates the short-term satiating effects of high-protein drinks. *British Journal of Nutrition, 110*, 578–586.
Biggs, L., Juravle, G., & Spence, C. (2016). Haptic exploration of plateware alters the perceived texture and taste of food. *Food Quality & Preference, 50*, 129–134.
Blumenthal, H. (2008). *The big Fat Duck cookbook*. London, UK: Bloomsbury.
Bourne, M. C. (1982). *Food texture and viscosity*. New York, NY: Academic Press.

Breslin, P. A. S., Gilmore, M. M., Beauchamp, G. K., & Green, B. G. (1993). Psychophysical evidence that oral astringency is a tactile sensation. *Chemical Senses, 18*, 405–417.

Brunstrom, J. M., Shakeshaft, N. G., & Scott-Samuel, N. E. (2008). Measuring expected satiety in a range of common foods using a method of constant stimuli. *Appetite, 51*, 604–614.

Bult, J. H. F., de Wijk, R. A., & Hummel, T. (2007). Investigations on multimodal sensory integration: Texture, taste, and ortho- and retronasal olfactory stimuli in concert. *Neuroscience Letters, 411*, 6–10.

Calvert, G. A., Spence, C., & Stein, B. E. (Eds.), (2004). *The handbook of multisensory processing*. Cambridge, MA: MIT Press.

Cerf-Ducastel, B., Van de Moortele, P.-F., Macleod, P., Le Bihan, D., & Faurion, A. (2001). Interaction of gustatory and lingual somatosensory perceptions at the cortical level in the human: A functional magnetic resonance imaging study. *Chemical Senses, 26*, 371–383.

Chandrashekar, J., Yarmolinsky, D., von Buchholtz, L., Oka, Y., Sly, W., Ryba, N. J. P., et al. (2009). The taste of carbonation. *Science, 326*, 443–445.

Christensen, C. M. (1980). Effects of solution viscosity on perceived saltiness and sweetness. *Perception & Psychophysics, 28*, 347–353.

Christensen, C. M. (1984). Food texture perception. In E. Mark (Ed.), *Advances in food research* (pp. 159–199). New York, NY: Academic Press.

Chylinski, M., Northey, G., & Ngo, L. V. (2015). Cross-modal interactions between color and texture of food. *Psychology & Marketing, 32*, 950–966.

Cruz, A., & Green, B. G. (2000). Thermal stimulation of taste. *Nature, 403*, 889–892.

David, E. (1987). *Italian food*. London, UK: Barrie & Jenkins.

De Araujo, I. E., & Rolls, E. T. (2004). Representation in the human brain of food texture and oral fat. *Journal of Neuroscience, 24*, 3086–3093.

de Wijk, R. A., Engelen, L., & Prinz, J. F. (2003). The role of intra-oral manipulation in the perception of sensory attributes. *Appetite, 40*, 1–7.

de Wijk, R. A., Polet, I. A., Engelen, L., et al. (2004). Amount of ingested custard dessert as affected by its color, odor, and texture. *Physiology & Behavior, 82*, 397–403.

de Wijk, R. A., Zijlstra, N., Mars, M., De Graaf, C., & Prinz, J. F. (2008). The effects of food viscosity on bite size, bite effort and food intake. *Physiology & Behavior, 95*, 527–532.

Delwiche, J. (2004). The impact of perceptual interactions on perceived flavor. *Food Quality and Preference, 15*, 137–146.

Delwiche, J. F. (2003). Attributes believed to impact flavor: An opinion survey. *Journal of Sensory Studies, 18*, 437–444.

Dematté, M. L., Pojer, N., Endrizzi, I., Corollaro, M. L., Betta, E., Aprea, E., et al. (2014). Effects of the sound of the bite on apple perceived crispness and hardness. *Food Quality and Preference, 38*, 58–64.

Diaz, M. E. (2004). Comparison between orthonasal and retronasal flavour perception at different concentrations. *Flavour and Fragrance Journal, 19*, 499–504.

Dornenburg, A., & Page, K. (1996). *Culinary artistry*. New York, NY: John Wiley & Sons.

Eldeghaidy, S., Marciani, L., McGlone, F., Hollowood, T., Hort, J., Head, K., et al. (2011). The cortical response to the oral perception of fat emulsions and the effect of taster status. *Journal of Neurophysiology, 105*, 2572–2581.

Engelen, L., De Wijk, R. A., Prinz, J. F., et al. (2002). The effect of oral temperature on the temperature perception of liquids and semisolids in the mouth. *European Journal of Oral Science, 110*, 412–416.

Francis, S., Rolls, E. T., Bowtell, R., McGlone, F., O'Doherty, J., Browning, A., et al. (1999). The representation of pleasant touch in the brain and its relationship with taste and olfactory areas. *Neuroreport, 10*, 453–459.

Frost, M. B., & Janhoj, T. (2007). Understanding creaminess. *International Dairy Journal, 17*, 1298–1311.

Furness, H. (2012). *How to eat with one's fingers: The Debrett's guide to very modern etiquette.* [Online], Available: <http://www.telegraph.co.uk/foodanddrink/foodanddrinknews/9696223/How-to-eat-with-ones-fingers-the-Debretts-guide-to-very-modern-etiquette.html> 18.06.13.

Gallace, A., & Spence, C. (2014). *In touch with the future: The sense of touch from cognitive neuroscience to virtual reality.* Oxford, UK: Oxford University Press.

Gawel, R. (2006). *Importance of "texture" to red wine quality acknowledged by the development of a red wine mouth-feel wheel.* Downloaded from <http://www.aromadictionary.com/articles/winemouthfeel_article.html> on 01.02.06.

Gawel, R., Oberholster, A., & Francis, I. L. (2000). A "mouth-feel wheel": Terminology for communicating the mouth-feel characteristics of red wine. *Australian Society of Viticulture and Oenology, 6*, 203–207.

Gill, A. (2009). Sea snot on the half shell. *The Sunday Times*, November 8.

Green, B. G. (2002). Studying taste as a cutaneous sense. *Food Quality and Preference, 14*, 99–109.

Green, B. G., & George, P. (2004). "Thermal taste" predicts higher responsiveness to chemical taste and flavor. *Chemical Senses, 29*, 617–628.

Green, B. G., & Lawless, H. T. (1991). The psychophysics of somatosensory chemoreception in the nose and mouth. In T. V. Getchell, R. L. Doty, L. M. Bartoshuk, & J. B. Snow (Eds.), *Smell and taste in health and disease* (pp. 235–253). New York, NY: Raven Press.

Guinard, J.-X., & Mazzucchelli, R. (1996). The sensory perception of texture and mouthfeel. *Trends in Food Science & Technology, 7*, 213–219.

Harrison, J. (2001). *Synaesthesia: The strangest thing.* Oxford, UK: Oxford University Press.

Hayes, J. E., & Duffy, V. B. (2007). Revisiting sugar-fat mixtures: Sweetness and creaminess vary with phenotypic markers of oral sensation. *Chemical Senses, 32*, 225–236.

Hayes, J. E., & Duffy, V. B. (2008). Oral sensory phenotype identifies level of sugar and fat required for maximal liking. *Physiology and Behaviour, 95*, 77–87.

Hogenkamp, P. S., Stafleu, A., Mars, M., Brunstrom, J. M., & de Graaf, C. (2011). Texture, not flavor, determines expected satiation of dairy products. *Appetite, 57*, 635–641.

Hollingworth, H. L., & Poffenberger, A. T. (1917). *The sense of taste.* New York, NY: Moffat Yard. Accessed from: <https://archive.org/stream/sensetaste01goog#page/n0/mode/2up>.

ISO (1992). *Standard 5492: Terms relating to sensory analysis.* International Organization for Standardization. Vienna: Austrian Standards Institute.

ISO (2008). *Standard 5492: Terms relating to sensory analysis.* International Organization for Standardization. Vienna: Austrian Standards Institute.

Jacobs, R., Serhal, C. B., & van Steenberghe, D. (1998). Oral stereognosis: A review of the literature. *Clinical Oral Investigations, 2*, 3–10.

Jowitt, R. (1974). The terminology of food texture. *Journal of Texture Studies, 5*, 351–358.

Kappes, S. M., Schmidt, S. J., & Lee, S.-Y. (2007). Relationship between physical properties and sensory attributes of carbonated beverages. *Journal of Food Science, 72*, S001–S011.

Keast, R. S. J., & Costanzo, A. (2015). Is fat the sixth taste primary? Evidence and implications. *Flavour, 4*, 5.

Kilcast, D. (2008). Creaminess. In H. Blumenthal (Ed.), *The big Fat Duck cookbook* (pp. 491–492). London, UK: Bloomsbury.

Kirkmeyer, S. V., & Tepper, B. J. (2005). Consumer reactions to creaminess and genetic sensitivity to 6-n-propylthiouracil: A multidimensional study. *Food Quality and Preference, 16*, 545–556.

Kramer, A. (1973). Food texture—Definition, measurement and relation to other food-quality attributes. In A. Kramer & A. S. Szczesniak (Eds.), *Texture measurements of foods* (pp. 55–64). Dordrecht, Holland: D. Reidel.

Kutter, A., Hanesch, C., Rauh, C., & Delgado, A. (2011). Impact of proprioception and tactile sensations in the mouth on the perceived thickness of semi-solid food. *Food Quality and Preference, 22*, 193–197.

Langstaff, S. A., Guinard, J.-X., & Lewis, M. J. (1991). Sensory evaluation of the mouthfeel of beer. *American Society of Brewing Chemists, 49*, 54–59.

Lim, J., & Green, B. G. (2008). Tactile interaction with taste localization: Influence of gustatory quality and intensity. *Chemical Senses, 33*, 137–143.

Lim, J., & Johnson, M. B. (2011). Potential mechanisms of retronasal odor referral to the mouth. *Chemical Senses, 36*, 283–289.

Lyman, B. (1989). *A psychology of food, more than a matter of taste.* New York, NY: Avi, van Nostrand Reinhold.

Marsilli, R. (1993). Texture and mouthfeel: Making rheology real. *Food Product Design*, August. Downloaded from http://www.foodproductdesign.com/archive/1993/0893QA.html

Masuda, M., & Okajima, K. (2011). *Added mastication sound affects food texture and pleasantness.* Poster presented at the 12th International Multisensory Research Forum meeting in Fukuoka, Japan, October 17–20.

Masuda, M., Yamaguchi, Y., Arai, K., & Okajima, K. (2008). Effect of auditory information on food recognition. *IEICE Technical Report, 108*(356), 123–126.

Mattes, R. D. (2009). Is there a fatty acid taste? *Annual Review of Nutrition, 29*, 305–327.

McBurney, D. H., Collings, V. B., & Glanz, L. M. (1973). Temperature dependence of human taste responses. *Physiology & Behavior, 11*, 89–94.

McCrickerd, K., Chambers, L., Brunstrom, J. M., & Yeomans, M. R. (2012). Subtle changes in the flavour and texture of a drink enhance expectations of satiety. *Flavour, 1*, 20.

Michel, C., Velasco, C., Salgado, A., & Spence, C. (2014). The butcher's tongue illusion. *Perception, 43*, 818–824.

Michel, C., Velasco, C., & Spence, C. (2015). Cutlery influences the perceived value of the food served in a realistic dining environment. *Flavour, 4*, 27.

Mony, P., Tokar, T., Pang, P., Fiegel, A., Meullenet, J.-F., & Seo, H.-S. (2013). Temperature of served water can modulate sensory perception and acceptance of food. *Food Quality and Preference, 28*, 449–455.

Mosca, A. C., van de Velde, F., Bult, J. H., van Boekel, M. A., & Stieger, M. (2012). Effect of gel texture and sucrose spatial distribution on sweetness perception. *LWT-Food Science and Technology, 46*, 183–188.

Munoz, A. M., & Civille, G. V. (1987). Factors affecting perception and acceptance of food texture by American consumers. *Food Reviews International, 3*, 285–322.

Murphy, C., & Cain, W. S. (1980). Taste and olfaction: Independence vs. interaction. *Physiology and Behavior, 24*, 601–605.

Nagata, H., Dalton, P., Doolittle, N., & Breslin, P. A. S. (2005). Psychophysical isolation of the modality responsible for detecting multimodal stimuli: A chemosensory example. *Journal of Experimental Psychology: Human Perception & Performance, 31*, 101–109.

Piqueras-Fiszman, B., & Spence, C. (2011). Do the material properties of cutlery affect the perception of the food you eat? An exploratory study. *Journal of Sensory Studies, 26*, 358–362.

Piqueras-Fiszman, B., & Spence, C. (2012). The influence of the feel of product packaging on the perception of the oral-somatosensory texture of food. *Food Quality & Preference, 26*, 67–73.

Piqueras-Fiszman, B., Harrar, V., Roura, E., & Spence, C. (2011). Does the weight of the dish influence our perception of food? *Food Quality & Preference*, *22*, 753–756.

Piqueras-Fiszman, B., & Spence, C. (2015). Sensory expectations based on product-extrinsic food cues: An interdisciplinary review of the empirical evidence and theoretical accounts. *Food Quality & Preference*, *40*, 165–179.

Powers, J. J., Howell, A. J., Lillard, D. A., & Vacinek, S. J. (1971). Effect of temperature on threshold values for citric acid, malic acid and quinine sulphate–energy of activation and extreme-value determination. *Journal of Science, Food, and Agriculture*, *22*, 543–547.

Prescott, J. (2012). *Taste matters: Why we like the foods we do*. London, UK: Reaktion Books.

Prinz, J. (2008). The role of the mouth in the appreciation of food. In H. Blumenthal (Ed.), *The big Fat Duck cookbook* (pp. 489–490). London, UK: Bloomsbury.

Roudnitzky, N., Bult, J. H. F., de Wijk, R. A., Reden, J., Schuster, B., & Hummel, T. (2011). Investigation of interactions between texture and ortho- and retronasal olfactory stimuli using psychophysical and electrophysiological approaches. *Behavioural Brain Research*, *216*, 109–115.

Rozin, P. (1982). "Taste-smell confusions" and the duality of the olfactory sense. *Perception & Psychophysics*, *31*, 397–401.

Sato, M. (1967). Gustatory response as a temperature-dependent process. *Contributions to Sensory Physiology*, *2*, 223–251.

Shepherd, G. M. (2012). *Neurogastronomy: How the brain creates flavor and why it matters*. New York, NY: Columbia University Press.

Simon, S. A., de Araujo, I. E., Gutierrez, R., & Nicolelis, M. A. L. (2006). The neural mechanisms of gustation: A distributed processing code. *Nature Reviews Neuroscience*, *7*, 890–901.

Slocombe, B. G., Carmichael, D. A., & Simner, J. (in press). Cross-modal tactile-taste interactions in food evaluations. *Neuropsychologia*.

Small, D. M. (2012). Flavor is in the brain. *Physiology and Behavior*, *107*, 540–552.

Small, D. M., Gerber, J. C., Mak, Y. E., & Hummel, T. (2005). Differential neural responses evoked by orthonasal versus retronasal odorant perception in humans. *Neuron*, *47*, 593–605.

Snyder, D. J., Prescott, J., & Bartoshuk, L. M. (2006). Modern psychophysics and the assessment of human oral sensation. *Acta Otorhinolaryngology*, *63*, 221–241.

Spence, C. (2011). Mouth-watering: The influence of environmental and cognitive factors on salivation and gustatory/flavour perception. *Journal of Texture Studies*, *42*, 157–171.

Spence, C. (2012). Multi-sensory integration & the psychophysics of flavour perception. In J. Chen & L. Engelen (Eds.), *Food oral processing–Fundamentals of eating and sensory perception* (pp. 203–219). Oxford, UK: Blackwell.

Spence, C. (2015a). Eating with our ears: Assessing the importance of the sounds of consumption to our perception and enjoyment of multisensory flavour experiences. *Flavour*, *4*, 3.

Spence, C. (2015b). On the psychological impact of food colour. *Flavour*, *4*, 21.

Spence, C. (2016). Oral referral: Mislocalizing odours to the mouth. *Food Quality & Preference*, *50*, 117–128.

Spence, C., Hobkinson, C., Gallace, A., & Piqueras-Fiszman, B. (2013). A touch of gastronomy. *Flavour*, *2*, 14.

Spence, C., & Piqueras-Fiszman, B. (2011). Multisensory design: Weight and multisensory product perception. In G. (2011). Hollington (Ed.), *Proceedings of right weight* (2, pp. 8–18). London: Materials KTN.

Spence, C., & Piqueras-Fiszman, B. (2014). *The perfect meal: The multisensory science of food and dining*. Oxford, UK: Wiley-Blackwell.

Spence, C., Shankar, M. U., & Blumenthal, H. (2011). Sound bites": Auditory contributions to the perception and consumption of food and drink. In F. Bacci & D. Melcher (Eds.), *Art and the senses* (pp. 207–238). Oxford, UK: Oxford University Press.

Spence, C., Smith, B., & Auvray, M. (2015). Confusing tastes and flavours. In D. Stokes, M. Matthen, & S. Biggs (Eds.), *Perception and its modalities* (pp. 247–274). Oxford, UK: Oxford University Press.

Spence, C., & Squire, S. B. (2003). Multisensory integration: Maintaining the perception of synchrony. *Current Biology, 13*, R519–R521.

Spence, C. & Wang, Q.(J.) (2015). *Krug champagne workshop*. Berlin, March.

Stevenson, R. J. (2009). *The psychology of flavour*. Oxford, UK: Oxford University Press.

Stevenson, R. J. (2012). Multisensory interactions in flavor perception. In B. E. Stein (Ed.), *The new handbook of multisensory processes* (pp. 283–299). Cambridge, MA: MIT Press.

Stevenson, R. J., & Mahmut, M. K. (2011). Experience dependent changes in odour-viscosity perception. *Acta Psychologica, 136*, 60–66.

Stevenson, R. J., Oaten, M. J., & Mahmut, M. K. (2011). The role of taste and oral somatosensation in olfactory localization. *Quarterly Journal of Experimental Psychology, 64*, 224–240.

Stieger, M., & van de Velde, F. (2013). Microstructure, texture and oral processing: New ways to reduce sugar and salt in foods. *Current Opinion in Colloid & Interface Science, 18*, 334–348.

Stuckey, B. (2012). *Taste what you're missing: The passionate eater's guide to why good food tastes good*. London, UK: Free Press.

Stummerer, S., & Hablesreiter, M. (2010). *Food design XL*. New York, NY: Springer.

Sundqvist, N. C., Stevenson, R. J., & Bishop, I. R. J. (2006). Can odours acquire fat-like properties? *Appetite, 47*, 91–99.

Szczesniak, A. S. (1979). Classification of mouthfeel characteristics of beverages. In P. Sherman (Ed.), *Food texture and rheology* (pp. 1–20). London, UK: Academic Press.

Szczesniak, A. S. (1988). The meaning of textural characteristics–crispness. *Journal of Texture Studies, 19*, 51–59.

Szczesniak, A. S. (1990). Psychorheology and texture as factors controlling consumer acceptance of food. *Cereal Foods World, 351*, 1201–1205.

Szczesniak, A. S. (2002). Texture is a sensory property. *Food Quality and Preference, 13*, 215–225.

Szczesniak, A. S., & Kahn, E. L. (1971). Consumer awareness of and attitudes to food texture. I: Adults. *Journal of Texture Studies, 2*, 280–295.

Szczesniak, A. S., & Kahn, E. L. (1984). Texture contrasts and combinations: A valued consumer attribute. *Journal of Texture Studies, 15*, 285–301.

Szczesniak, A. S., & Kleyn, D. H. (1963). Consumer awareness of texture and other food attributes. *Food Technology, 17*, 74–77.

Talavera, K., Ninomiya, Y., Winkel, C., Voets, T., & Nilius, B. (2007). Influence of temperature on taste perception. *Cellular, Molecular, & Life Sciences, 64*, 377–381.

Tepper, B. J., & Nurse, R. J. (1997). Fat perception is related to PROP taster status. *Physiology & Behavior, 61*, 949–954.

Todrank, J., & Bartoshuk, L. M. (1991). A taste illusion: Taste sensation localized by touch. *Physiology & Behavior, 50*, 1027–1031.

Topolinski, S., & Türk Pereira, P. (2012). Mapping the tip of the tongue—deprivation, sensory sensitisation, and oral haptics. *Perception, 41*, 71–92.

Tournier, C., Sulmont-Rossé, C., & Guichard, E. (2007). Flavour perception: Aroma, taste and texture interactions. *Food, 1*, 246–257.

Tournier, C., Sulmont-Rossé, C., Sémon, E., Vignon, A., Issanchou, S., & Guichard, E. (2009). A study on texture-taste-aroma interactions: Physico-chemical and cognitive mechanisms. *International Dairy Journal, 19*, 450–458.

Trivedi, B. (2012). Hardwired for taste: Research into human taste receptors extends beyond the tongue to some unexpected places. *Nature, 486*, S7.
van Aken, G.A. (2013a). *Listening to what the tongue feels*. Downloaded from <http://www.nizo.com/news/latest-news/67/listening-to-what-the-tongue-feels/> on 01.08.14.
van Aken, G. A. (2013b). Acoustic emission measurement of rubbing and tapping contacts of skin and tongue surface in relation to tactile perception. *Food Hydrocolloids, 31*, 325–331.
Velasco, C., Jones, R., King, S., & Spence, C. (2013a). "Hot or cold?" On the informative value of auditory cues in the perception of the temperature of a beverage. In K. Bronner, R. Hirt, & C. Ringe (Eds.), *(((ABA))) Audio Branding Academy Yearbook 2012/2013* (pp. 177–187). Baden-Baden: Nomos.
Velasco, C., Jones, R., King, S., & Spence, C. (2013b). The sound of temperature: What information do pouring sounds convey concerning the temperature of a beverage. *Journal of Sensory Studies, 28*, 335–345.
Verhagen, J. V., & Engelen, L. (2006). The neurocognitive bases of human multimodal food perception: Sensory integration. *Neuroscience and Biobehavioral Reviews, 30*, 613–650.
Wang, G.-J., Volkow, N. D., Felder, C., Fowler, J. S., Levy, A. V., Pappas, N. R., et al. (2002). Enhanced resting state activity of the oral somatosensory cortex in obese subjects. *Neuroreport, 13*, 1151–1155.
Weel, K. G. C., Boelrijk, A. C., Alting, P. J. J. M., van Mil, J. J., Burger, H., Gruppen, H., et al. (2002). Flavor release and perception of flavored whey protein gels: Perception is determined by texture rather than by release. *Journal of Agricultural and Food Chemistry, 50*, 5149–5155.
Woods, A. T., Poliakoff, E., Lloyd, D. M., Dijksterhuis, G. B., & Thomas, A. (2010). Flavor expectation: The effects of assuming homogeneity on drink perception. *Chemosensory Perception, 3*, 174–181.
Woods, S. C. (2009). The control of food intake: Behavioral versus molecular perspectives. *Cell Metabolism, 9*, 489–498.
Yackinous, C., & Guinard, J.-X. (2001). Relation between PROP taster status and fat perception, touch, and olfaction. *Physiology & Behavior, 72*, 427–437.
Zampini, M., & Spence, C. (2004). The role of auditory cues in modulating the perceived crispness and staleness of potato chips. *Journal of Sensory Science, 19*, 347–363.
Zampini, M., & Spence, C. (2005). Modifying the multisensory perception of a carbonated beverage using auditory cues. *Food Quality and Preference, 16*, 632–641.

Sound:
The Forgotten Flavor Sense

Charles Spence
Crossmodal Research Laboratory, Department of Experimental Psychology,
University of Oxford, Oxford, United Kingdom

1 Introduction

Sound definitely takes a back seat as far as research into the field of multisensory flavor perception is concerned (see Spence, 2012, 2015a, for reviews). In fact, when quizzed a few years ago, the majority of sensory scientists appeared to share the view that what we hear simply is not an important contributor to our tasting experiences (Delwiche, 2003). As Spence (2015a) noted recently, it is hard to find a review paper or book on multisensory flavor perception that devotes more than 3% of the text to sound.[1] Consumers apparently share the same view, as they too failed to rate sound as a particularly important aspect of their interaction with food and drink products in another study (Schifferstein, 2006). In this chapter, though, I want to argue (and hopefully convince you) that sound, in all its guises, is actually much more important than most of us realize to the ways in which we experience food and drink. My view is very much that sound is "the forgotten flavor sense."[2]

At the outset, though, it is crucial to highlight the broad range of different ways in which what we hear while, or prior to, eating and drinking can affect our sensory-discriminative and hedonic responses to that which we are tasting. There are, for instance, often sounds associated with the preparation of foods and beverages (think here only of the sizzle of the steak or the ding of the microwave). Then there are sounds of the packaging (think of the pop of the champagne cork or the distinctive sound of someone opening a bottle of Snapple). Such food-related sounds can undoubtedly set our taste and flavor expectations, and, by so doing, anchor and thus modulate our subsequent taste and flavor experiences (see Piqueras-Fiszman & Spence, 2015, for a review). At the same time, however, our physical interactions with food and drink while biting, chewing, masticating, sucking, and slurping all generate potentially informative auditory cues that may influence our perception of the textural properties of the food, and even the oral-somatosensory perception of mouthfeel. It can be argued that the crunchy, crackly, crispy, carbonated, creamy, and squeaky sounds of food and drink consumption all have value to the consumer.

---

[1] If the senses were all given equal weighting, one might have expected a figure closer to 20%, assuming that there are five main senses that contribute to flavor experiences.
[2] I will, though, leave aside the question of whether that means that sound should be considered as constitutive of multisensory flavor perception, or merely a modulatory factor (see Spence, Smith, & Auvray, 2015, for a review).

Beyond these informative sounds, though, one also needs to consider the impact of any ambient noise and/or background music. Once again, both have been shown to have a much more pronounced effect (both positive and negative) on the tasting experience than most people probably realize (see Spence, 2014a; Spence, Michel, & Smith, 2014, for reviews). Taken together, then, there is now every reason to believe that what we hear should be considered as an important factor that can modulate our taste and flavor expectations, as well as the ensuing perceptual experience when eating and drinking. I would argue that if all of the evidence that has been published over the last 70 years is put together, then it is, in fact, easy to support the conclusion that sound is one of the forgotten flavor senses.

Perhaps unsurprisingly, given the burgeoning body of evidence in the area of sound and food, a growing number of companies and brands are now starting to take these emerging insights and growing awareness concerning the importance of audition on board in the design of their food and beverage products, the packaging in which those products are presented to the consumer, not to mention the sound of both product and pack as presented auditorily in their advertising (see Spence, 2014b; Spence & Wang, 2015a, for reviews). Furthermore, a few of the most innovative companies are now starting to pay far more attention to the role of background noise, music, and even specially designed soundscapes in the consumers' (or, in some cases, diners') encounters with their food and beverage products/offerings as well (see Spence, 2014b; Spence & Wang, 2015b–d, for reviews). Intriguingly, a number of the developments in this area have been facilitated by the recent emergence of sensory apps (Crisinel, Jacquier, Deroy, & Spence, 2013), and even by digitally augmented glassware (Baker, 2015).

2 The sounds of food and beverage preparation

Marketers have known about the importance of food and beverage preparation sounds for a very long time now. In fact, one of the first great marketers, the legendary North American, Elmer Wheeler, became famous, at least in part, for his ubiquitous strapline that: "You sell the sizzle not the steak." (see Wheeler, 1938). One can perhaps think of this as one of the original examples of auditory sensory marketing (eg, Krishna, 2012, 2013; Spence, 2014b). It hints, I think, at the importance that the sounds of food preparation can have. It is, after all, the sizzle that helps set your expectations about the sensory qualities of the steak that is about to come (Stuckey, 2012). In the 1970s and 1980s (if memory serves me correctly), there was a long-running series of adverts here in the United Kingdom (and who knows, perhaps elsewhere too) revolving around a central character who, on having a surprise visit from some guests, would disappear off into the kitchen, only to find that there was no fresh coffee. He would then make a cup of instant while imitating the sounds of the coffee machine gurgling and spluttering out of sight of the guests sitting none the wiser in the living room. The advert was presumably so successful, at least in part, because it played on the importance of the sensory-discriminative and hedonic expectations that are set by the sounds of preparation. These preparation sounds are not part of the taste (or better said flavor) of the steak, nor of the delicious orthonasal aroma of the coffee, yet they very definitely do help to set the expectations (both sensory and hedonic) that anchor our subsequent tasting experiences (see Piqueras-Fiszman & Spence, 2015, for a recent review of the literature on sensory expectations).[3] And while it is certainly true to say that the majority of the literature in this area has tended to focus on the expectations that are set by the consumer's eyes, I would argue that the expectations set by their ears are oftentimes just as important (Spence & Wang, 2015a).

In other words, the marketers and advertisers would appear to have known long before the sensory scientists and gastrophysicists that the sounds of food and beverage preparation are key to setting the right expectations in the mind of the consumer. As an everyday example, think only of what runs through your mind when you hear the "ding" of the microwave when sitting at the dining table of some fancy restaurant or other. More recently, product designers have also started to take a growing interest in the sounds of preparation elicited by various kitchen gadgets. In one intriguing series of experiments, Knöferle (2011, 2012; see also Knöferle, Sprott, Landwehr, & Herrmann, 2014), a former postdoc here at the Crossmodal Research Laboratory in Oxford, used to bring participants into the lab, and make them a Nespresso. The participants had to rate the taste of the drink and how much they liked the experience. Klemens would make the pod coffee fresh for each participant with a standard Nespresso Lattissima coffee machine. The participants were able to hear, but not to see, their coffee being prepared. However, unbeknownst to them, the coffee machine had been modified so as to change the frequency distribution of the sounds that it made.[4] Despite the fact that exactly the same coffee was presented in both conditions, the results revealed that the participants rated the coffee as tasting around 10% better, and as being of significantly higher quality, having listened to the more pleasant sounding (ie, the less sharp) coffee machine soundtrack. A similar pattern of results was also documented in a second experiment in which a coffee machine was wired up to make one or other of the two sounds. The effect was particularly noticeable amongst those consumers who expressed a general interest in, and enjoyment of, product sounds.

It is interesting to note that 2014 saw the release of the Marlow (Houghton, 2014). This hot beverage vending machine would actually play the sounds of the coffee shop, not to mention squirting out the smell of coffee, when the customer made the appropriate drinks choice. As one writer put it: "Marlow's advanced multi-sensory coffee shop experience has the ability to captivate all five senses. Consumers will be able to hear, see, touch, smell and taste all the different elements that make Costa coffee shops so warm and welcoming." (Anon., 2014).[5] Given the importance of sound to the coffee-drinking experience, it would not seem like too much of a stretch

---

[3] And one shouldn't forget the classic black-and-white Maxwell House coffee ads featuring the "singing" coffee pot, with a voiceover encouraging the viewer to "listen to it perk" and to "listen to good coffee."
[4] Specifically, the sounds associated with the operation of the coffee machine were varied. Knöferle either boosted or cut the spectral contents by 20 dB between the frequencies of 2.5 and 6.5 kHz.
[5] Notice here that sound is the first sense that is mentioned! In this case, Costa apparently worked with eMixpro (who have worked with the likes of Coldplay, U2, and the Rolling Stones) in order to try to imitate the sounds of a bustling Costa coffee shop (Anon., 2014).

of the imagination to think that the manufacturers of kitchen appliances, such as coffee makers, but also mixers, liquidizers, etc., will increasingly start to design their appliances to give off a certain sound, much like those working in the car industry have been doing for years when they modify the sound of the engine and car door when it is closed to appeal to the ears of the driver (see Byron, 2012; Spence & Zampini, 2006, for reviews).

> *Sounds are associated with food preparation, such as the popping of corn, perking of coffee, simmering of liquids, bubbling of syrups, broiling of meats, frying of eggs, and cracking of nuts.*
> Amerine, Pangborn, and Roessler (1965, p. 227).

2.1 Interim summary

As the above quote makes clear, sensory scientists have been thinking about the sounds of preparation for half a century or more. However, I would argue that the last decade has seen an explosive growth of interest in modifying and enhancing the sounds of food and beverage preparation in order to set better taste and flavor expectations in the mind of the consumer. What is more, there is growing awareness amongst both sensory scientists and various commercial interests that congruent sounds can enhance the consumer's experience of a range of different food and beverage products (eg, see Seo & Hummel, 2011; Seo, Lohse, Luckett, & Hummel, 2014; and see Seo & Hummel, 2015, for a review).

3 The sounds of packaging

While rarely given the consideration it deserves, any sounds that are made by the packaging of a food or beverage product prior to the consumption of the contents can also help to set sensory and hedonic expectations in the mind of the consumer. Here, one could think of the distinctive (or signature) sound of opening of certain packaging, for example, the Snapple pop (Byron, 2012; see Spence & Wang, 2015a, for a review). A more generic sense of expectation might be elicited by the sound of the pop on the bottle of sparkling wine, say. But, the question that we really want an answer to here is "Can packaging sounds actually influence the consumer's perception of the contents?" While I am not aware of anyone having conducted a study in which branding information was provided solely by the distinctive sound of the packaging being opened or handled, there is no obvious reason to think that such signature auditory cues should be any less effective than a brand name or logo, or image mold presented visually (see Spence, in press), in terms of cuing consumers to the brand. Of course, that said, in everyday life, we nearly always see brand-relevant visual cues before we hear any possibly distinctive auditory packaging cues, and hence the former may well play a dominant role in identification, as they so often do in other aspects of everyday life (see Gates, Copeland, Stevenson, & Dillon, 2007; Posner, Nissen, & Klein, 1976; Schifferstein & Cleiren, 2005; cf. Fenko, Schifferstein, & Hekkert, 2011).

Relevant here, Spence and his colleagues (2011) reported on the results of a small pilot study in which their participants were shown to rate potato chips as tasting significantly crisper when the sound of a noisy crisps packet was heard rattling in the background, as compared to when the rattling of a quieter packaging material was heard instead (Spence et al., 2011). One can only imagine, then, at the effect on crispiness perception that would have been elicited by the now-infamous packaging of the Sun Chip potato chip (Horovitz, 2010; Vranica, 2010a,b). This packaging came in at an incredible 100 decibels when gently held and rattled in the hands. Unsurprisingly, the packaging was soon withdrawn from the shelves because of the racket it made.[6]

The last few years have seen a great deal of innovation in the design of food and beverage packaging. While much of that development has tended to focus around the visual and haptic/tactile aspects of packaging design (eg, Gallace & Spence, 2014; Spence, in press; Spence & Gallace, 2011), there seems no good reason to doubt the suggestion that in the coming years we are going to see further development in the area of distinctive signature sensory design in terms of the sounds of opening (and possibly also of pouring; Spence & Wang, 2015a). I believe that the growing body of research demonstrating the functional benefits to the product experience that can be induced via food and beverage packaging, will also lead to a growing emphasis on the more functional aspects of the auditory design of product packaging (see Spence & Zampini, 2006, 2007; Spence, in press). Finally here, it is worth noting how informative pouring (and possibly also opening) sounds can be in terms of the likely temperature, level and type of carbonation, and possibly even viscosity of a drink (Velasco, Jones, King, & Spence, 2013a,c; see Spence & Wang, 2015a, for a review). In terms of recent commercial interest in this space, think here only of Krug champagne. The company introduced the limited edition Krug shell that sits on top of the champagne glass in order to accentuate the sound of the bubbles (see King, 2014).[7]

4 The sound of food and drink and our interaction with them

You can tell a lot about the texture of a food—think crispy, crunchy, and crackly—from the mastication sounds heard while biting and chewing. The latest techniques from the field of cognitive neuroscience have started to revolutionize our understanding of just how important what we hear is to our experience and enjoyment of food and drink. A growing body of research now shows that by synchronizing eating

---

[6] Interestingly, it was, once again, the early marketers who first suggested that a noisy snack like crisps should probably be packaged in a noisy packaging format (see Smith, 2011).
[7] This piece of Bernardaud Limoges porcelain sits on top of a custom Riedel "Joseph" glass. According to Krug: *"listening to their champagne via the 'Krug Shell,' connects you to the 'eternal resonance of the seashore'."* They further claim that the first time you went to the beach and lifted a shell to your ear, that "rushing" sound you heard brought you back to your *"first acoustic experience within the womb"* (Antin, 2014). As one commentator put it, the Krug Sound campaign (of which the shell was just a part) is all about making: *"a physical and sentimental connection to the brand via sound."* (King, 2014).

sounds with the act of consumption, one can change a person's experience of what they think they are eating and how much they enjoy the experience.

4.1 The sound of crispness

Zampini and Spence (2004) took a crossmodal interaction that had originally been discovered in the psychophysics laboratory—namely, "the parchment skin illusion"—and applied it to the world of food. In this perceptual illusion, the dryness/texture of a person's hands can be changed simply by changing the sound that the person hears when they rub their palms together (Guest, Catmur, Lloyd, & Spence, 2002; Jousmäki & Hari, 1998). Zampini and Spence wanted to know whether a similar auditory modulation of oral-somatosensory perception and/or of crispness perception would be experienced when their participants bit into a noisy food product while the interaction sounds were modified. The 20 participants who took part in this study were given 180 potato chips (Pringles) to evaluate over the course of an hour-long experimental testing session. The participants had to bite each potato chip between their front teeth and rate it in terms of its "freshness" or "crispness" using an anchored visual analog scale. During each trial, the participants received real-time auditory feedback of the sounds associated with their own biting action over closed-ear headphones.[8] On a crisp-by-crisp basis, this auditory feedback could be manipulated by the computer controlling the experiment in terms of its overall loudness and/or frequency composition. Consequently, on some proportion of the trials, the participants heard the veridical sounds that they were actually making while biting into the crisp. On the majority of the trials, though, the overall volume of their crisp-biting sounds was attenuated by either 20 or 40 dB. The higher-frequency components of the sound (>2 kHz) could also either be boosted or attenuated (by 12 dB) on a trial-by-trial basis. Interestingly, on debriefing, many of the participants thought that the crisps had been taken from different packets during the course of the experiment, hinting at the power of this particular crossmodal illusion.

The key result to emerge from Zampini and Spence's (2004) study was that participants rated the potato chips as tasting both significantly crisper and significantly fresher when the overall sound level was higher and/or when just the high-frequency sounds were boosted (see Fig. 5.1). By contrast, the crisps were rated as both staler and softer when the overall sound intensity was reduced and/or when the high-frequency sounds associated with their biting into the potato chips were attenuated instead.

Recently, a group of Italian scientists has extended this cognitive neuroscience-inspired approach to studying the role of sound in the perception of the crispness and hardness of apples (Dematté et al., 2014). Once again, reducing the volume of the auditory feedback on biting into the apples was shown to lead to a reduction in the

---

[8] Interestingly though, while the sounds were delivered over headphones, the participants typically reported perceiving the sounds as originating from the potato chip in their mouth, rather than from the headphones, due to the well-known ventriloquism illusion (eg, Alais & Burr, 2004; Caclin, Soto-Faraco, Kingstone, & Spence, 2002; Jackson, 1953).

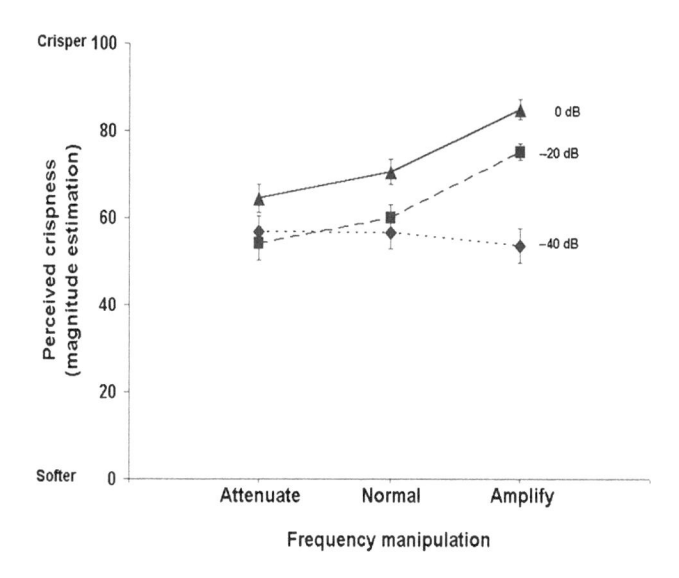

Figure 5.1 Results of a study by Zampini and Spence (2004) demonstrating that the real-time modification of biting sounds can significantly influence the perceived crispness of potato chips/crisps.
Source: Zampini and Spence (2004).

perceived crispness of the "Renetta Canada," "Golden Delicious," and "Fuji" apples that the participants were given to evaluate. Specifically, a small but significant reduction in mean crispness and hardness ratings was observed for this moist food product, when the participants' high-frequency biting sounds were attenuated by 24 dB and/or when there was an absolute reduction in the overall sound level. Thus, it would appear that people's perception of the textural properties of both dry and moist food products can be changed simply by modifying the sounds that they hear.

It is, though, perhaps worth bearing in mind here that while the two studies that have just been reviewed (Demattè et al., 2014; Zampini & Spence, 2004) clearly demonstrate that the sounds of our interaction with food play an important role in the perception of the oral-somatosensory aspects of food *texture*, this should not necessarily be taken to show an effect of sound on the *flavor* of food itself (see Spence, 2012; Spence, Smith, & Auvray, 2015). What is more, it should also be remembered that the magnitude of the auditory manipulations that were introduced in these studies was pretty substantial, and hence unlikely to be captured in full by the change in the design or manufacture of any real food product. Nevertheless, these studies do serve a function in terms of "proof-of-principle" and in helping to emphasize the role of sound in our perception of food. What is more, they provide one of the first examples where paradigms/approaches from the cognitive neuroscience study of multisensory perception has been extended to the study of multisensory flavor perception (Spence, 2013). The approach outlined here can be contrasted with previous research where the participants would either have to rate crispness while listening to white noise, or else try to rate prerecorded biting/crunching sounds (see Spence, 2015a, for a review). While the two studies reviewed here focused on the perception of crispness/freshness, it should be borne in mind that our perceptions of crunchiness, crackliness, and creaminess are likely similarly influenced by the sounds we hear when biting into and masticating food (see Spence, 2015a, for a review).

4.2 The sounds of carbonation

It turns out that our perception of the carbonation in a beverage is based partly on the sounds of effervescence and popping that we hear when the drink is poured and/or while holding the drink in our hand(s): Make the carbonation sounds louder, or else make the bubbles pop more frequently, and people judge the drink to be more carbonated (Zampini & Spence, 2005). That said, such crossmodal effects of sonic cues on the perception of carbonation tend to dissipate just as soon as people take a drink. One reason for this may be that the sour-sensing cells that act as the taste sensors for carbonation (Chandrashekar et al., 2009), and/or the associated oral-somatosensory cues (van Aken, 2013) associated with the bubbles popping dominate the multisensory tasting experience once we take the carbonated beverage into our mouths. Auditory cues also become less salient. Alternatively, however, it could simply be argued that the auditory manipulations to the sounds of carbonation that were introduced by Zampini and Spence were simply not ecologically valid enough to override the other sensory cues when they were put into conflict. Given how desirable a characteristic level of carbonation is amongst consumers, it would certainly merit further research to better understand how the design of cans, bottles, cups, and glasses could be redesigned in order to help amplify, or otherwise enhance, such sounds. As noted by McMains (2015), the advertising agencies have certainly not been slow to build on the growing interest in this area.

4.3 Interim summary

Given the various results that have been reported in this section, and in the absence of further research, the most appropriate conclusion to draw at the present time would appear to be that oral-somatosensory and auditory cues play somewhat different roles in the perception of different food attributes such as crispy, crunchy, crackly, creamy, carbonated, etc. (see Spence, 2015a, for a review). Taken together, though, the research that has been published to date is at least consistent with the view that people rely more on their sense of touch when judging the hardness of foods and the carbonation of drinks in-mouth. By contrast, the two senses (of hearing and oral-somatosensation) would appear to make a much more balanced contribution to our judgments when it comes to evaluating the crispiness of foods, say. And, while more research is most definitely needed, crackly may turn out to be a percept that is a little more auditory dominant than the others. One other desirable attribute of our interaction with at least certain foods, is squeakiness, as when we bite into halloumi or the Finnish Leipäjuusto cheese (Anon, 2011). While squeakiness has not always been thought of as a desirable food characteristic (see Amerine et al., 1965, p. 228), its absence would, I think, likely cause a negative disconfirmation of expectation in many consumers nowadays (Spence, 2015a).

In terms of the importance of sound, I will end this section with the following quote from a consumer describing their experience of a Magnum ice cream (a product that first appeared on the shelves in Sweden back in 1989): "I experienced the crack of the chocolate while biting into it and the "mmmmm" sound in my mind while eating the ice-cream. I was lost into it:) It was pure pleasure indeed." (http://rakshaskitchen. blogspot.com/2014/02/magnum-masterclass-with-kunal-kapur.html). In fact, there is a lot of talk of "cracking chocolate" in online product descriptions (http://www. mymagnum.co.uk/products/). Getting the sound of the crack right is obviously critical, given that Unilever sells somewhere in the region of 2 billion of these ice creams per year (http://alvinology.com/2014/05/25/magnum-celebrates-25-years-of-pleasure/). However, it is important to note that consumer reports concerning what they want in their ice cream can't always be relied on.

Some years ago, researchers working on behalf of Unilever asked their brand-loyal consumers what they would change about the chocolate-covered Magnum ice. A frequent complaint that came back concerned all of those bits of chocolate falling onto the floor and staining one's clothes when biting into one of these iconic choc ices. This feedback was promptly passed back to the product development team who, or so the rumor goes, set about trying to alter the formulation so as to make the chocolate coating adhere better to the ice cream. In so doing, the distinctive cracking sound of the chocolate coating was lost. When the enhanced product offering was launched, consumers once again complained. It turned out that they didn't like the new formulation. The developers were understandably confused. Hadn't they fixed the problem that their consumers had been complaining about? Subsequent analysis, though, revealed that it was that distinctive cracking sound that consumers were missing. It turned out that this was a signature feature of the product experience even though the consumers (not to mention the market researchers) didn't necessarily realize it. Since then, Unilever have returned to the original formulation, thus ensuring a solid cracking sound every time someone bites into one, regardless of the consequences for your clothes and carpet.

5 Background noise

In recent years, a growing number of commentators have started to highlight the exceedingly high levels of background noise that one finds in many restaurants nowadays (Sietsema, 2008a,b; see Spence, 2014a, for a review). There are likely a variety of reasons for the increasing decibel count, including the removal of sound-absorbing materials in many contemporary restaurants, through to the tendency of a growing number of chefs and restaurateurs to subject diners to whatever they happened to have been listening to in the kitchens (see Spence, 2015b). Intriguingly, the evidence from a number of recent laboratory-based studies now demonstrates that loud background noise can indeed suppress our ability to taste food and drink (eg, Woods et al., 2011; Yan & Dando, 2015). So, for instance, Woods et al. (2011) conducted a couple of studies to assess the impact of presenting loud versus quiet background noise (75–85 vs 45–55 dB, respectively) over headphones on people's perception of the sweetness, sourness, and liking of a variety of foodstuffs. In this repeated measures experimental design, the participants rated Pringles potato chips, cheese, biscuits/crackers, and flapjack on a number of labeled magnitude scales. In one study, nearly 50 people rated salty foods (crisps and cheese) as significantly less salty, and sweet foods (biscuits and flapjack) as tasting less sweet, under conditions of loud background noise (see Fig. 5.2). By contrast, there was no such effect on the participants' liking (ie, hedonic) ratings.

Intriguingly though, the results of Woods et al.'s (2011) second experiment demonstrated that people's ratings of the crunchiness of rice cakes was actually significantly increased under conditions of loud background noise (see Fig. 5.2). These results therefore highlight the significant effect that loud noise can have on the sensory-discriminative aspects of taste perception. It is, though, perhaps worth noting that somewhat different results were reported by Stafford, Fernandes, and Agobiani (2012). The latter researchers observing that their participants rated alcoholic beverages as tasting sweeter when listening to loud background music than in its absence (see also Ferber & Cabanac, 1987; Stafford, Agobiani & Fernandes, 2013).

Yan and Dando (2015) recently conducted an intriguing study in which their participants were required to rate the intensity of solutions containing one of the five basic tastants. The results showing that the perception of sweetness was suppressed by loud background noise (actually simulated airplane noise presented at 80–85 dB). By contrast, the taste of umami appeared to be enhanced by this loud noise (see Fig. 5.3).[9] Perception of the other three tastants was unaffected by the simulated airplane noise. The differing crossmodal effect of background noise on the different basic taste qualities argues against a simple account of such findings in terms of crossmodal sensory masking (or attentional distraction), say, as that would presumably predict that the perception of all tastants should be affected equally. Yan and Dando themselves were left to speculate as to what the cause of such a particular pattern of results might be. As such, further research replicating these results, and perhaps using more complex food stimuli would likely be beneficial here.[10]

---

[9] Intriguingly, Spence et al. (2014) hypothesized that such an effect might have occurred, based on observations concerning the number of people who order a bloody Mary, or tomato juice, while on the plane. Many of these people, or so it turns out, never drink tomato juice while on the ground. Knowing about the asymmetric influence of noise on different tastants also makes sense of the umami-rich menus that some airlines have been promoting in recent years.

[10] Intriguingly, early research by Ferber and Cabanac (1987) led to the suggestion that loud noise (a stressor) might be expected to have a particular influence on the perception of sweetness (a nutritive taste).

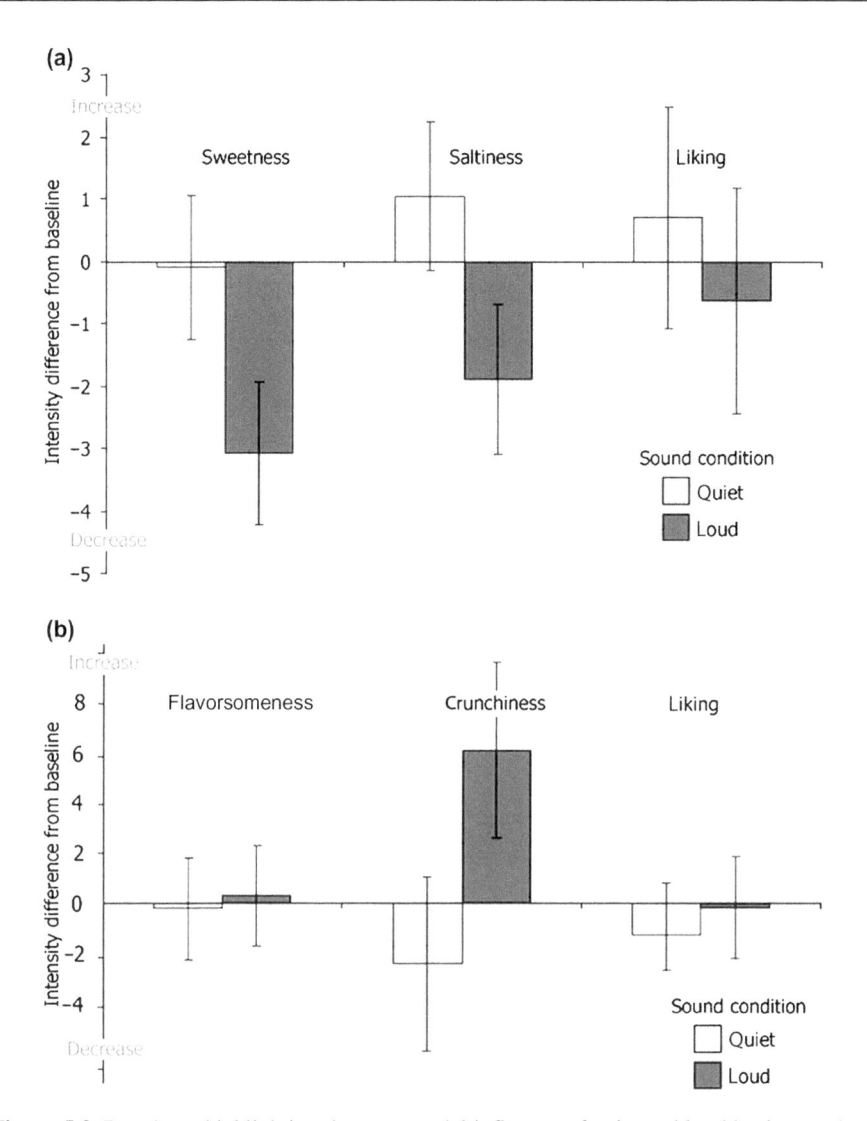

Figure 5.2 Bar charts highlighting the crossmodal influence of quiet and loud background noise on the perception of sugar, salt, flavorsomeness, crunchiness, and liking. (a) The effect of background noise levels on sweetness, saltiness, and liking intensity ratings for a range of foods in Woods et al.'s (2011) Experiment 1, calculated relative to a baseline silent condition. Negative values indicate a lower rating than the baseline while positive values indicate a higher level than baseline. Error bars = 2 SEM. (b) The effect of quiet and loud background sounds on the perception of flavorsomeness, crunchiness, and liking intensity ratings for rice cakes reported in Woods et al.'s (2011) Experiment 2, again relative to the silent baseline condition.
Source: Adapted from Woods et al. (2011).

Figure 5.3 Salty (a: NaCl, sodium chloride), sour (b: citric acid), and bitter (c: quinine hydrochloride) stimuli were delivered to panelists at three concentrations in Yan and Dando's (2015) study. Panelists rated the intensity of taste on the general labeled magnitude scale. No significant differences were observed between sound (80–85 dB of simulated airplane noise) and silent condition. (d) Intriguingly, participants' responses to sweet (sucrose) solution showed a noticeable and significant decrement, when delivered in the simulated cabin noise condition, across all three concentration levels. (e) By contrast, umami (MSG, monosodium glutamate) taste appeared to be significantly augmented by the noise condition, particularly at higher concentrations. Errors represent mean ± SEM; ns, nonsignificant. *$P < 0.05$; **$P < 0.01$; ***$P < 0.001$; ****$P < 0.0001$.
Source: Adapted from Yan and Dando (2015).

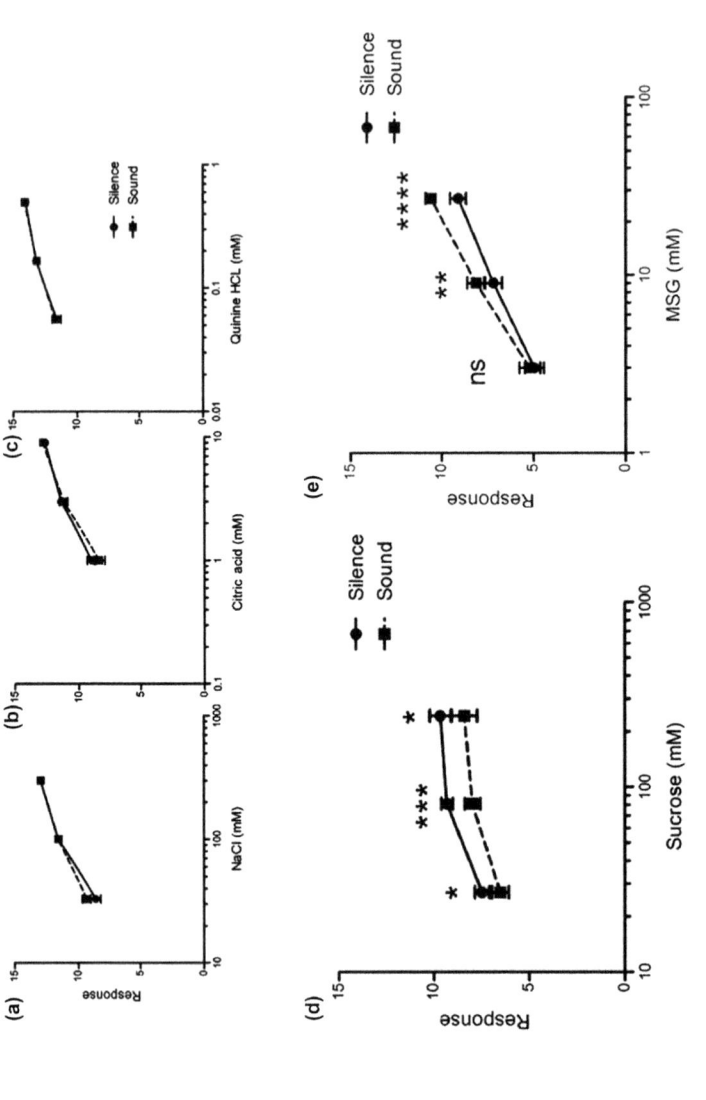

5.1 Interim summary

In conclusion, it would appear that background noise can indeed suppress (and, on occasion, enhance) taste, and presumably also flavor perception. There are likely a number of routes by which such crossmodal effects might occur (see Spence, 2014a). The recent discovery of direct connections between the nose and the ear are potentially interesting in this regard (Wesson & Wilson, 2010, 2011). The latest research has highlighted the specificity of some of these effects, in terms of certain tastants, oral sensations, or flavors being affected more than others by sound/noise. The only study to have investigated whether loud background noise would influence how people mixed drinks failed to demonstrate any such influence (see Ferber & Cabanac, 1987; though see also Kontukoski et al., 2015). It will be interesting to see whether the campaign for more dining and drinking establishments where customers can eat and drink without being bombarded by music (ie, noise) will gain in popularity in the coming years (eg, Kogan, 1991; Moir, 2015; Spence, 2014a).

6 Background music

Background music has a dramatic role on our purchasing decisions. So, for example, marketing research shows that the ethnicity of the music playing in the background can have a dramatic effect on the type of wine that people buy: Play French music in the supermarket wine section and the majority of people apparently buy French wine, play distinctively German music, and the majority of wines sold are German instead (North, Hargreaves, & McKendrick, 1997, 1999). Intriguingly, most people deny that they would be so easily influenced by background music when told about the results of this study, including those customers who agreed to be questioned when they came away from the till whether the music had influenced their purchasing decisions. This despite the fact that the sales figures told a very different story (see Table 5.1)! There

Table 5.1 **Number (and % in brackets) of bottles of French versus German wine sold as a function of the type of background music played in North et al.'s (1997) oft-cited marketing study**

	Background music	
	French accordian music	German Bierkeller music
Bottles of French wine sold	40 (77)	12 (23)
Bottles of German wine sold	8 (27)	22 (73)

is also evidence to suggest that playing classical music in the wine store leads customers to spend more than when top-40 tracks are played instead (Areni & Kim, 1993). Results such as these are undoubtedly intriguing, and often cited on the sensory marketing conference circuit. However, given the number of years that have passed since they were first published, it would, I think, undoubtedly be worthwhile to determine how well they replicate in the contemporary retail landscape (see Open Science Collaboration, 2015). Who can know whether similar results would still be obtained today (see Spence, Puccinelli, Grewal, & Roggeveen, 2014)?

Related evidence concerning the effect of the ethnicity of the music on people's (diners') food choices comes from a study published by Yeoh and North (2010). These researchers served people either Indian or Malay food while the music of either country could be played in the background. Once again, the results revealed that the ethnicity of the background music biased people's (meal) choices. That said, Yeoh and North's results also suggested that the crossmodal effects of the ethnicity of the music may be more likely to influence food choice under those conditions where the diner concerned does not have a strong preference for a particular type of cuisine.

In terms of the effects of classical music on people's food behavior, North and Hargreaves (1998) observed that people reported being willing to pay more when classical music was played in a student cafeteria than when easy listening music was played instead. Classical music also led to people being willing to pay more than either pop music or a no music baseline. Meanwhile, both North, Shilcock, and Hargreaves (2003) and Wilson (2003) have demonstrated that diners do indeed actually spend significantly more when classical music is played in a restaurant setting. A 10% increase in spending is not unheard of in research in this area. That said, it is important to bear in mind the congruency between the restaurant concept, the clientele, and the type of music that is played (see Lammers, 2003; Spence & Piqueras-Fiszman, 2014; see also Wansink & Van Ittersum, 2012).

Meanwhile, playing faster music has been shown to result in people eating and drinking more rapidly. In what is perhaps the classic study to have been published in this area, Milliman (1986) manipulated the tempo (that is, the number of beats per minute, or bpm) of the music playing in a medium-sized restaurant in North America, and assessed its impact on the behavior of 1400 diners. Intriguingly, the diners ate significantly more rapidly when fast (as compared to slow) instrumental music was played. In the slow music condition, the diners spent more than 10 min longer eating (bringing the total duration of their restaurant stay up to almost an hour). Although there was no effect of musical tempo on how much people spent on their food, there was a marked difference on the final bar bill, with those exposed to the slow music spending around a third more! In other words, simply by slowing down the music it was possible to increase the gross margin at the restaurant by almost 15%.

Elsewhere, playing fast-tempo instrumental, nonclassical background music ($M = 122$ bpm) has been shown to result in the diners in a cafeteria taking significantly more bites per minute than when a slow-tempo piece of music ($M = 56$ bpm) was played instead (4.4 vs 3.2 bites/min, respectively; Roballey et al., 1985). Finally,

McElrea and Standing (1992) reported that doubling the bpm of the background music increased the rate at which students drank soda in the lab. Taken together, then, the results from both the restaurant and the psychology lab demonstrate the profound effect that the tempo of the background music can have on the behavior of diners and drinkers.

Furthermore, both the style and loudness of the background music also appear to influence how long people choose to stay in a given venue eating and/or drinking (Jacob, 2006). Playing loud music during a meal can be used to accelerate the rate at which people eat their food (Stroebele & de Castro, 2004). No surprise for guessing, though, that the more discomforting the music, the less time diners want to spend in a restaurant, while the more they like the music, the longer they tend to stay (North & Hargreaves, 1996).

6.1 Interim summary

The evidence reported in this section clearly demonstrates that the music playing in the background in those places in which we choose to eat and drink can have a major impact on both our food and beverage choices and on our consumption behaviors (Stroebele & de Castro, 2006). It can affect everything from how long we choose to stay in a restaurant through to how much we end up eating, from how rapidly we bring the fork or spoon up to our mouths through to how much we end up spending, and from how ethnic we rate the dish as being through to how acceptable we rate the overall multisensory dining experience (Fiegel, Meullenet, Harrington, Humble, & Seo, 2014; see Spence, 2012, for a review). Ultimately, of course, the majority of restaurateurs and bar managers are going to be most interested in whether they can increase their takings by playing a certain type of music. The results reported here are certainly consistent with such a suggestion, though when clients ask, I always suggest that they should probably conduct research in the relevant venue to know what works best for them.

Taken together then, the findings reported in the last couple of sessions are broadly consistent with Crocker's (1950, p. 7) early suggestion that: "A loud noise, for instance, may prevent entirely our ability to smell or taste, yet softly played dinner music can create an environment favorable for elegant dining." While there is little sign that restaurants are turning down the volume yet (see Spence, 2015b), there is undoubtedly growing interest, at least from some of the larger chain restaurants in optimizing the sonic environments in their outlets and restaurants. For example, as Chris Golub, the man responsible for selecting the music that plays in all 1500 Chipotle restaurants put it: "The lunch and dinner rush have songs with higher BPMs because they need to keep the customers moving." (Suddath, 2013).[11] However, beyond changing what people order, how ethnic they rate it, and how much they are

---

[11] As Chipotle's spokesman, Chris Arnold, says: *"We could've just piped in some satellite radio. It would've been cheaper...But if you're creating a restaurant, you need the right atmosphere. We're sticklers for detail. We wanted something that belonged just to us."* (quoted in Suddath, 2013).

willing to pay, the latest evidence from the field of sonic seasoning suggests that one can actually use off-the-shelf and specially composed music and soundscapes to accentuate certain components of the tasting experience.

7 Sonic seasoning

For almost half a century now, researchers have been reporting that people experience a surprising crossmodal correspondence between sounds (eg, pure tones and music) and the taste and flavor of certain foods and drinks (see Holt-Hansen, 1968, 1976; Rudmin & Cappelli, 1983, for early research; and Spence & Wang, 2015b–d, for reviews of the literature). While the original work in this area was conducted with the participants tasting beer, there is no doubt that by far the majority of the research in the intervening years has involved the study of the interaction between music and wine. Seminal empirical work on the impact of music on the wine-tasting experience comes from North (2012). In this study, 250 university students were offered a glass of red or white wine, which they drank while listening to one of four pieces of specially selected music, or else they drank in silence. The musical selections had been chosen to convey different meanings: *Carmina Burana* by Orff—"powerful and heavy"; *Waltz of the Flowers* from The Nutcracker by Tchaikovsky—"subtle and refined"; *Just Can't Get Enough* by Nouvelle Vague—"zingy and refreshing"; and *Slow Breakdown* by Michael Brook—"mellow and soft."

The students who took part in this study gave the wine a score from 0 to 10 on each of four rating scales: "powerful and heavy," "subtle and refined," "zingy and refreshing," and "mellow and soft." A score of 0 indicating that "The wine definitely does not have this characteristic" while a score of 10 indicated that "The wine definitely does have this characteristic." The results revealed that the two wines were judged as tasting significantly more powerful/heavy, by those who had listened to the "heavy" music. Meanwhile, the wines were rated as tasting 40% more zingy/fresh by those listening to the Nouvelle Vague track. A similar pattern of results was also reported for those listening to the mellow/soft and subtle/refined music selections. However, while the results of North's study are clear enough, the most appropriate interpretation of the underlying cause of these crossmodal effects has been debated by researchers (Spence & Deroy, 2013). One possibility here is in terms of the notion of sensation transference or a "halo effect" (see Lawless & Heymann, 1997; Thorndike, 1920). The idea here being that what a diner or drinker thinks about what they are listening to, and how much they like the experience, may influence what they say about what they are tasting (see Spence et al., 2013).

Ideas around sonic seasoning have been developed most extensively in the world of wine (see Spence & Wang, 2015b–d, for reviews). By now, many studies show that what people say about a wine can be systematically altered as a function of the sonic properties of the music that happens to be playing in the background (see Wang & Spence, 2015, 2016).

In the world of sonic seasoning as it applies to food, the classic research was conducted by Crisinel et al. (2012). Here researchers gave their participants a series of pieces of bittersweet cinder toffee. The participants in this study had to rate the perceived taste on a scale anchored by bitterness at one end and by sweetness on the other. The participants had to rate the food while listening to one of two specially composed soundscapes designed to bring out the sweetness and bitterness, respectively. The results showed a 5–10% change in people's rating of the taste (see Fig. 5.4). That is, sweetness was rated 5–10% higher while listening to the putatively "sweet" soundscape than when listening to the putatively "bitter" soundscape instead (though see Wang & Spence, submitted). Importantly, one generalization that would appear to emerge from the growing body of research that has been published in this area is that, while sonic seasoning can be used to draw a taster's attention to some aspect of their tasting experience, and hence make it more salient, it has not, as yet proved possible to use sonic stimulation to create a percept that is not actually there. It is for this reason that an attentional account of these findings is preferred (see Spence & Wang, 2015c, 2015d). Subsequently, researchers have extended the sonic seasoning approach, based on the growing literature demonstrating a range of crossmodal correspondences between what we hear and what we taste/smell to a number of other foods, such as chocolates (see Reinoso Carvalho et al., 2015a, 2015b; see also Knöferle & Spence, 2012; Wang & Spence, submitted).

Typical of experiments in this area, Knöferle, Woods, Käppler, and Spence (2015) recently conducted a series of three studies in which they demonstrated that even with a very restricted sonic variation, they were able to create four brief pieces of music that people associated with different tastes. In the first part of their study, the participants were asked to imagine each of the four basic tastes and then move a scale to change

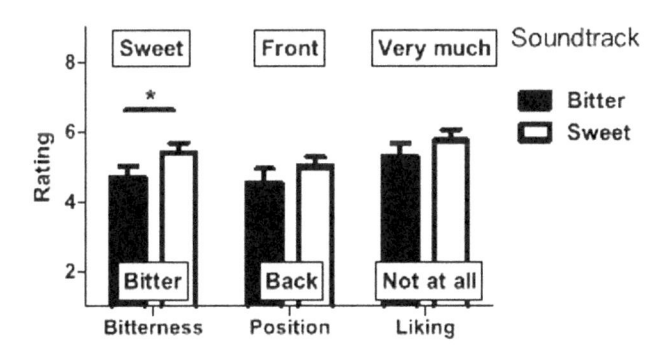

Figure 5.4 The results of the experiment by Crisinel et al. (2012) highlighting the effect of the auditory soundtrack ("bitter" vs "sweet") on participants' ratings on the bitter–sweet scale, the back–front (of mouth) scale, and the hedonic rating scale. The asterisk highlights a significant difference between the conditions ($P<0.05$). The error bars represent the SEM.
Source: Adapted from Crisinel et al. (2012).

some attribute of a sound. By moving a slider, a short chord progression could be modified in one of six ways (ie, to increase or decrease the sound's attack, discontinuity, pitch, roughness, sharpness, and speed). In terms of pitch, the sweet taste was found to be the highest, then sour, then salty, and finally bitter was the lowest in pitch. Sour was sharpest relative to other tastes. Meanwhile, sweet was least rough, bitter most rough. Bitter, salty, and sour were all high in discontinuity, whereas sweet was not (ie, it was low in discontinuity). Based on these results, a German sound design agency came up with four soundtracks. These soundtracks were tested in subsequent research. In a forced choice task, nearly one-third of the participants assigned the music to the four basic tastes perfectly. While the main study was conducted in North American participants, a subsequent experiment revealed similar results, with just a small reduction in the accuracy of participants' responses when tested on those living in India.

Over the last few years, a number of composers/designers have come up with music/soundscapes designed to match each of the basic tastes. In fact, this field of research has now developed to such a point that it is possible to compare the many different pieces of music to determine how well they match a specific taste (see Wang, Woods, & Spence, 2015). Moving forward, the hope is that the results of such comparative studies may help to further our understanding in the field. And while the majority of the latest research has tended to focus on the musical matching to, and modification of, taste, there is a large body of evidence on crossmodal correspondences between auditory and olfactory stimuli (see Deroy, Crisinel, & Spence, 2013, for a review), thus suggesting a number of lines for additional research moving forward. In the years to come, one should expect the arrival of more sensory apps designed to provide musical accompaniment for branded tasting experiences (see Crisinel et al., 2013; Spence, 2014b). I would not be surprised if we see more digital artifacts, that is glassware and cutlery, that has been designed specifically to provide some sort of sonic accompaniment (eg, see Anon., 2013; Baker, 2015; Spence & Piqueras-Fiszman, 2014, for a number of intriguing examples).

Relevant here is the Bittersweet Symphony dish, presented by Caroline Hobkinson at The House of Wolf restaurant in London (as well as at The Experimental Food Society annual dinner; http://www.stirringwithknives.com/2012/09/caroline-hobkinson-debuts-as-house-of-wolf%E2%80%99s-inaugural-guest-chef/; http://houseofwolf.co.uk/). For dessert, a bittersweet chocolate lollipop would be brought to the table. There was a telephone number printed on the menu for the diner to call on their mobile device. Choose one option on being connected and the diner would hear a soundtrack that has been designed to enhance the dish's sweetness, dial the other number and the bitter notes in the dish should become a little more prominent—a kind of digital seasoning if you will (based on Crisinel et al.'s (2012) findings). But notice here how the technology has moved from something brought out by the chef in one of the world's top restaurants (in the case of the sound of the sea dish) through to the diners using their own technology in a more mainstream restaurant.

Although beyond the scope of the present chapter, it is also worth noting that soundscapes have also been shown to influence the tasting experience. Everything from the sounds of the sea, as served at Heston Blumenthal's *The Fat Duck* restaurant in Bray (see Spence & Piqueras-Fiszman, 2014; Spence et al., 2011),[12] through to a growing number of chefs offering multicourse, multisensory, dining experiences with different atmospheric soundscapes and music often synchronized with each of the courses (eg, see Moore, 2015; Pigott, 2015). Sonic backdrops are likely to play an increasingly important role in our dining experiences (at least at the high end) in the future. Welcome to the new world of "Sensploration" (Leow, 2015; Velasco et al., 2013b).

8 Conclusions

In conclusion, what we hear when eating and drinking, and even when deciding what to eat and drink, actually plays a much more important role in our food and beverage choices, as well as in our taste and flavor experiences, than many people realize (eg, see Delwiche, 2003; Schifferstein, 2006; Shepherd, 2012). Everything from the sounds of food preparation (Spence, 2015b) and food packaging (Spence, in press; Spence & Wang, 2015a) through to the sounds of consumption (Spence, 2015a), and from background noise (Spence, 2014a), through background music (Spence & Piqueras-Fiszman, 2014) and increasingly sonic soundscapes. Get the background music (or soundscape) right, and one can significantly enhance the perceived acceptability, quality, and ethnicity of the food. The music can make the food and drink appear more valuable. It certainly has an impact on how much we end up consuming, and even on how much we spend. Music can be chosen to arouse or relax us, and even to increase the amount of time we spend in a given venue. Some restaurant chains know this only too well, and carefully pick the music playing in their stores (Suddath, 2013). Research on the emerging science of digital/sonic seasoning (see Spence, 2014b; Spence & Wang, 2015b–d) is providing novel avenues for multisensory experience design. Given the multiple routes by which what we hear influences what we taste, it should come as little surprise that a number of different mechanisms, at both the psychological and neurological levels, have been put forward to account for such findings over the years (see Piqueras-Fiszman & Spence, 2015; Spence, 2014a, 2015a).

---

[12] In this study, the delegates at the Art and the Senses conference that was hosted in Science Oxford in 2007 tasted and rated two oysters in terms of their pleasantness and the intensity of their flavors using pencil-and-paper-based response scales. One oyster was served in the shell from a wooden basket (of the type that one commonly sees at the seaside, well in the UK at least). The other oyster had been removed from its shell and was served in a petri-dish instead. The first oyster was served while people listened to the "sounds of the sea" soundtrack (this consisted of the sound of seagulls squawking and waves crashing gently on the beach), the second while they listened to the noise of farmyard animals (ie, an incongruent soundtrack). The participants rated the oyster tasted while listening to the "sound of the sea" as tasting significantly more pleasant than the oyster that had been tasted while they listened to the sounds of the farmyard instead. Interestingly, though, no such effect was found for participants' intensity ratings. That is, changing the sound playing in the background had no effect on people's perception of the intensity of the flavor of the oysters (see Spence et al., 2011).

References

Alais, D., & Burr, D. (2004). The ventriloquist effect results from near-optimal bimodal integration. *Current Biology, 14*, 257–262.

Amerine, M. A., Pangborn, R. M., & Roessler, E. B. (1965). *Principles of sensory evaluation of Food*. New York, NY: Academic Press.

Anon. (2011). Squeaky cheese. *The New Scientist*, December 7 (2842). Downloaded from http://www.newscientist.com/article/mg21228421.800-squeaky-cheese.html on February 1, 2016.

Anon. (2013). Musical spoons to go with your Heinz beans. *Advertising Age*, March 28. Downloaded from http://adage.com/article/creativity-pick-of-the-day/bompas-parr-design-musical-spoons-heinz-beans/240605/ on February 1, 2016.

Anon. (2014). Costa Express revolutionises self-service coffee with launch of Marlow. *Hospitality & Catering News*, April 9. Downloaded from http://www.hospitalityandcateringnews.com/2014/04/costa-express-revolutionises-self-serve-coffee-with-launch-of-marlow/ on May 25, 2014.

Antin, C. (2014). What does wine sound like? *Punch*, January 28. Downloaded from http://punchdrink.com/articles/what-does-wine-sound-like/ on August 6, 2014.

Areni, C. S., & Kim, D. (1993). The influence of background music on shopping behavior: Classical versus top-forty music in a wine store. *Advances in Consumer Research, 20*, 336–340.

Baker, N. (2015). Johnnie Walker reveals futuristic glass. *The Drinks Business*, April 10. Downloaded from http://www.thedrinksbusiness.com/2015/04/johnnie-walker-reveals-futuristic-glass/ on July 2, 2015.

Byron, E. (2012). The search for sweet sounds that sell: Household products' clicks and hums are no accident; Light piano music when the dishwasher is done? *The Wall Street Journal*, October 23. Downloaded from http://online.wsj.com/article/SB10001424052970203406404578074671598804116.html?mod=googlenews_wsj#articleTabs%3Darticle on February 1, 2016.

Caclin, A., Soto-Faraco, S., Kingstone, A., & Spence, C. (2002). Tactile 'capture' of audition. *Perception & Psychophysics, 64*, 616–630.

Chandrashekar, J., Yarmolinsky, D., von Buchholtz, L., Oka, Y., Sly, W., Ryba, N. J. P., et al. (2009). The taste of carbonation. *Science, 326*, 443–445.

Crisinel, A.-S., Cosser, S., King, S., Jones, R., Petrie, J., & Spence, C. (2012). A bittersweet symphony: Systematically modulating the taste of food by changing the sonic properties of the soundtrack playing in the background. *Food Quality and Preference, 24*, 201–204.

Crisinel, A.-S., Jacquier, C., Deroy, O., & Spence, C. (2013). Composing with cross-modal correspondences: Music and smells in concert. *Chemosensory Perception, 6*, 45–52.

Crocker, E. C. (1950). The technology of flavors and odors. *Confectioner, 34*(January), 7–8, 36–37.

Delwiche, J. F. (2003). Attributes believed to impact flavor: An opinion survey. *Journal of Sensory Studies, 18*, 437–444.

Dematté, M. L., Pojer, N., Endrizzi, I., Corollaro, M. L., Betta, E., Aprea, E., et al. (2014). Effects of the sound of the bite on apple perceived crispness and hardness. *Food Quality and Preference, 38*, 58–64.

Deroy, O., Crisinel, A.-S., & Spence, C. (2013). Crossmodal correspondences between odors and contingent features: Odors, musical notes, and geometrical shapes. *Psychonomic Bulletin & Review, 20*, 878–896.

Fenko, A., Schifferstein, H. N. J., & Hekkert, P. (2011). Noisy products: Does appearance matter? *International Journal of Design*, *5*(3), 77–87.
Ferber, C., & Cabanac, M. (1987). Influence of noise on gustatory affective ratings and preference for sweet or salt. *Appetite*, *8*, 229–235.
Fiegel, A., Meullenet, J.-F., Harrington, R. J., Humble, R., & Seo, H.-S. (2014). Background music genre can modulate flavor pleasantness and overall impression of food stimuli. *Appetite*, *76*, 144–152.
Gallace, A., & Spence, C. (2014). *In touch with the future: The sense of touch from cognitive neuroscience to virtual reality*. Oxford, UK: Oxford University Press.
Gates, P. W., Copeland, J., Stevenson, R. J., & Dillon, P. (2007). The influence of product packaging on young people's palatability ratings for RTDs and other alcoholic beverages. *Alcohol and Alcoholism*, *42*, 138–142.
Guest, S., Catmur, C., Lloyd, D., & Spence, C. (2002). Audiotactile interactions in roughness perception. *Experimental Brain Research*, *146*, 161–171.
Holt-Hansen, K. (1968). Taste and pitch. *Perceptual and Motor Skills*, *27*, 59–68.
Holt-Hansen, K. (1976). Extraordinary experiences during cross-modal perception. *Perceptual and Motor Skills*, *43*, 1023–1027.
Horovitz, B. (2010). Frito-Lay sends noisy, 'green' SunChips bag to the dump. *USA Today*, May 10. Downloaded from http://www.usatoday.com/money/industries/food/2010-10-05-sunchips05_ST_N.htm on December 22, 2015.
Houghton, L. (2014). Costa Express launches 'Marlow' self-serve coffee. Downloaded from http://www.bighospitality.co.uk/content/view/print/910569 on May 15, 2014.
Jackson, C. V. (1953). Visual factors in auditory localization. *Quarterly Journal of Experimental Psychology*, *5*, 52–65.
Jacob, C. (2006). Styles of background music and consumption in a bar: An empirical evaluation. *International Journal of Hospitality Management*, *25*, 716–720.
Jousmäki, V., & Hari, R. (1998). Parchment-skin illusion: Sound-biased touch. *Current Biology*, *8*, 869–872.
King, J. (2014). Krug heightens sensory experience with listening device. *Luxury Daily*, January 2. Downloaded from http://www.luxurydaily.com/krug-heightens-sensory-experience-with-listening-device/ on February 1, 2016.
Knöferle, K. M. (2011). *It's the sizzle that sells: Crossmodal influences of acoustic product cues varying in auditory pleasantness on taste perception*. Unpublished manuscript.
Knöferle, K. M. (2012). Using customer insights to improve product sound design. *Marketing Review St. Gallen*, *29*(2), 47–53.
Knöferle, K. M., & Spence, C. (2012). Crossmodal correspondences between sounds and tastes. *Psychonomic Bulletin & Review*, *19*, 992–1006.
Knöferle, K. M., Sprott, D. E., Landwehr, J. R., & Herrmann, A. (2014). *Crossmodal influences of acoustic product cues varying in auditory pleasantness on consumer taste perceptions*. Unpublished manuscript.
Knöferle, K. M., Woods, A., Käppler, F., & Spence, C. (2015). That sounds sweet: Using crossmodal correspondences to communicate gustatory attributes. *Psychology & Marketing*, *32*, 107–120.
Kogan, P. (1991). *Muzak-free London: A guide to eating and drinking and shopping in peace*. London, UK: Kogan Page.
Kontukoski, M., Luomala, H., Mesz, B., Sigman, M., Trevisan, M., Rotola-Pukkila, M., et al. (2015). Sweet and sour: Music and taste associations. *Nutrition and Food Science*, *45*, 357–376.

Krishna, A. (2012). An integrative review of sensory marketing: Engaging the senses to affect perception, judgment and behavior. *Journal of Consumer Psychology, 22*, 332–351.

Krishna, A. (2013). *Customer sense: How the 5 senses influence buying behaviour*. New York, NY: Palgrave Macmillan.

Lammers, H. B. (2003). An oceanside field experiment on background music effects on the restaurant tab. *Perceptual and Motor Skills, 96*, 1025–1026.

Lawless, H. T., & Heymann, H. (1997). *Sensory evaluation of food: Principles and practices*. Gaithersburg, MD: Chapman & Hall.

Leow, H. C. (2015). Never heard of Sensploration? Time to study up on epicure's biggest high-end pattern. *The Veox*, December 22. Downloaded from http://www.theveox.com/never-heard-of-sensploration-time-to-study-up-on-epicures-biggest-high-end-pattern/ on January 31, 2016.

McElrea, H., & Standing, L. (1992). Fast music causes fast drinking. *Perceptual and Motor Skills, 75*, 362.

McMains, A. (2015). How JWT Brazil and Dolby captured the iconic sound of Coke being poured over ice. *Adweek*, May 21. Downloaded from http://www.adweek.com/news/advertising-branding/how-jwt-brazil-and-dolby-captured-iconic-sound-experience-coke-being-poured-over-ice-164920 on July 8, 2015.

Milliman, R. E. (1986). The influence of background music on the behavior of restaurant patrons. *Journal of Consumer Research, 13*, 286–289.

Moir, J. (2015). Why are restaurants so noisy? Can't hear a word your other half says when you dine out? Our test proves restaurants can be a loud as rock concerts. *Daily Mail Online*, December 5. Downloaded from http://www.dailymail.co.uk/news/article-3346929/Why-restaurants-noisy-t-hear-word-half-says-dine-test-proves-restaurants-loud-rock-concerts.html on February 1, 2016.

Moore, M. (2015). Taste the difference. *The Financial Times Weekend*, August 30 (Life & Arts), 1–2.

North, A. C. (2012). The effect of background music on the taste of wine. *British Journal of Psychology, 103*, 293–301.

North, A. C., & Hargreaves, D. J. (1996). The effects of music on responses to a dining area. *Journal of Environmental Psychology, 16*, 55–64.

North, A. C., & Hargreaves, D. J. (1998). The effects of music on atmosphere and purchase intentions in a cafeteria. *Journal of Applied Social Psychology, 28*, 2254–2273.

North, A. C., Hargreaves, D. J., & McKendrick, J. (1997). In-store music affects product choice. *Nature, 390*, 132.

North, A. C., Hargreaves, D. J., & McKendrick, J. (1999). The influence of in-store music on wine selections. *Journal of Applied Psychology, 84*, 271–276.

North, A. C., Shilcock, A., & Hargreaves, D. J. (2003). The effect of musical style on restaurant customers' spending. *Environment and Behavior, 35*, 712–718.

Open Science Collaboration, (2015). Estimating the reproducibility of psychological science. *Science, 349*(6251), aac4716. http://dx.doi.org/10.1126/science.aac4716.

Pigott, S. (2015). Appetite for invention. *Robb Report, May*, 98–101.

Piqueras-Fiszman, B., & Spence, C. (2015). Sensory expectations based on product-extrinsic food cues: An interdisciplinary review of the empirical evidence and theoretical accounts. *Food Quality & Preference, 40*, 165–179.

Posner, M. I., Nissen, M. J., & Klein, R. M. (1976). Visual dominance: An information-processing account of its origins and significance. *Psychological Review, 83*, 157–171.

Reinoso Carvalho, F., Van Ee, R., Rychtarikova, M., Touhafi, A., Steenhaut, K., Persoone, D., et al. (2015a). Using sound-taste correspondences to enhance the subjective value of tasting experiences. *Frontiers in Psychology: Eating Behaviour, 6*, 1309.

Reinoso Carvalho, F., Van Ee, R., Rychtarikova, M., Touhafi, A., Steenhaut, K., Persoone, D., et al. (2015b). Does music influence the multisensory tasting experience? *Journal of Sensory Science*, *30*, 404–412.

Roballey, T. C., McGreevy, C., Rongo, R. R., Schwantes, M. L., Steger, P. J., Wininger, M. A., et al. (1985). The effect of music on eating behavior. *Bulletin of the Psychonomic Society*, *23*, 221–222.

Rudmin, F., & Cappelli, M. (1983). Tone-taste synesthesia: A replication. *Perceptual & Motor Skills*, *56*, 118.

Schifferstein, H. N. J. (2006). The perceived importance of sensory modalities in product usage: A study of self-reports. *Acta Psychologica*, *121*, 41–64.

Schifferstein, H. N. J., & Cleiren, M. P. H. D. (2005). Capturing product experiences: A split-modality approach. *Acta Psychologica*, *118*, 293–318.

Seo, H.-S., & Hummel, T. (2011). Auditory-olfactory integration: Congruent or pleasant sounds amplify odor pleasantness. *Chemical Senses*, *36*, 301–309.

Seo, H.-S., & Hummel, T. (2015). Influence of auditory cues on chemosensory perception Guthrie (Ed.), *The chemical sensory informatics of food: Measurement, analysis, integration* (pp. 41–56). Washington, DC: ACS Symposium Series; American Chemical Society.

Seo, H.-S., Lohse, F., Luckett, C. R., & Hummel, T. (2014). Congruent sound can modulate odor pleasantness. *Chemical Senses*, *39*, 215–228.

Shepherd, G. M. (2012). *Neurogastronomy: How the brain creates flavor and why it matters*. New York, NY: Columbia University Press.

Sietsema, T. (2008a). No appetite for noise. *The Washington Post*, April 6. Downloaded from http://www.washingtonpost.com/wp-dyn/content/article/2008/04/01/AR2008040102210_pf.html.

Sietsema, T. (2008b). Revealing raucous restaurants. *The Washington Post*, April 6. Downloaded from http://www.washingtonpost.com/wp-dyn/content/article/2008/04/04/AR20080404022735_pf.html.

Smith, P. (2011). Watch your mouth: The sounds of snacking. *Good*, August 27. Downloaded from http://magazine.good.is/articles/watch-your-mouth-the-sounds-of-snacking on August 2, 2014.

Spence, C. (2012). Auditory contributions to flavour perception and feeding behaviour. *Physiology & Behaviour*, *107*, 505–515.

Spence, C. (2013). Multisensory flavour perception. *Current Biology*, *23*, R365–R369.

Spence, C. (2014a). Noise and its impact on the perception of food and drink. *Flavour*, *3*, 9.

Spence, C. (2014b). Multisensory advertising & design. In B. Flath & E. Klein (Eds.), *Advertising and design. Interdisciplinary perspectives on a cultural field* (pp. 15–27). Bielefeld: Verlag.

Spence, C. (2015a). Eating with our ears: Assessing the importance of the sounds of consumption to our perception and enjoyment of multisensory flavour experiences. *Flavour*, *4*, 3.

Spence, C. (2015b). Music from the kitchen. *Flavour*, *4*, 25.

Spence, C. (in press). Multisensory packaging design: Colour, shape, texture, sound, and smell. In M. Chen & P. Burgess (Eds.), *Integrating the packaging and product experience: A route to consumer satisfaction*. Oxford, UK: Elsevier.

Spence, C., & Deroy, O. (2013). On why music changes what (we think) we taste. *i-Perception*, *4*, 137–140.

Spence, C., & Gallace, A. (2011). Multisensory design: Reaching out to touch the consumer. *Psychology & Marketing*, *28*, 267–308.

Spence, C., Michel, C., & Smith, B. (2014). Airplane noise and the taste of umami. *Flavour*, *3*, 2.

Spence, C., & Piqueras-Fiszman, B. (2014). *The perfect meal: The multisensory science of food and dining*. Oxford, UK: Wiley-Blackwell.

Spence, C., Puccinelli, N., Grewal, D., & Roggeveen, A. L. (2014). Store atmospherics: A multisensory perspective. *Psychology & Marketing, 31*, 472–488.

Spence, C., Richards, L., Kjellin, E., Huhnt, A.-M., Daskal, V., Scheybeler, A., et al. (2013). Looking for crossmodal correspondences between classical music & fine wine. *Flavour, 2*, 29.

Spence, C., Shankar, M. U., & Blumenthal, H. (2011). 'Sound bites': Auditory contributions to the perception and consumption of food and drink. In F. Bacci & D. Melcher (Eds.), *Art and the senses* (pp. 207–238). Oxford, UK: Oxford University Press.

Spence, C., Smith, B., & Auvray, M. (2015). Confusing tastes and flavours. In D. Stokes, M. Matthen, & S. Biggs (Eds.), *Perception and its modalities* (pp. 247–274). Oxford, UK: Oxford University Press.

Spence, C., & Wang, Q. (J.) (2015a). Sonic expectations: On the sounds of opening and pouring. *Flavour, 4*, 35.

Spence, C., & Wang, Q. (J.) (2015b). Wine & music (I): On the crossmodal matching of wine & music. *Flavour, 4*, 34.

Spence, C., & Wang, Q. (J.) (2015c). Wine & music (II): Can you taste the music? Modulating the experience of wine through music and sound. *Flavour, 4*, 33.

Spence, C., & Wang, Q. (J.) (2015d). Wine & music (III): So what if music influences taste? *Flavour, 4*, 35.

Spence, C., & Zampini, M. (2006). Auditory contributions to multisensory product perception. *Acta Acustica united with Acustica, 92*, 1009–1025.

Spence, C., & Zampini, M. (2007). Affective design: Modulating the pleasantness and forcefulness of aerosol sprays by manipulating aerosol spraying sounds. *CoDesign, 3*(Suppl. 1), 109–123.

Stafford, L. D., Agobiani, E., & Fernandes, M. (2013). Perception of alcohol strength impaired by low and high volume distraction. *Food Quality and Preference, 28*, 470–474.

Stafford, L. D., Fernandes, M., & Agobiani, E. (2012). Effects of noise and distraction on alcohol perception. *Food Quality & Preference, 24*, 218–224.

Stroebele, N., & de Castro, J. M. (2004). Effects of ambience on food intake and food choice. *Nutrition, 20*, 821–838.

Stroebele, N., & de Castro, J. M. (2006). Listening to music while eating is related to increases in people's food intake and meal duration. *Appetite, 47*, 285–289.

Stuckey, B. (2012). *Taste what you're missing: The passionate eater's guide to why good food tastes good*. London, UK: Free Press.

Suddath, C. (2013). How Chipotle's DJ, Chris Golub, creates his playlists. *Businessweek*, October 17. Downloaded from http://www.businessweek.com/articles/2013-10-17/chipotles-music-playlists-created-by-chris-golub-of-studio-orca on August 27, 2014.

Thorndike, E. L. (1920). A constant error in psychological ratings. *Journal of Applied Psychology, 4*, 25–29.

van Aken, G. (2013). *Listening to what the tongue feels*. Downloaded from http://www.nizo.com/news/latest-news/67/listening-to-what-the-tonguefeels/ on August 1, 2014.

Velasco, C., Jones, R., King, S., & Spence, C. (2013a). "Hot or cold?" On the informative value of auditory cues in the perception of the temperature of a beverage. In K. Bronner, R. Hirt, & C. Ringe (Eds.), *(((ABA))) Audio branding academy yearbook 2012/2013* (pp. 177–187). Baden-Baden: Nomos.

Velasco, C., Jones, R., King, S., & Spence, C. (2013b). Assessing the influence of the multisensory environment on the whisky drinking experience. *Flavour, 2*, 23.

Velasco, C., Jones, R., King, S., & Spence, C. (2013c). The sound of temperature: What information do pouring sounds convey concerning the temperature of a beverage. *Journal of Sensory Studies, 28*, 335–345.

Vranica, S. (2010a). Snack attack: Chip eaters make noise about a crunchy bag green initiative has unintended fallout: A snack as loud as 'the cockpit of my jet'. *Wall Street Journal*, August 10. Downloaded from http://online.wsj.com/news/articles/SB10001424052748703 960004575427150103293906 on July 24, 2014.

Vranica, S. (2010b). Sun Chips bag to lose its crunch. *The Wall Street Journal*, October 6. Downloaded from http://online.wsj.com/article/SB10001424052748703843804575534182403878708.html on December 2, 2015.

Wang, Q. (J.), & Spence, C. (2015). Assessing the effect of musical congruency on wine tasting in a live performance setting. *i-Perception*, *6*(3), 1–13.

Wang, Q. (J.), & Spence, C. (2016). "Striking a sour note": Assessing the influence of consonant and dissonant music on taste perception. *Multisensory Research*, *30*, 195–208.

Wang Q. (J.), & Spence, C. (submitted). Unravelling the bittersweet symphony: Assessing the influence of crossmodally congruent soundtracks on bitter and sweet taste evaluation of taste solutions and complex foods. *Food Quality & Preference*.

Wang, Q. (J.), Woods, A., & Spence, C. (2015). "What's your taste in music?" A comparison of the effectiveness of various soundscapes in evoking specific tastes. *i-Perception*, *6*(6), 1–23.

Wansink, B., & Van Ittersum, K. (2012). Fast food restaurant lighting and music can reduce calorie intake and increase satisfaction. *Psychological Reports: Human Resources & Marketing*, *111*(1), 1–5.

Wesson, D. W., & Wilson, D. A. (2010). Smelling sounds: Olfactory-auditory sensory convergence in the olfactory tubercle. *Journal of Neuroscience*, *30*, 3013–3021.

Wesson, D. W., & Wilson, D. A. (2011). Sniffing out the contributions of the olfactory tubercle to the sense of smell: Hedonics, sensory integration, and more? *Neuroscience and Biobehavioral Reviews*, *35*, 655–668.

Wheeler, E. (1938). *Tested sentences that sell*. New York, NY: Prentice & Co. Hall.

Wilson, S. (2003). The effect of music on perceived atmosphere and purchase intentions in a restaurant. *Psychology of Music*, *31*, 93–112.

Woods, A. T., Poliakoff, E., Lloyd, D. M., Kuenzel, J., Hodson, R., Gonda, H., et al. (2011). Effect of background noise on food perception. *Food Quality & Preference*, *22*, 42–47.

Yan, K. S., & Dando, R. (2015). A crossmodal role for audition in taste perception. *Journal of Experimental Psychology: Human Perception & Performance*, *41*, 590–596.

Yeoh, J. P. S., & North, A. C. (2010). The effects of musical fit on choice between two competing foods. *Musicae Scientiae*, *14*, 127–138.

Zampini, M., & Spence, C. (2004). The role of auditory cues in modulating the perceived crispness and staleness of potato chips. *Journal of Sensory Science*, *19*, 347–363.

Zampini, M., & Spence, C. (2005). Modifying the multisensory perception of a carbonated beverage using auditory cues. *Food Quality and Preference*, *16*, 632–641.

Food Color and Its Impact on Taste/Flavor Perception

Charles Spence[1] and Betina Piqueras-Fiszman[2]
[1]Crossmodal Research Laboratory, Department of Experimental Psychology, University of Oxford, Oxford, United Kingdom, [2]Marketing and Consumer Behavior Group, Wageningen University, Wageningen, The Netherlands

1 Introduction

Under most everyday conditions, consumers have the opportunity to inspect food and drink visually before deciding on whether or not to purchase or taste it (Cardello, 1996). Color constitutes one of the most salient of visual cues concerning the likely sensory properties (eg, taste/flavor) of food and drink (eg, Chylinski, Northey, & Ngo, 2015; Clydesdale, 1984, 1991, 1993; Delwiche, 2012; Hall, 1958; Kanig, 1955; Kostyla & Clydesdale, 1978; Valentin, Parr, Peyron, Grose, & Ballester, 2016; Watson, 2013). Indeed, people have been coloring food and drink for millennia now (eg, Downham & Collins, 2000; Tannahill, 1973; Walford, 1980). Furthermore, although little studied to date, those colors that we take to suggest that a food may have gone off can exert a particularly powerful effect on our food avoidance behaviors (eg, Crumpacker, 2006; Wheatley, 1973). As such, food color can legitimately be considered as perhaps the single most important product-intrinsic[1] cue governing the sensory and hedonic expectations that consumers hold concerning the foods and drinks they search for, purchase, and which they may subsequently consume.

A growing body of scientific research now suggests that our experience of flavor is to a large degree determined by the expectations that we generate (often automatically) prior to tasting (see Deliza & MacFie, 1997; Hutchings, 2003; Piqueras-Fiszman & Spence, 2015). Such expectations can result from branding, labeling, packaging, and other contextual effects (ie, from a host of product-extrinsic cues; Spence & Piqueras-Fiszman, 2011, 2012), but also from a variety of product-intrinsic cues. It is vision then, and most often color, that our brains use to help identify sources of food and make predictions about their likely taste and flavor (Cardello, 1994; Hutchings, 2003). As a spokesperson for the Institute of Food Technologists put it a few years back: "*Color creates a psychological expectation for a certain flavor that is often impossible to dislodge.*" (Harris, 2011).

---

[1]While the term "product-intrinsic" is widely used in the literature when talking about the color of a food or beverage, the appropriateness of this notion can be questioned from the perspective of holistic perception. Strictly speaking, color is not a property of a material but rather a percept (or construct) in an observer that originates from an interaction with a material, under the influence of many other cues that are external to the colored surface, but certainly internal to the person who is observing it (see, eg, Shepherd, 2012, p. 5).

The focus in this article will be on the psychological effect(s) that food color exerts over the mind and behavior of the consumer. We start by looking at the effect of food coloring on sensory expectations, and hence on our judgments of taste/flavor intensity and flavor identity (see Spence, Levitan, Shankar, & Zampini, 2010, for a review). The literature on off-colors in foods and drinks will be briefly reviewed, and the debate concerning natural versus artificial food coloring touched upon. Attention will be drawn to the growing body of research that is now starting to highlight some of the important individual differences that have been identified in terms of the meaning and influence of color in food as a function of a person's age, culture, or even their genetic make-up. Along the way, we will highlight some of the problems associated with the interpretation of much of the laboratory research that has been conducted to date, and take a speculative look at the future of color in food.

2 Psychological effects of food color: setting sensory expectations

Over the last 80 years or so, researchers have investigated the psychological impact of food color in everything from noodles (Zhou, Wan, Mu, & Du, 2015), vegetables (Urbányi, 1982; Yang, Cho, & Seo, 2015), meat (Carpenter, Conforth, & Whittier, 2001), salsas (Shermer & Levitan, 2014), and cheese (Wadhwani & McMahon, 2012) through to yoghurt (Calvo, Salvador, & Fiszman, 2001; Dolnick, 2008), cake (DuBose, Cardello, & Maller, 1980), jams, jellies, and sherbets (eg, Hall, 1958; Moir, 1936). That said, the majority of studies in this area have been conducted with beverages, presumably because it is simply much easier to manipulate the level of color in solution (see Spence, 2015a). In this section, we describe a number of representative studies that have focused on the investigation of the impact of color on flavor/taste identity, on the one hand, and on its perceived intensity on the other.

2.1 Flavor identity

Perhaps the most robustly demonstrated effect of adding (or changing) food coloring has been on the ability of people to identify the flavor of food or, more commonly, drink (see Spence, 2015b, for a review). For instance, classic research by DuBose et al. (1980) demonstrated that the addition of food coloring (green, red, or orange) biased participants' judgments concerning the identity of the cherry-flavored solution. Nearly 20% of the participants reported that the drink tasted of orange when the solution was colored orange compared to no such responses when the same drink was colored red, green, or remained colorless. Meanwhile, coloring the same drink green led to 26% lime-flavored responses as compared to no such responses when the drink was colored red or orange (see also Hyman, 1983; Stillman, 1993; and Watson, 2013, for similar results).

Oram et al. (1995) gave more than 300 participants of various ages four drinks to taste. Four possible flavors (chocolate, orange, pineapple, and strawberry) were presented in four different colors (brown, yellow, orange, and red) thus giving rise to a total of 16 possible drinks. The participant had to try to discriminate the flavor of the drinks. The results highlighted a clear developmental trend toward an increased ability to correctly report the flavor of the drinks, regardless of the drink's color (see Fig. 6.1). That is, the crossmodal modulation of flavor perception by color apparently decreased from 2 years of age up.

Importantly, in this and the majority of other published studies that have been reported so far, the participants were given no information about the possibility that the solutions might have been incongruently colored (a point to which we return later). Subsequent research by Zampini, Sanabria, Phillips, and Spence (2007) and Zampini, Wantling, Phillips, and Spence (2008) demonstrated that adult participants were often confused by the addition of inappropriate color to a range of fruit-flavored soft drinks (see Fig. 6.2). Zampini et al. demonstrated that such a crossmodal influence of beverage color on flavor identification could not be circumvented even if the participants had been explicitly informed that they should try to ignore the potentially misleading food coloring as much as possible. Such results therefore hint at the automaticity of these kinds of crossmodal effects. Although beyond the scope of the present chapter, it is perhaps also worth noting that food coloring can influence the perceived thirst-quenching (or refreshing) properties of drinks as well (eg, Clydesdale, Gover,

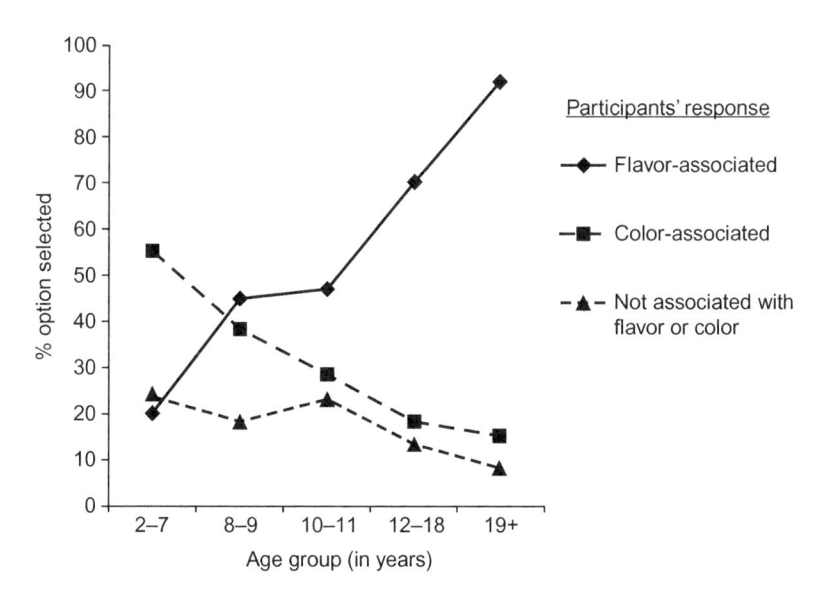

Figure 6.1 Graph highlighting the percentage of trials in which the participants' flavor discrimination response matched the color of the drink, the actual flavor of the drink, or matched neither the color nor flavor of the drink as a function of the age of the participants in a developmental study of the psychological impact of color on people's flavor discrimination responses.
Source: Redrawn from Oram et al. (1995).

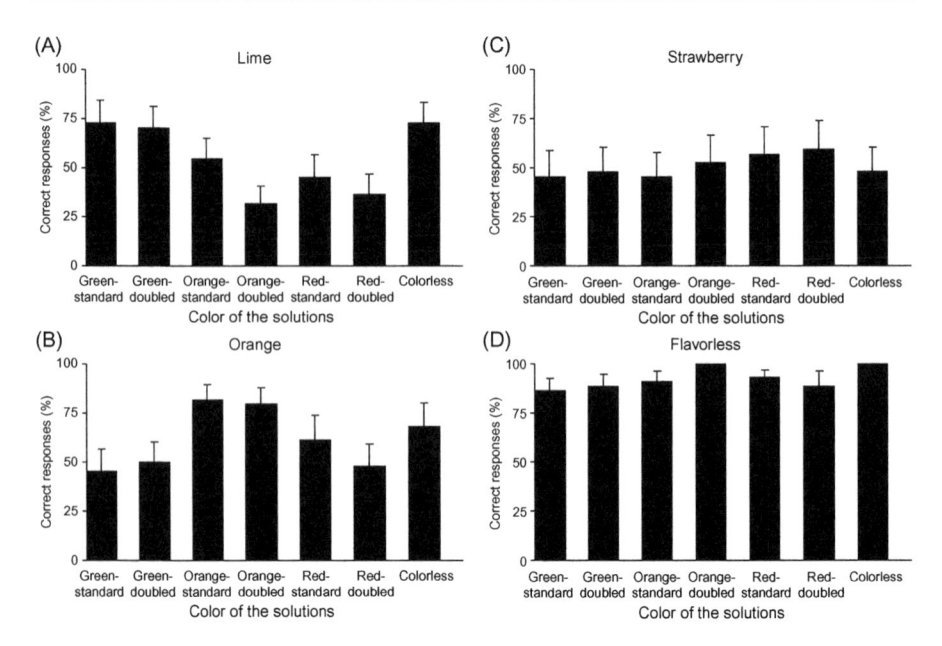

Figure 6.2 Mean percentage of correct flavor discrimination responses for the lime (A), orange (B), strawberry (C), and flavorless (D) solutions presented in Zampini et al. (2007; Experiment 2). The error bars represent the between-participants standard errors of the means. These results clearly show the deleterious effect of adding inappropriate food color on participants' flavor identification responses, at least for the lime- and orange-flavored drinks. (Somewhat surprisingly, the addition of food coloring had little effect on the accuracy of participants' flavor discrimination responses for the strawberry-flavored solution.) Critical to the present discussion, increasing the intensity of food coloring had no effect on flavor identification, nor on judgments of flavor intensity.
Source: Figure reprinted with permission from Zampini et al. (2007).

Philipsen, & Fugardi, 1992; Guinard, Souchard, Picot, Rogeaux, & Siefferman, 1998; Zellner & Durlach, 2002, 2003).

2.2 Taste/flavor intensity

The majority of consumers expect more intensely colored foods and beverages (not to mention the packaging in which such products come) to have a more intense taste/flavor. Over the last 50 years or so, a large body of laboratory research has demonstrated that adding more coloring to a food, or more often, to a beverage (see Spence, 2015b; Spence et al., 2010, for reviews), can lead the participants in laboratory research to rate the taste and/or flavor as more intense (eg, Calvo et al., 2001; Johnson & Clydesdale, 1982; Johnson, Dzendolet, & Clydesdale, 1983; Johnson, Dzendolet, Damon, Sawyer, & Clydesdale, 1982; Norton & Johnson, 1987).

The addition of food coloring has also been shown to influence sensory thresholds for certain basic tastes. For example, in one classic study, Maga (1974) demonstrated that adding food coloring (red, green, or yellow) to an otherwise clear solution exerted a significant effect on thresholds for the detection of certain of the basic tastes when they were presented in solution: The addition of green food coloring decreased people's detection threshold for sourness, while simultaneously increasing their sensitivity to sweetness. The addition of yellow coloring reduced people's detection threshold to sourness and sweetness, while the addition of red food coloring reduced people's sensitivity to bitterness.[2] By contrast, the threshold for salt was unaffected by the addition of coloring.

Maga's (1974) suggestion here is that salty foods come in many different colors, and so are not associated especially with any particular color. However, the latest research on the color associations that people hold to basic tastes clearly shows that saltiness is strongly associated with white (and to a lesser extent blue). As it happens, neither of these colors were included in Maga's study. Thus, it could be argued that the appropriate color was simply not tested, and that is why no effect of color on the saltiness threshold was observed (see Spence et al., 2015).

Perhaps the most convincing evidence published to date concerning the influence of food coloring on ratings of taste intensity comes from Clydesdale et al. (1992). These researchers conducted a number of psychophysical studies showing that the addition of the appropriate food coloring (in this case, red) increased perceived sweetness by as much as 10%. Indeed, such results have led some to wonder whether food coloring could be used as an effective means of reducing the sugar content of foods. While this is certainly a theoretical possibility it is worth bearing in mind that the majority of the studies that have been published to date have involved fairly short-term exposure to particular combinations of color–taste/flavor. And while it is one thing to demonstrate that food coloring has a short-term impact on sweetness perception, it is quite another to demonstrate that it will necessarily have psychological effects over the long term (cf. Levitan, Zampini, Li, & Spence, 2008). As such, it can be argued that longer-term follow-ups are urgently needed.

At this point, it is, however, important to note that changing the color of a food (or more often, beverage) has not always influenced people's ratings of perceived taste/flavor intensity. For instance, Norton and Johnson (1987) reported one such null result. Although these researchers manipulated the intensity of four typical drink colors, they were unable to find any meaningful relationship between the intensity of the color and participants' flavor ratings on either a sweet–sour scale, or on a distinct–indistinct flavor scale. Elsewhere, Zampini et al. (2007) conducted an experiment in which a group of adults had to try to identify the flavor of a variety of drinks and rate the perceived intensity of the flavor using a Labeled Magnitude Scale. The drinks were flavorless, or else had an orange, lime, or strawberry flavor added. The solutions were artificially colored with either a standard or double concentration of

---

[2] Note that the participants in this study only ever had to report whether or not the solution had a taste. That is, they never had to identify the tastant. In fact, somewhat surprisingly, the question of whether color influences people's ability to *identify/discriminate* the basic tastes has not been studied to date (see Spence et al., 2010, for a review). This is despite the fact that extensive evidence has been collected concerning the colors that people in different cultures associate with each of the basic tastes (see Spence et al., 2015; Wan, Woods et al., 2014b, for reviews and cross-cultural evidence).

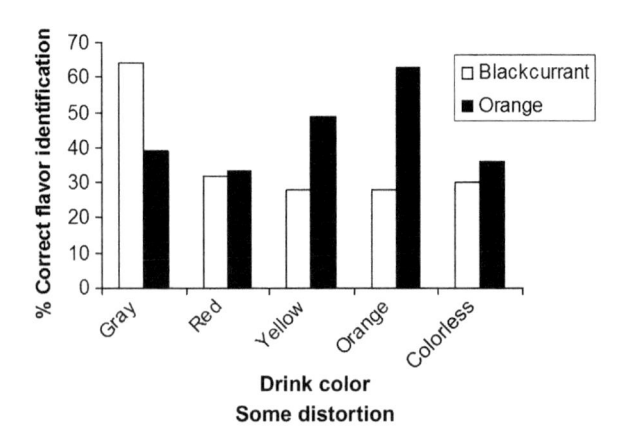

Figure 6.3 Summary of the results of Zampini et al.'s (2008) study highlighting the influence of color on people's ability to correctly identify orange- and blackcurrant-flavored solutions. *Source*: Figure reprinted with permission from Spence et al. (2010).

red, green, or orange coloring, or else were left colorless. Once again, however, variations in the intensity of the food coloring that was added (no matter whether that color was appropriate or inappropriate to the flavor of the drink) did not influence the perceived intensity of a drink's flavor. That said, the addition of inappropriate food coloring significantly impaired the participants' flavor identification responses (see Fig. 6.3), thus suggesting that the participants were unable to ignore the color of the drinks completely, as they had been encouraged to do by the experimenter.

Meanwhile, Lavin and Lawless (1998) investigated the influence of varying the intensity of food coloring on people's ratings of sweetness intensity for two pairs of strawberry-flavored drinks. One pair light- and dark-red, the other pair light- and dark-green. The drinks were equally physically sweet and varied only in terms of their color intensity. The adults who took part in this study rated the dark-red and light-green drinks as tasting sweeter than the light-red and dark-green samples, respectively. By contrast, the intensity of the color did not exert any effect on the responses of children aged between 5 and 14 years. Elsewhere, Alley and Alley (1998) also failed to demonstrate any effect of the addition of food coloring (red, blue, yellow, or green) to otherwise colorless sweet solutions in a group of 11–13-year-olds. Philipsen, Clydesdale, Griffin, and Stern (1995) had a group of young adults, aged 18–22 years, and a group of older participants, aged 60–75 years, rate a number of attributes (eg, sweetness, flavor intensity, flavor quality, flavor identification, etc.) of 15 samples of an artificially flavored cherry beverage that varied in terms of their sucrose content, flavor, and color. Interestingly, variations in color intensity had no effect on sweetness ratings in either age group. However, they did influence flavor intensity ratings in the older group.

Chan and Kane-Martinelli (1997) examined the effect of food coloring on perceived flavor intensity and acceptability ratings for chicken bouillon and chocolate pudding. The foods were presented with no color added, with the normal (ie, commercial) level of food coloring, or else with double the normal amount of coloring added. The participants tasted and then evaluated the three samples of either food using visual analog scales. The results of this study revealed that the younger adults (20–35 years of age) were more influenced by the presence of food coloring than were the older adults (60–90 years of age). In particular, the younger group's judgment of the overall flavor intensity of the chicken bouillon was influenced by the amount of coloring that had been added to the sample.

2.3 Interim summary

Taken together, the results that have been reviewed in this section suggest that the psychological effects of adding or changing the intensity of food coloring on the *intensity* of taste/flavor perception are not altogether clear. Null results have been reported in a number of seemingly well-conducted studies (eg, Alley & Alley, 1998; Frank, Ducheny, & Mize, 1989), while even those who have obtained significant effects of color on taste/flavor intensity ratings/perception have tended to do so only under a subset of experimental conditions, or else in a subset of those individuals whom they have chosen to test (Bayarri, Calvo, Costell, & Duran, 2001; Chan & Kane-Martinelli, 1997; DuBose et al., 1980; Fernández-Vázquez et al., 2014; Gifford & Clydesdale, 1986; Gifford, Clydesdale, & Damon, 1987; Lavin & Lawless, 1998; Maga, 1974; McCullough, Martinsen, & Moinpour, 1978; Pangborn, 1960; Pangborn & Hansen, 1963; Philipsen et al., 1995; Roth, Radle, Gifford, & Clydesdale, 1988; Strugnell, 1997; Zampini et al., 2007). Hence, it is difficult to draw any clear conclusions from the range of results that have been published to date as to when exactly the addition of food coloring will or will not influence ratings of taste/flavor intensity. That the addition of food coloring *can* influence thresholds and ratings of stimulus intensity is not in doubt. However, one of the questions that has been left unresolved by the research that has been published over the last 80 years or so concerns why it is that these seemingly inconsistent results have been obtained.

According to Shankar, Simons, Levitan et al. (2010), Shankar, Simons, Shiv et al. (2010), and Shankar, Simons, Shiv, McClure, and Spence (2010), one way to resolve this inconsistency is by considering the "*degree of discrepancy*" between the expected flavor set by color, and the actual flavor when experienced by the participant (or consumer). Under low discrepancy conditions, the perceived disparity between the expected and actual flavor of a drink (or food) is small. Low discrepancy color–flavor combinations might, for example, consist of cranberry- or blueberry-flavored drinks colored purple (purple being associated with grape flavor), whereas high discrepancy combinations might include banana- or vanilla-flavored drinks that had been colored purple. Across several experiments, when a particular color—identified by participants as one that generated a strong flavor expectation—was added to the drinks that the participants had been given to sniff (as compared to when no such color was added), a significantly greater proportion of participants' identification responses were consistent with this expectation. By contrast, under conditions of high discrepancy, adding the same colors to the drinks no longer affected their identification responses as much. That is, there was a significant difference in the proportion of responses that were consistent with participants' color-based expectations in

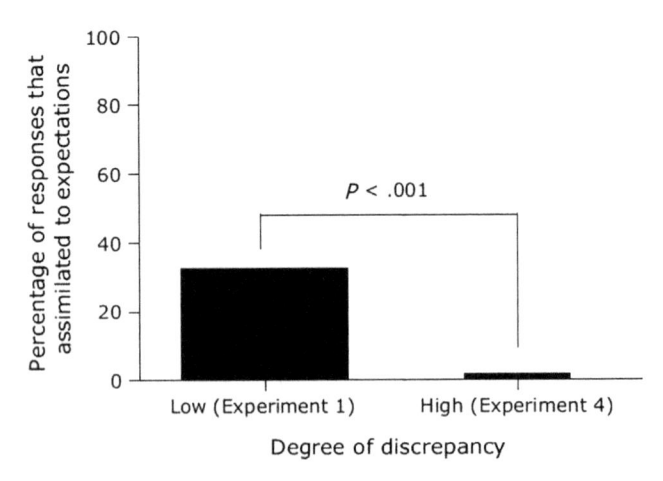

Figure 6.4 Summary results from two of the experiments (conducted with the same participants) reported by Shankar et al. (2010c) showing how the addition of food coloring to an otherwise colorless flavored solution led to assimilation when the "degree of discrepancy" between the flavor expected by the color and the actual flavor of the drink when sniffed orthonasally was small (left), but not when the degree of discrepancy was large (right). *Source*: Figures reprinted with permission from Shankar et al. (2010c).

conditions of low as compared with high discrepancy (see Fig. 6.4). Shankar et al.'s results therefore indicate how the degree of discrepancy between an individual's expected experience (on seeing a drink) and their actual experience (on sniffing, or tasting, it) can significantly affect the extent to which color crossmodally influences judgments of flavor identity.

That said, one thing to bear in mind about the studies of Shankar, Simons, Levitan et al. (2010), Shankar, Simons, Shiv et al. (2010), and Shankar, Simons, Shiv, McClure, et al. (2010) is that the participants never got to taste the flavored drinks that they were judging. That is, all their judgments/ratings were made on the basis of orthonasal olfaction. One might not expect this to matter all that much, given the extensive literature showing that color cues modulate orthonasal olfactory discrimination/identification responses across a wide range of experimental conditions (eg, Blackwell, 1995; Davis, 1981; Michael, Galich, Relland, & Prud'hon, 2010; Petit, Hollowood, Wulfert, & Hort, 2007; Stevenson & Oaten, 2008; Zellner, Bartoli, & Eckard, 1991; Zellner & Kautz, 1990; Zellner & Whitten, 1999; see Zellner, 2013, for a review).[3] However, Koza, Cilmi, Dolese, and Zellner (2005) have shown differing effects of food color on orthonasal and retronasal judgments of a commercial fruit-flavored water drink. Coloring a tangerine–pineapple–guava-flavored solution red led to odor enhancement in those participants who sniffed the odor orthonasally, while leading to a reduction in perceived odor intensity when the same olfactory stimulus was presented retronasally instead!

---

[3] That said, Koza et al.'s (2005) results concerning the differing effect of color on orthonasal and retronasal olfactory intensity judgments needs to be borne in mind here.

Koza and colleagues attempted to account for this admittedly surprising pattern of results by suggesting that it is more important for us to correctly evaluate foods once they have entered our mouths, since that is when they pose a greater risk of poisoning. By contrast, foods that are located outside the mouth are simply less likely to poison us. Whatever the most appropriate explanation turns out to be, the main point to note is that one cannot simply assume that a color's effect on orthonasal olfactory judgments of a food or drink's flavor (ie, when sniffing) will necessarily be the same when people come to actually taste the food or drink item (and retronasal olfaction comes into play). Another point to note here is that a lack of precise color measurement has made it difficult for those waiting to replicate many of the studies that have been published to date (cf. Clydesdale, 1991; Francis & Clydesdale, 1975).

However, here, perhaps, one also needs to take a step back and consider what happens if the sensory expectations set by the intensity of food coloring fail to match the experience when a food or beverage item is actually tasted by the participant/consumer. In much of the laboratory research, there is a very real question as to whether the participants actually believed that the colors of the foods or drinks that they were tasting had meaning—that is, to what extent did they really think that the food coloring they saw was linked to the actual taste/flavor of the drinks that they were tasting? It can be wondered whether or not participants noticed any discrepancy between what they saw and what they tasted (Oram et al., 1995). While such an assumption is presumably likely out there in the real world, it is not so clearly the case for those participants in laboratory research who may well have been exposed to a whole series of inappropriately colored samples to taste and evaluate. What is more, the studies that have been published to date vary between those where the researchers have been very explicit about the fact that the color cues were designed to be misleading (Zampini et al., 2007, 2008; see also Parr, White, & Heatherbell, 2003), and those where the researchers have done everything that they can to hide the true purpose of their study from their participants (Garber, Hyatt, & Starr, 2000).

3 Names, brands, and colors

Given the ambiguity in the meaning of color in foods and beverages, it can sometimes be important that the name/description of a food or beverage sets the right sensory or hedonic expectations, or else help to disambiguate between the different possible meanings of a given color. Perhaps the classic example here comes from the work of Yeomans, Chambers, Blumenthal, and Blake (2008). These researchers demonstrated that when the meaning of the color of food is misinterpreted (ie, when it sets the wrong sensory expectations) this can have an adverse effect on people's subsequent taste ratings. Yeoman et al.'s participants tasted a bright pink ice-cream. One group of participants was given no information about the dish, another group were informed that the food was called "Food 386," and a third group was told that what they were about to eat was a "frozen savory mousse." The participants in the former group who were led by their eyes into expecting that they would taste a strawberry-flavored ice-cream (which has the same pinkish-red color) did not like the dish when they tried it. Specifically, they rated the savory smoked salmon ice-cream as tasting too salty. By contrast, the participants in the other two groups were more likely to rate the seasoning of the dish as being about right, and, what is more, liked the savory ice-cream more as well (see also Hoegg & Alba, 2007; Miller & Kahn, 2005; Shankar, Levitan, Prescott, & Spence, 2009). Results such as these demonstrate that the meaning of color in food and drink can be altered simply by the description that is given to a product or dish (Spence & Piqueras-Fiszman, 2014). Generally speaking (ie, in all environments except perhaps the modernist restaurant), it is important to avoid disconfirmed expectations (Carlsmith & Aronson, 1963; Lelièvre, Chollet, Abdi, & Valentin, 2009; Mace & Enzie, 1970; Piqueras-Fiszman & Spence, 2012).

Indeed, the typical laboratory situation can be contrasted with that of everyday consumption episodes when a food or drink will most likely be encountered in the context of branding/packaging information, or may well have been described by whoever has prepared or happens to be serving it. In other words, it can be argued that the situation that is typically studied in the laboratory is quite unlike that of everyday life (see also Garber, Hyatt, & Starr, 2001, 2003a,b, on this point). Hence, one concern here is that the results of much of the laboratory research that has been conducted to date may end up giving a biased view of the importance of color in multisensory flavor perception. Under typical laboratory testing conditions, color is the only cue that participants have normally had to go on. By contrast, in the majority of real-world consumption situations, color is but one of many cues (including branding, pricing, labeling, etc.) that consumers can, and most certainly do, use (Di Monaco, Cavella, Iaccarino, Mincione, & Masi, 2003).

One other product-extrinsic cue that modulates the meaning of color in beverages is the nature of the glass or receptacle in which a drink happens to be presented (see Wan, Velasco et al., 2014; Wan, Woods, Seoul, Butcher, & Spence, 2015). On occasion, exactly the same drink color may have a very different meaning if shown in a plastic bathroom cup than in a cocktail glass say. The interested reader will find this topic covered in the chapter "Assessing the Influence of the Drinking Receptacle on the Perception of the Contents."

4 Psychological effects on food color on behavior

It is important to realize that the psychological effects of food coloring are not restricted to the sensory-discriminative domain. It has often been suggested that food coloring modulates certain of our food-related behaviors as well (eg, Birren, 1963; Piqueras-Fiszman & Spence, 2014). Certainly, getting the color right plays an important role in food acceptance, liking, and hence, ultimately, intake (eg, de Wijk, Polet, & Engelen, 2004; Gossinger et al., 2009; Harris, 2011; Imram, 1999; Jantathai, Danner, Joechl, & Dürrschmid, 2013; Schutz, 1954; Wei, Ou, Luo, & Hutchings, 2012). However, as Garber et al. (2001) point out, while the claim that color influences food preferences is often made by researchers, good marketing-relevant insights tend to be a little harder to come by. Color cues can play an important role in modulating a consumer's affective expectations (eg, Wilson & Klaaren, 1992; Zellner, Strickhouser, & Tornow, 2004). And just as there can be a sensory disconfirmation of expectation (as outlined above), there can also be a negative disconfirmation of hedonic expectation—that is, when a consumer does not like a food or beverage as much as they were expecting to.

4.1 On the impact of color variety

People have been shown to consume more candy if it comes in a variety of colors than if offered just a single color of candy (Rolls, Rowe, & Rolls, 1982; see also Weir, 2009), even if that color happens to be their favorite. Whether sensory-specific satiety or boredom is the most appropriate explanation for such results is still being deliberated by researchers (see Piqueras-Fiszman & Spence, 2014, for a review). Interestingly, though, while the use of color (specifically increasing the variety of color) is usually portrayed as a means by which food companies can get consumers to consume more, there is some evidence to suggest that color cues can also be used to help control intake, by providing an effective cue to portion control (Geier, Wansink, & Rozin, 2012; see also Kahn & Wansink, 2004; Redden & Hoch, 2009). So, for example, Geier et al. (2012) conducted a study showing that people eat less potato chips from a tube if every seventh chip in the stack was colored red.[4]

4.2 Off coloring in food

Color is a key parameter to determine food quality and it serves as an indicator for consumers to predict when a food is becoming rotten or spoiled. Researchers have been interested in the response of consumers to food coloring that they associate with products that have been in some way spoiled. Anecdotal support for the suggestion that such off-colors have a profound effect on people's food behaviors was suggested by the response of consumers to a batch of Tropicana grapefruit juice that was donated to a food bank back in 1981. According to Crumpacker (2006, p. 6), nobody wanted to drink it because of its abnormal brown color. This was despite the fact that those who tried it reported that it tasted just fine (see also Fernández-Vázquez et al., 2014; Hoegg & Alba, 2007; Tepper, 1993).

Meanwhile, the only thing that may have struck the guests at Wheatley's (1973) classic steak, chips, and peas dinner as odd was how dim the ambient lighting was. However, this hid the food's true color. Part-way through the meal, the lighting was returned to normal, revealing that the steak was colored blue, the chips green, and the peas red. A number of Wheatley's guests suddenly felt ill, with several of them apparently heading straight for the bathroom (cf. Moir, 1936). Here, it is noticeable how the majority of the research on the psychological impact of off-color in food is rather anecdotal in nature (presumably because it can be difficult to get ethical approval to present food to participants and have them believe that the color indicates that it has gone off). Nevertheless, the evidence such as it is, does at least hint at the strong avoidance responses that such food coloring can induce, especially in the case of meats and fish that happen to look off (see also Profet, 1992).

---

[4] Elsewhere, researchers have also shown that serving food off red plates can be used to reduce people's consumption (Bruno, Martani, Corsini & Oleari, 2013; Genschow, Reutner, & Wanke, 2012).

4.3 Artificial/natural

Over the years, there have been frequent concerns expressed about the negative health and well-being consequences associated with certain artificial food colorings (see Accum, 1820; Anon., 1979a,b, 1980; Goldenberg, 1977; Harris, 2011; Kramer, 1978; Lucas, Hallagan, & Taylor, 2001; Meggos, 1995; Sinclair, 1906, p. 93; Stevens et al., 2013; Tuorila-Ollikainen, 1982; Weiss et al., 1980; Whitehill, 1980; Wilson, 2009). This had led some to search out those foods that are free from all artificial food coloring. However, many commercial foods and drinks are disappointing, as they are perceived by many as lacking in taste/flavor if served in a colorless (ie, clear or white) format (Harris, 2011). Of course, that the food coloring used is natural does not in-and-of-itself necessarily make it appealing in the mind of the consumer. Here, one only needs to think of the red coloring of, for example, Smarties (the candy-covered chocolate) that used to be made from the carminic acid that was extracted from a particular variety of scaly insect. Unappealing to most consumers, one imagines; nowadays, the red coloring comes from red cabbage instead (Wilson, 2009). And what, exactly, constitutes natural is not obvious: so, for example, the vibrant orange-colored carrots that we are all familiar with nowadays are actually the result of selective breeding. Once upon a time, the majority of carrots were purple. The selective breeding designed to deliver the orange color of the Dutch royal family back in the 17th century (Anon., 2005; Dalby 2003; Macrae 2011).[5]

A number of the modernist chefs we have been fortunate enough to work with over the years have been particularly interested in surprising their diners by presenting foods that have one color (and hence set a particular taste/flavor expectation) while delivering another unexpected flavor instead. Note that while under the majority of everyday conditions, people prefer foods and beverages that taste as they expect them to taste (ie, people don't like surprises, especially when it comes to the stimuli that enter the mouth, and hence have the potential to poison them), there are occasions, in particular, while sitting at the table in a modernist restaurant where diners seem to positively relish having their expectations played with (see Piqueras-Fiszman & Spence, 2012; Spence & Piqueras-Fiszman, 2014). However, the chefs typically do not want to achieve such results by means of artificial food colorings for fear of the negative reaction of their diners. One elegant example, then, of the use of natural coloring to create surprise and delight in the mind of the diner comes from the beetroot and orange jelly dish that was served as one of the opening courses on the menu at The Fat Duck restaurant in Bray a few years ago (http://www.thefatduck.co.uk/). This dish would be presented as two blocks of jelly, one bright orange, the other a dark purple, placed side-by-side on the plate.

And where the modernist chefs lead, the market often follows. Such unusually colored fruits and vegetables have apparently been selling well in the supermarkets in recent years (Macrae, 2011). More generally, there would appear to be renewed interest in surprisingly colored foods for the mass market as well. For example, a few years

---

[5] Although another, perhaps more plausible, explanation for why orange carrots may have been favored over the original purple variety was because the latter would color the soups, stews, etc. in which they were used.

back, one well-known burger chain launched a pitch black bamboo and squid ink burger in Japan. It was seasoned with black squid ink ketchup and served in a black bun (Cook, 2012).[6] As a group, children seem to be particularly fond of such miscolored foods (think confused Skittles) and beverages (Anon., 2007; Garber, Hyatt, & Boya, 2008; see also Piqueras-Fiszman & Spence, 2012; Walsh, Toma, Tuveson, & Sondhi, 1990).

5 Marketing color

Adding color to food, or else changing the color of a food or beverage (or its packaging) has long been used as an effective marketing tool (eg, Favre & November, 1979; Gimba, 1998; Hicks, 1979; Jantathaia, Danner, Joechl, & Dürrschmid, 2013; Singh, 2006; see also Stummerer & Hablesreiter, 2010; http://www.ddwcolor.com/hue/why-color/). In fact, according to an informal store audit reported by Garber et al. (2001), 97% of all food brands displayed (in all categories), used color to signify flavor. Food color is used in marketing for a number of reasons: everything from increasing shelf stand-out through to blurring the distinction between competing products. Indeed, in the early days, there was quite a fight by the butter lobby in order to try to prevent the makers of margarine from adding a golden yellow hue to their product to give it the appearance of its better established rival (eg, see Masurovsky, 1939).

More recently, the potential role in marketing of adding food coloring was amply demonstrated by the dramatic rise in sales of tomato ketchup when Heinz decided to add a tiny amount of food coloring and turn this staple of the dining table green (Farrell, 2000). However, it is important to note that not every attempt by marketers to use color to boost sales has been so successful. Clear cola drinks have, for example, generally failed in the marketplace (see Triplett, 1994), and it could be this very reason why nowadays sales of Cordial Campari, a clear version of the classic Italian red-colored bitter aperitif, remain disappointing (http://www.saveur.com/article/Wine-and-Drink/Campari-Good-and-Bitter). There are certainly a number of theories out there in the marketing literature about what may have gone wrong in the case of clear cola. Our guess is that when such drinks were tasted away from their packaging then the likely disconfirmation of expectation that results from experiencing a cola flavor when the visual appearance of the drink led the consumer to expect lemonade or soda water may have been especially problematic.

6 Individual differences in the psychological effects of color

One thing that is noticeable about much of the early research on the psychological effects of food coloring is how little attention was paid to the profiles of the

---

[6] Obviously not content with that, a red burger was recently launched (see Anon., 2015).

participants themselves. This is turning out to be an important caveat since the latest research is now starting to show that exactly the same food color can elicit qualitatively different expectations concerning the likely taste/flavor of food and drink in different groups of consumers.

6.1 Cross-cultural differences

Researchers have started to study how exactly the same color (eg, in a beverage) can set up qualitatively different expectations in the minds of different groups of consumers. Just take the drinks shown in Fig. 6.5: when these were shown to young adults in Taiwan and the United Kingdom, the former expected them to taste of cranberry and mint (perhaps mouthwash), respectively, whereas the latter expected cherry/strawberry and raspberry, instead (see Shankar, Levitan, & Spence, 2010). Wan, Velasco et al. (2014) and Wan et al. (2015) conducted a number of internet-based studies designed to assess which food colors have a similar meaning in terms of expected flavor across culture and which differ markedly in terms of the expectations that they set. Jantathaia, Sungsri-in, Mukprasirt, and Dürrschmid (2014) conducted a cross-cultural study of food color in Austria and Thailand. These researchers demonstrated significant differences in terms of how much consumers from the two countries expected to like the various brightly colored Thai desserts, as well as how intense they expected the taste to be. Furthermore, when other consumers actually got to taste the desserts, significant differences in liking between the two cultural groups were observed. Food marketers working in the global marketplace obviously need to be aware of such cultural differences in the meaning of food color (see also Garber, Hyatt, & Nafees, 2015).

Figure 6.5 Two of the six colored drinks shown to the participants from Taiwan and the United Kingdom in a study by Shankar et al. (2010a). The results of this cross-cultural study demonstrated that exactly the same food color (red on the left and blue on the right) can elicit qualitatively sensory different expectations as far as the likely flavor of a drink might be in consumers from different countries. The most frequently expected flavors for drinks of these colors are shown at the bottom.

6.2 Developmental differences

Developmental differences in the meaning, and influence, of food color have occasionally been observed. As noted already, young children seem to be more drawn to brightly (some would say artificially) colored foods than are adults. However, as noted by Spence (2014), adults are usually quite willing to accept unusually colored drinks when it comes to the case of cocktails. In terms of changes in the psychological influence of food coloring across the lifespan, then, on the basis of the evidence published to date (eg, see Christensen, 1985; Oram et al., 1995; Philipsen et al., 1995), it would appear as though, if anything, visual cues may exert a somewhat greater influence on flavor identification early in development (see Fig. 6.1), and in old age, than in adulthood (see Spence, 2012, for a review).

One reason as to why children might show more visual sensory dominance (ie, simply relying on what they see) than adults is because they have not yet learned to integrate their senses in an adult-like manner (cf. Gori, Del Viva, Sandini, & Burr, 2008). At the other end of the age spectrum, the well-documented decline of taste and smell sensitivity may mean that the residual senses (especially those where prostheses, such as glasses or hearing aids, are available) take on a far more important role in terms of determining the final taste/flavor experience (eg, Christensen, 1985; Clydesdale, 1984). However, it has to be said that the available evidence on this topic (some of which was reviewed earlier) is rather mixed. While some researchers have been able to demonstrate more pronounced psychological effects of food coloring in, say, older adults (eg, Philipsen et al., 1995; Tepper, 1993), such differences have not always been found. Here, it is perhaps also worth bearing in mind any changes in the meaning and acceptability of color over time. One only needs to be reminded that blue foods were traditionally considered unacceptable to consumers (eg, Cheskin, 1957; Hine, 1995; Tysoe, 1985). Nowadays, many foods are blue, although in this case note that they are primarily marketed at the younger consumer (Garber et al., 2001). One could even think over a much longer timescale of how the flavor of carrots has switched its color association from purple to orange (see above).

6.3 Expertise

Expertise has been shown to modulate the psychological effect of food coloring on flavor perception. Some of the most impressive studies in this area have come from the world of wine (see Pangborn, Berg, & Hansen, 1963, for early research; and Spence, 2010, for a review). In one oft-cited study, Morrot, Brochet, and Dubourdieu (2001) fooled a group of students on a university wine course in Bordeaux, France into choosing red wine aroma descriptors when given a white wine that had been artificially colored red with odorless, tasteless food dye to evaluate. Parr et al. (2003) conducted a follow-up in New Zealand in which they tested both experts (including professional wine tasters and wine makers) and "social" drinkers. The descriptions of the aroma of a Chardonnay wine given by the experts when it had been artificially colored red were more accurate when the wine was served in an opaque glass than when served in a clear glass instead. Interestingly, this color-induced biasing of flavor judgments occurred despite the fact that the experts had been explicitly instructed to ignore the color of the wines they were tasting (thus suggesting that the crossmodal effect of vision is not under cognitive control). In this case, the social drinkers were so bad at reliably identifying the aromas present in the wine that it was difficult for Parr et al. to discern any pattern in the data when an inappropriate color was added to the wine.

Taken together, the evidence that has been published to date is consistent with the view that expert wine tasters differ from social drinkers in the degree to which visual (color) cues influence their orthonasal perception of flavor (Parr et al., 2003) and their perception of sweetness (Pangborn et al., 1963; see also Lelièvre et al., 2009). That said, not all food/flavor experts necessarily exhibit the same increased responsiveness to color cues when evaluating the taste and flavor of food and drink. Shankar, Simons, Shiv, McClure et al. (2010), for example, reported that the flavor experts working on a descriptive panel at an international flavor house (who all had more than 3 years of experience flavor profiling food and drink products) exhibited just as much visual capture (or assimilation) of their orthonasal olfactory flavor judgments as did the nonexperts they tested. Thus, based on the range of research that has been published to date, the most appropriate conclusion regarding flavor experts would appear to be that while some (specifically those with an expertise in wine) show an enhanced susceptibility to the crossmodal influence of color on judgments within their area of expertise (Pangborn et al., 1963; Parr et al., 2003), there is not enough evidence to know whether the same holds true of other groups of flavor experts (Shankar, Simons, Shiv, McClure et al., 2010).

6.4 Genetic differences

There is also some evidence to suggest that genetic differences modulate the psychological effect of color. Here, for example, one might think both of those individuals who are born color blind (primarily males, and constituting approximately 6% of the population; Broackes, 2010). Presumably such differences in color perception ought to modulate the psychological effect of food color, though, that said, it is surprisingly hard to find any published research on the topic (see http://www.colourblindawareness.org/colour-blindness/living-with-colour-vision-deficiency/food/). However, just as important as any deficits in color perception may be an individual's taster status (Bartoshuk, 2000). Genetic differences in taster status may play a role in determining just how much of an influence color has over multisensory flavor perception. Some people simply have far more taste buds than others (the former are known as supertasters, the latter, non-tasters) (Miller & Reedy, 1990). There is also an intermediate group known as medium tasters, with roughly a third of consumers falling into each group. In at least one study (Zampini et al., 2008), supertasters were significantly less affected by the color of a drink than were medium tasters, who, in turn, were less affected than were non-tasters (see Fig. 6.6). It is somewhat surprising to find that this is the only study of the impact of the psychological impact of food color to have assessed the taster status of their participants.[7] And, although beyond the scope of the

---

[7] Indeed, given the relatively small sample size, and the post-hoc nature of Zampini et al.'s (2008) finding, replication in a larger sample would undoubtedly be desirable to check on the generalizability of this potentially important result.

Figure 6.6 Mean percentage of correct flavor identification responses for the three groups of participants (non-tasters, medium tasters, and supertasters) for the blackcurrant, orange, and flavorless solutions. The black columns represent solutions where fruit acids had been added and the white columns solutions without fruit acids. The error bars represent the between-participants standard errors of the means. The results highlight the fact that genetic differences in taster status may determine just how much of a psychological effect on flavor identification color cues can have. *Source*: Figure reprinted with permission from Zampini et al. (2008).

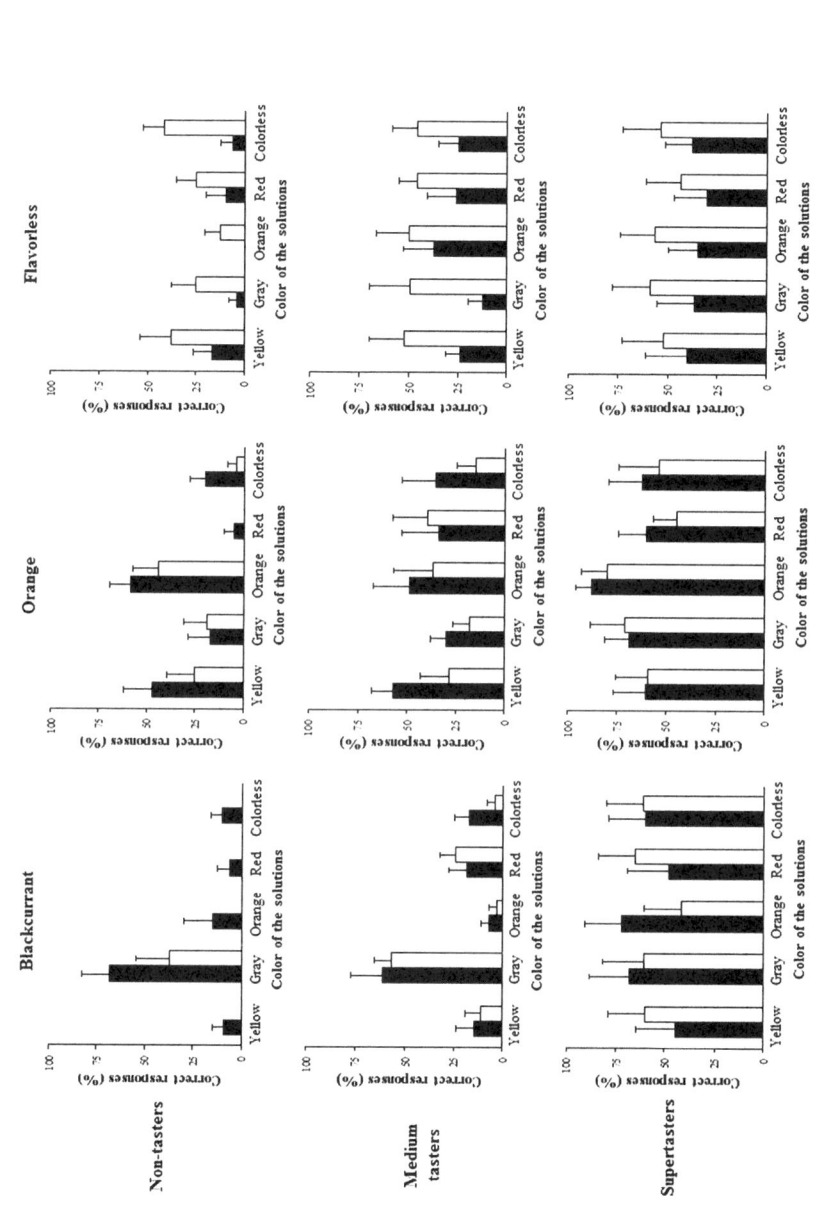

present article, it has also been suggested that there may be racial differences in terms of color preferences as well (Scanlon, 1985).

We would like to argue that one cannot hope to attain a comprehensive understanding of the psychological impact of food color on taste/flavor perception without taking into account the individual differences relevant to this topic that have been identified to date. These include genetic differences in terms of taster status and color perception, as well as cross-cultural and age-related differences. What is clear from the research that has been published to date is that these individual differences can influence both the meaning of color and its influence on the consumer. Having established the importance of such individual differences, the question becomes one of how best, moving forward, to assess the psychological impact of food color experimentally.

7 Conclusions

Since the first scientific reports that changing the color of a food could change the taste/flavor were published (see Duncker, 1939; Moir, 1936), somewhere in the region of 150 other papers have investigated the impact of food coloring on the perception and behavior of participants/consumers. While the majority of those studies have tended to focus on color's effect on taste/flavor identification (see Spence, 2015a, 2015b; Spence et al., 2010, for reviews), it is important to note that color cues can also play a number of different roles in modulating our food- and drink-related behavior (see Piqueras-Fiszman & Spence, 2012, 2014; Spence & Piqueras-Fiszman, 2014). Food coloring undoubtedly plays an important role in driving liking and the consumer acceptability of a variety of food and beverage products. And while increasing color variety in food can lead to enhanced consumption (see Piqueras-Fiszman & Spence, 2014), what we see can also lead to a suppression of our appetitive behaviors when associated with off-colors (or, at least, coloration that is interpreted by the consumer as such).

Finally, given the practical difficulties associated with delivering flavors while a participant lies in a brain scanner (see Spence & Piqueras-Fiszman, 2014), it is perhaps understandable that there has not been a great deal of neuroimaging research that has looked at the influence of color on flavor perception (Skrandies & Reuther, 2008; Verhagen & Engelen, 2006; see also Österbauer et al., 2005). Whether or not as the result of further neuroimaging, it is clear that further research is most definitely needed in order to develop a better understanding of the psychological mechanisms underlying the various effects of color on our perception of, and behaviors toward, food (see Kappes, Schmidt, & Lee, 2006; Spence, 2015b). And, finally, while the focus in this chapter has very much been on the influence of color, it is important to remember that consumers also use other visual cues in order to evaluate the quality properties of foods. However, reviewing those influences will have to remain the subject of another review.

References

Accum, F. (1820). *A treatise on adulteration of food and culinary poisons*. Available from <http://www.gutenberg.org/files/19031/19031-h/19031-h.htm>

Alley, R. L., & Alley, T. R. (1998). The influence of physical state and color on perceived sweetness. *Journal of Psychology: Interdisciplinary and Applied, 132*, 561–568.

Anon. (1979a). Colourings—An interim review. *International Flavours and Food Additives, 10*(3), 96–97.

Anon. (1979b). Additive use triggers consumer food concerns. *Food Product Development, 13*(8), 8.

Anon. (1980). Food colors. A scientific status summary by the Institute of Food Technologists' Expert Panel on Food Safety & Nutrition and the Committee on Public Information. *Food Technology, 34*(7), 77–84.

Anon. (2005). Orange is not the only carrot. *Yeahbaby*, April, 16.

Anon. (2007). "Anything" and "Whatever" beverages promise a surprise, every time. *Press release*, 17th May.

Anon. (2015). Burger King Japan to sell "red" burgers. *The Wall Street Journal*, June 17th. Downloaded from <http://blogs.wsj.com/japanrealtime/2015/06/17/burger-king-japan-to-sell-red-burgers/> on 27.06.2015.

Bartoshuk, L. M. (2000). Comparing sensory experiences across individuals: Recent psychophysical advances illuminate genetic variation in taste perception. *Chemical Senses, 25*, 447–460.

Bayarri, S., Calvo, C., Costell, E., & Duran, L. (2001). Influence of color on perception of sweetness and fruit flavor of fruit drinks. *Food Science and Technology International, 7*, 399–404.

Birren, F. (1963). Color and human appetite. *Food Technology, 17*(May), 45–47.

Blackwell, L. (1995). Visual clues and their effects on odour assessment. *Nutrition and Food Science, 5*, 24–28.

Broackes, J. (2010). What do the color-blind see? In J. Cohen & M. Matthen (Eds.), *Color ontology and color science* (pp. 291–389). Cambridge, MA: MIT Press.

Bruno, N., Martani, M., Corsini, C., & Oleari, C. (2013). The effect of the color red on consuming food does not depend on achromatic (Michelson) contrast and extends to rubbing cream on the skin. *Appetite, 71*, 307–313.

Calvo, C., Salvador, A., & Fiszman, S. (2001). Influence of colour intensity on the perception of colour and sweetness in various fruit-flavoured yoghurts. *European Food Research and Technology, 213*, 99–103.

Cardello, A. V. (1994). Consumer expectations and their role in food acceptance. In H. J. H. MacFie & D. M. H. Thomson (Eds.), *Measurement of food preferences* (pp. 253–297). London: Blackie Academic & Professional.

Cardello, A. V. (1996). The role of the human senses in food acceptance. In H. L. Meiselman & H. J. H. MacFie (Eds.), *Food choice, acceptance and consumption* (pp. 1–82). New York, NY: Blackie Academic and Professional.

Carlsmith, J. M., & Aronson, E. (1963). Some hedonic consequences of the confirmation and disconfirmation of expectancies. *Journal of Abnormal and Social Psychology, 66*, 151–156.

Carpenter, C. E., Conforth, D. P., & Whittier, D. (2001). Consumer preference for beef colour and packaging did not affect eating satisfaction. *Meat Science, 57*, 359–363.

Chan, M. M., & Kane-Martinelli, C. (1997). The effect of color on perceived flavor intensity and acceptance of foods by young adults and elderly adults. *Journal of the American Dietetic Association, 97*, 657–659.

Cheskin, L. (1957). *How to predict what people will buy*. New York, NY: Liveright.
Christensen, C. (1985). Effect of color on judgments of food aroma and food intensity in young and elderly adults. *Perception, 14*, 755–762.
Chylinski, M., Northey, G., & Ngo, L. V. (2015). Cross-modal interactions between color and texture of food. *Psychology & Marketing, 32*, 950–966.
Clydesdale, F. M. (1984). The influence of colour on sensory perception and food choices. In J. Walford (Ed.), *Developments in food colours–2* (pp. 75–112). London: Elsevier Applied Science.
Clydesdale, F. M. (1991). Color perception and food quality. *Journal of Food Quality, 14*, 61–74.
Clydesdale, F. M. (1993). Color as a factor in food choice. *Critical Reviews in Food Science and Nutrition, 33*, 83–101.
Clydesdale, F. M., Gover, R., Philipsen, D. H., & Fugardi, C. (1992). The effect of color on thirst quenching, sweetness, acceptability and flavor intensity in fruit punch flavored beverages. *Journal of Food Quality, 15*, 19–38.
Cook, W. (2012). Would you eat a "gourmet" burger made with charred bamboo and squid ink? *Daily Mail Online*, 25th September. Downloaded from: http://www.dailymail.co.uk/news/article-2208321/Burger-King-black-burger-Japan-bamboo-charcoal-squid-ink.html on 26.12.14.
Crumpacker, B. (2006). *The sex life of food: When body and soul meet to eat*. New York, NY: Thomas Dunne Books.
Dalby, A. (2003). *Food in the ancient world from A to Z*. London: Routledge.
Davis, R. G. (1981). The role of nonolfactory context cues in odor identification. *Perception & Psychophysics, 30*, 83–89.
Deliza, R., & MacFie, H. J. H. (1997). The generation of sensory expectation by external cues and its effect on sensory perception and hedonic ratings: A review. *Journal of Sensory Studies, 2*, 103–128.
Delwiche, J. F. (2012). You eat with your eyes first. *Physiology & Behavior, 107*, 502–504.
de Wijk, R. A., Polet, I. A., Engelen, L., et al. (2004). Amount of ingested custard dessert as affected by its color, odor, and texture. *Physiology & Behavior, 82*, 397–403.
Di Monaco, R., Cavella, S., Iaccarino, T., Mincione, A., & Masi, P. (2003). The role of the knowledge of color and brand name on the consumer's hedonic ratings of tomato purees. *Journal of Sensory Studies, 18*, 391–408.
Dolnick, E. (2008). Fish or foul? *The New York Times*, September 2. Downloaded from http://www.nytimes.com/2008/09/02/opinion/02dolnick.html?_r=1&scp=1&sq=chocolate%20strawberry%20yogurt&st=cse on 26.12.14.
Downham, A., & Collins, P. (2000). Coloring our foods in the last and next millennium. *International Journal of Food Science and Technology, 35*, 5–22.
DuBose, C. N., Cardello, A. V., & Maller, O. (1980). Effects of colorants and flavorants on identification, perceived flavor intensity, and hedonic quality of fruit-flavored beverages and cake. *Journal of Food Science, 45*, 1393–1399, 1415.
Duncker, K. (1939). The influence of past experience upon perceptual properties. *American Journal of Psychology, 52*, 255–265.
Farrell, G. (2000). What's green. Easy to squirt? Ketchup! *USA Today*, July 10th, 2b.
Favre, J. P., & November, A. (1979). *Color and communication*. Zurich: ABC-Verlag.
Fernández-Vázquez, R., Hewson, L., Fisk, I., Vila, D., Mira, F., Vicario, I. M., et al. (2014). Color influences sensory perception and liking of orange juice. *Flavor, 3*, 1.
Francis, F. J., & Clydesdale, F. M. (1975). *Food colorimetry: Theory and applications*. New York, NY: Van Nostrand Reinhold/AVI.

Frank, R. A., Ducheny, K., & Mize, S. J. S. (1989). Strawberry odor, but not red color, enhances the sweetness of sucrose solutions. *Chemical Senses, 14*, 371–377.

Garber, L. L., Jr., Hyatt, E. M., & Boya, Ü. Ö. (2008). The mediating effects of the appearance of nondurable consumer goods and their packaging on consumer behavior. In H. N. J. Schifferstein & P. Hekkert (Eds.), *Product experience* (pp. 581–602). London: Elsevier.

Garber, L. L., Hyatt, E. M., & Nafees, L. (2015). The effects of food color on perceived flavor: A factorial investigation in India. *Journal of Food Products Marketing, 21*, 1–20.

Garber, L. L., Jr., Hyatt, E. M., & Starr, R. G., Jr. (2000). The effects of food color on perceived flavor. *Journal of Marketing Theory and Practice, 8*(4), 59–72.

Garber, L. L., Jr., Hyatt, E. M., & Starr, R. G., Jr. (2001). Placing food color experimentation into a valid consumer context. *Journal of Food Products Marketing, 7*(3), 3–24.

Garber, L. L., Jr., Hyatt, E. M., & Starr, R. G., Jr. (2003). Measuring consumer response to food products. *Food Quality and Preference, 14*, 3–15.

Garber, L. L., Jr., Hyatt, E. M., & Starr, R. G., Jr. (2003). Reply to commentaries on: "Placing food color experimentation into a valid consumer context". *Food Quality and Preference, 14*, 41–43.

Geier, A., Wansink, B., & Rozin, P. (2012). Red potato chips: Segmentation cues can substantially decrease food intake. *Health Psychology, 31*, 398–401.

Genschow, O., Reutner, L., & Wanke, M. (2012). The color red reduces snack food and soft drink intake. *Appetite, 58*, 699–702.

Gifford, S. R., & Clydesdale, F. M. (1986). The psychophysical relationship between color and sodium chloride concentrations in model systems. *Journal of Food Protection, 49*, 977–982.

Gifford, S. R., Clydesdale, F. M., & Damon, R. A., Jr. (1987). The psychophysical relationship between color and salt concentrations in chicken flavored broths. *Journal of Sensory Studies, 2*, 137–147.

Gimba, J. G. (1998). Color in marketing: Shades of meaning. *Marketing News, 32*(6), 16.

Goldenberg, N. (1977). Colours—Do we need them? British Nutrition Foundation (Ed.), *Why food additives? The safety of foods* (pp. 22–24). London: Forbes Publications.

Gori, M., Del Viva, M., Sandini, G., & Burr, D. C. (2008). Young children do not integrate visual and haptic information. *Current Biology, 18*, 694–698.

Gossinger, M., Mayer, F., Radochan, N., Höfler, M., Boner, A., Grolle, E., et al. (2009). Consumer's color acceptance of strawberry nectars from puree. *Journal of Sensory Studies, 24*, 78–92.

Guinard, J.-X., Souchard, A., Picot, M., Rogeaux, M., & Siefferman, J.-M. (1998). Sensory determinants of the thirst-quenching character of beer. *Appetite, 31*, 101–115.

Hall, R. L. (1958). Flavor study approaches at McCormick and Company, Inc A. D. Little, Inc (Ed.), *Flavor research and food acceptance: A survey of the scope of flavor and associated research, compiled from papers presented in a series of symposia given in 1956–1957* (pp. 224–240). New York, NY: Reinhold.

Harris, G. (2011). Colorless food? We blanch. *The New York Times*, April 3, 3. Downloaded from http://www.nytimes.com/2011/04/03/weekinreview/03harris.html?_r=0 on 21.12.2014.

Hicks, D. (1979). Benefits of added colourings in food and drinks. *International Flavours and Food Additives, 10*(1), 31–32.

Hine, T. (1995). *The total package: The secret history and hidden meanings of boxes, bottles, cans, and other persuasive containers*. New York, NY: Little Brown.

Hoegg, J., & Alba, J. W. (2007). Taste perception: More than meets the tongue. *Journal of Consumer Research, 33*, 490–498.

Hutchings, J. B. (2003). *Expectations and the food industry: The impact of color and appearance*. New York, NY: Plenum Publishers.

Hyman, A. (1983). The influence of color on the taste perception of carbonated water preparations. *Bulletin of the Psychonomic Society, 21*, 145–148.

Imram, N. (1999). The role of visual cues in consumer perception and acceptance of a food product. *Nutrition & Food Science, 99*, 224–230.

Jantathaia, S., Danner, L., Joechl, M., & Dürrschmid, K. (2013). Gazing behaviour, choice and color of food: Does gazing behaviour predict choice? *Food Research International, 54*, 1621–1626.

Jantathaia, S., Sungsri-in, M., Mukprasirt, A., & Dürrschmid, K. (2014). Sensory expectations and perceptions of Austrian and Thai consumers: A case study with six colored Thai desserts. *Food Research International, 64*, 65–73.

Johnson, J., & Clydesdale, F. M. (1982). Perceived sweetness and redness in colored sucrose solutions. *Journal of Food Science, 47*, 747–752.

Johnson, J. L., Dzendolet, E., & Clydesdale, F. M. (1983). Psychophysical relationships between perceived sweetness and redness in strawberry-flavored beverages. *Journal of Food Protection, 46*, 21–25, 28.

Johnson, J. L., Dzendolet, E., Damon, R., Sawyer, M., & Clydesdale, F. M. (1982). Psychophysical relationships between perceived sweetness and color in cherry-flavored beverages. *Journal of Food Protection, 45*, 601–606.

Kahn, B. E., & Wansink, B. (2004). The influence of assortment structure on perceived variety and consumption quantities. *Journal of Consumer Research, 30*, 519–533.

Kanig, J. L. (1955). Mental impact of colors in foods studied. *Food Field Reporter, 23*, 57.

Kappes, S. M., Schmidt, S. J., & Lee, S.-Y. (2006). Color halo/horns and halo-attribute dumping effects within descriptive analysis of carbonated beverages. *Journal of Food Science, 71*, S590–S595.

Kostyla, A. S., & Clydesdale, F. M. (1978). The psychophysical relationships between color and flavor. *CRC Critical Reviews in Food Science & Nutrition, 10*, 303–319.

Koza, B. J., Cilmi, A., Dolese, M., & Zellner, D. A. (2005). Color enhances orthonasal olfactory intensity and reduces retronasal olfactory intensity. *Chemical Senses, 30*, 643–649.

Kramer, A. (1978). Benefits and risks of color additives. *Food Technology, 32*(8), 65–67.

Lavin, J. G., & Lawless, H. T. (1998). Effects of color and odor on judgments of sweetness among children and adults. *Food Quality and Preference, 9*, 283–289.

Lelièvre, M., Chollet, S., Abdi, H., & Valentin, D. (2009). Beer-trained and untrained assessors rely more on vision than on taste when they categorize beers. *Chemosensory Perception, 2*, 143–153.

Levitan, C., Zampini, M., Li, R., & Spence, C. (2008). Assessing the role of color cues and people's beliefs about color-flavor associations on the discrimination of the flavor of sugar-coated chocolates. *Chemical Senses, 33*, 415–423.

Lucas, C. D., Hallagan, J. B., & Taylor, S. L. (2001). The role of natural color additives in food allergy. *Advances in Food and Nutrition Research, 43*, 195–216.

Mace, K. C., & Enzie, R. F. (1970). Dissonance versus contrast in an ego-involved situation with disconfirmed expectancies. *The Journal of Psychology, 75*, 107–121.

Macrae, F. (2011). What's for dinner? Rainbow coloured carrots and super broccoli that's healthier and sweeter. *DailyMail Online*, 15 October. Available at http://www.dailymail.co.uk/health/article-2044695/Purple-carrots-sale-Tescosupermarket-Orange-year.html (accessed January 2014).

Maga, J. A. (1974). Influence of color on taste thresholds. *Chemical Senses and Flavor, 1*, 115–119.

Masurovsky, B. I. (1939). How to obtain the right food color. *Food Industries, 11(Jan.)*13, 55–56.

McCullough, J. M., Martinsen, C. S., & Moinpour, R. (1978). Application of multidimensional scaling to the analysis of sensory evaluations of stimuli with known attribute structures. *Journal of Applied Psychology, 65*, 103–109.

Meggos, H. (1995). Food colours: An international perspective. *The Manufacturing Confectioner, 75*, 59–65.

Michael, G. A., Galich, H., Relland, S., & Prud'hon, S. (2010). Hot colors: The nature and specificity of color-induced nasal thermal sensations. *Behavioural Brain Research, 207*, 418–428.

Miller, E. G., & Kahn, B. E. (2005). Shades of meaning: The effect of color and flavor names on consumer choice. *Journal of Consumer Research, 32*, 86–92.

Miller, I. J., & Reedy, D. P. (1990). Variations in human taste bud density and taste intensity perception. *Physiology and Behavior, 47*, 1213–1219.

Moir, H. C. (1936). Some observations on the appreciation of flavour in foodstuffs. *Journal of the Society of Chemical Industry: Chemistry & Industry Review, 14*, 145–148.

Morrot, G., Brochet, F., & Dubourdieu, D. (2001). The color of odors. *Brain and Language, 79*, 309–320.

Norton, W. E., & Johnson, F. N. (1987). The influence of intensity of colour on perceived flavour characteristics. *Medical Science Research, 15*, 329–330.

Oram, N., Laing, D. G., Hutchinson, I., Owen, J., Rose, G., Freeman, M., et al. (1995). The influence of flavor and color on drink identification by children and adults. *Developmental Psychobiology, 28*, 239–246.

Österbauer, R. A., Matthews, P. M., Jenkinson, M., Beckmann, C. F., Hansen, P. C., & Calvert, G. A. (2005). Color of scents: Chromatic stimuli modulate odor responses in the human brain. *Journal of Neurophysiology, 93*, 3434–3441.

Pangborn, R. M. (1960). Influence of color on the discrimination of sweetness. *American Journal of Psychology, 73*, 229–238.

Pangborn, R. M., Berg, H. W., & Hansen, B. (1963). The influence of color on discrimination of sweetness in dry table-wine. *American Journal of Psychology, 76*, 492–495.

Pangborn, R. M., & Hansen, B. (1963). The influence of color on discrimination of sweetness and sourness in pear-nectar. *American Journal of Psychology, 76*, 315–317.

Parr, W. V., White, K. G., & Heatherbell, D. (2003). The nose knows: Influence of colour on perception of wine aroma. *Journal of Wine Research, 14*, 79–101.

Petit, C. E. F., Hollowood, T. A., Wulfert, F., & Hort, J. (2007). Colour-coolant-aroma interactions and the impact of congruency and exposure on flavour perception. *Food Quality and Preference, 18*, 880–889.

Philipsen, D. H., Clydesdale, F. M., Griffin, R. W., & Stern, P. (1995). Consumer age affects response to sensory characteristics of a cherry flavored beverage. *Journal of Food Science, 60*, 364–368.

Piqueras-Fiszman, B., & Spence, C. (2011). Crossmodal correspondences in product packaging: Assessing color-flavor correspondences for potato chips (crisps). *Appetite, 57*, 753–757.

Piqueras-Fiszman, B., & Spence, C. (2012). Sensory incongruity in the food and beverage sector: Art, science, and commercialization. *Petits Propos Culinaires, 95*, 74–118.

Piqueras-Fiszman, B., & Spence, C. (2014). Colour, pleasantness, and consumption behaviour within a meal. *Appetite, 75*, 165–172.

Piqueras-Fiszman, B., & Spence, C. (2015). Sensory expectations based on product-extrinsic food cues: An interdisciplinary review of the empirical evidence and theoretical accounts. *Food Quality & Preference, 40*, 165–179.

Profet, M. (1992). Pregnancy sickness as adaptation: A deterrant to maternal ingestion of teratogens. In J. H. Barkow, L. Cosmides, & J. Tooby (Eds.), *The adapted mind: Evolutionary psychology and the generation of culture* (pp. 327–365). Oxford: Oxford University Press.

Redden, J. P., & Hoch, S. J. (2009). The presence of variety reduces perceived quantity. *Journal of Consumer Research, 36*, 406–417.

Rolls, B. J., Rowe, E. A., & Rolls, E. T. (1982). How sensory properties of foods affect human feeding behaviour. *Physiology & Behavior, 29*, 409–417.

Roth, H. A., Radle, L. J., Gifford, S. R., & Clydesdale, F. M. (1988). Psychophysical relationships between perceived sweetness and color in lemon- and lime-flavored drinks. *Journal of Food Science, 53*, 1116–1119, 1162.

Scanlon, B. A. (1985). Race differences in selection of cheese color. *Perceptual and Motor Skills, 61*, 314.

Schutz, H. G. (1954). Color in relation to food preference. In K. T. Farrell, J. R. Wagner, M. S. Peterson, & G. MacKinney (Eds.), *Color in foods: A symposium sponsored by the Quartermaster Food and Container Institute for the Armed Forces Quartermaster Research and Development Command U. S. Army Quartermaster Corps* (pp. 16–23). Washington, DC: National Academy of Sciences – National Research Council.

Shankar, M. U., Levitan, C. A., Prescott, J., & Spence, C. (2009). The influence of color and label information on flavor perception. *Chemosensory Perception, 2*, 53–58.

Shankar, M. U., Levitan, C., & Spence, C. (2010). Grape expectations: The role of cognitive influences in color-flavor interactions. *Consciousness & Cognition, 19*, 380–390.

Shankar, M., Simons, C., Levitan, C., Shiv, B., McClure, S., & Spence, C. (2010). An expectations-based approach to explaining the crossmodal influence of color on odor identification: The influence of temporal and spatial factors. *Journal of Sensory Studies, 25*, 791–803.

Shankar, M., Simons, C., Shiv, B., Levitan, C., McClure, S., & Spence, C. (2010). An expectations-based approach to explaining the influence of color on odor identification: The influence of degree of discrepancy. *Attention, Perception, & Psychophysics, 72*, 1981–1993.

Shankar, M., Simons, C., Shiv, B., McClure, S., & Spence, C. (2010). An expectation-based approach to explaining the crossmodal influence of color on odor identification: The influence of expertise. *Chemosensory Perception, 3*, 167–173.

Shepherd, G. M. (2012). *Neurogastronomy: How the brain creates flavor and why it matters*. New York: Columbia University Press.

Shermer, D. Z., & Levitan, C. A. (2014). Red hot: The crossmodal effect of color intensity on piquancy. *Multisensory Research, 27*, 207–223.

Sinclair, U. (1906). *The jungle*. New York, NY: Penguin.

Singh, S. (2006). Impact of color on marketing. *Management Decision, 44*, 783–789.

Skrandies, W., & Reuther, N. (2008). Match and mismatch of taste, odor, and color is reflected by electrical activity in the human brain. *Journal of Psychophysiology, 22*, 175–184.

Spence, C. (2010). The color of wine—Part 1. *The World of Fine Wine, 28*, 122–129.

Spence, C. (2012). The development and decline of multisensory flavour perception. In A. J. Bremner, D. Lewkowicz, & C. Spence (Eds.), *Multisensory development* (pp. 63–87). Oxford: Oxford University Press.

Spence, C. (2014). Drinking in colour. *The Cocktail Lovers, 13*, 28–29.

Spence, C. (2015a). Visual contributions to taste and flavour perception. In M. Scotter (Ed.), *Colour additives for food and beverages* (pp. 189–210). Cambridge: Woodhead Publishing.

Spence, C. (2015b). On the psychological impact of food colour. *Flavour, 4*, 21.

Spence, C., Levitan, C., Shankar, M. U., & Zampini, M. (2010). Does food color influence taste and flavor perception in humans? *Chemosensory Perception, 3*, 68–84.

Spence, C., & Piqueras-Fiszman, B. (2012). The multisensory packaging of beverages. In M. G. Kontominas (Ed.), *Food packaging: Procedures, management and trends* (pp. 187–233). Hauppauge, NY: Nova Publishers.

Spence, C., & Piqueras-Fiszman, B. (2014). *The perfect meal: The multisensory science of food and dining*. Oxford: Wiley-Blackwell.

Spence, C., Wan, X., Woods, A., Velasco, C., Deng, J., Youssef, J., et al. (2015). On tasty colours and colourful tastes? Assessing, explaining, and utilizing crossmodal correspondences between colours and basic tastes. *Flavour, 4*, 23.

Stevens, L. J., Kuczek, T., Burgess, J. R., Stochelski, M. A., Arnold, L. E., & Galland, L. (2013). Mechanisms of behavioral, atopic, and other reactions to artificial food colors in children. *Nutrition Reviews, 71*, 268–281.

Stevenson, R. J., & Oaten, M. (2008). The effect of appropriate and inappropriate stimulus color on odor discrimination. *Perception & Psychophysics, 70*, 640–646.

Stillman, J. (1993). Color influences flavor identification in fruit-flavored beverages. *Journal of Food Science, 58*, 810–812.

Strugnell, C. (1997). Colour and its role in sweetness perception. *Appetite, 28*, 85.

Stummerer, S., & Hablesreiter, M. (2010). *Food design XL*. New York, NY: Springer.

Tannahill, R. (1973). *Food in history*. New York, NY: Stein and Day.

Tepper, B. J. (1993). Effects of a slight color variation on consumer acceptance of orange juice. *Journal of Sensory Studies, 8*, 145–154.

Triplett, T. (1994). Consumers show little taste for clear beverages. *Marketing News, 28*(11) 2, 11.

Tuorila-Ollikainen, H. (1982). Pleasantness of colourless and coloured soft drinks and consumer attitudes to artificial food colours. *Appetite, 3*, 369–376.

Tysoe, M. (1985). What's wrong with blue potatoes? *Psychology Today, 19*(12), 6–8.

Urbányi, G. (1982). Investigation into the interaction of different properties in the course of sensory evaluation. I. The effect of colour upon the evaluation of taste in fruit and vegetable products. *Acta Alimentaria, 11*, 233–243.

Valentin, D., Parr, W. V., Peyron, D., Grose, C., & Ballester, J. (2016). Colour as a driver of Pinot noir wine quality judgments: An investigation involving French and New Zealand wine professionals. *Food Quality and Preference, 48*, 251–261.

Verhagen, J. V., & Engelen, L. (2006). The neurocognitive bases of human multimodal food perception: Sensory integration. *Neuroscience and Biobehavioral Reviews, 30*, 613–650.

Wadhwani, R., & McMahon, D. J. (2012). Color of low-fat cheese influences flavor perception and consumer liking. *Journal of Dairy Science, 95*, 2336–2346.

Walford, J. (1980). Historical development of food coloration. In J. Walford (Ed.), *Developments in food colours* (pp. 1–26). London: Applied Science.

Walsh, L. M., Toma, R. B., Tuveson, R. V., & Sondhi, L. (1990). Color preference and food choice among children. *Journal of Psychology, 124*, 645–653.

Wan, X., Velasco, C., Michel, C., Mu, B., Woods, A. T., & Spence, C. (2014). Does the shape of the glass influence the crossmodal association between colour and flavour? A cross-cultural comparison. *Flavour, 3*, 3.

Wan, X., Woods, A. T., Seoul, K.-H., Butcher, N., & Spence, C. (2015). When the shape of the glass influences the flavour associated with a coloured beverage: Evidence from consumers in three countries. *Food Quality & Preference, 39*, 109–116.

Wan, X., Woods, A. T., van den Bosch, J., Mckenzie, K. J., Velasco, C., & Spence, C. (2014). Cross-cultural differences in crossmodal correspondences between tastes and visual features. *Frontiers in Psychology: Cognition, 5*, 1365.

Watson, E. (2013). *We eat with our eyes: Flavor perception strongly influenced by food color, says DDW*. Downloaded from <http://www.foodnavigator-usa.com/Science/We-eat-with-our-eyes-Flavor-perception-strongly-influenced-by-food-color-says-DDW> on 19.12.14.

Wei, S., Ou, L. -C., Luo, M. R., & Hutchings, J. B. (2012). Optimization of food expectations using product color and appearance. *Food Quality & Preference, 23*, 49–62.

Weir, K. (2009). Taste the rainbow. *Psychology Today, 42*, 26.

Weiss, B., Williams, J. H., Margen, S., Abrams, B., Caan, B., Citron, L. J., et al. (1980). Behavioral responses to artificial food colors. *Science, 207*, 1487–1489.

Wheatley, J. (1973). Putting colour into marketing. *Marketing, October*, 24–29, 67.

Whitehill, I. (1980). Human idiosyncratic responses to food colours. *Food Flavourings, Ingredients, Packaging & Processing, 1*(7), 23–27, 37.

Wilson, B. (2009). *Swindled: From poison sweets to counterfeit coffee—The dark history of the food cheats*. London: John Murray.

Wilson, T., & Klaaren, K. (1992). Expectation whirls me round: The role of affective expectations on affective experiences. In M. S. Clear (Ed.), *Review of personality and social psychology: Emotion and social behavior* (pp. 1–31). Newbury Park, CA: Sage.

Yang, F. L., Cho, S., & Seo, H.-S. (2015). Effects of light color on consumers' acceptability and willingness to eat apples and bell peppers. *Journal of Sensory Studies*. http://dx.doi.org/10.1111/joss.12183.

Yeomans, M., Chambers, L., Blumenthal, H., & Blake, A. (2008). The role of expectancy in sensory and hedonic evaluation: The case of smoked salmon ice-cream. *Food Quality and Preference, 19*, 565–573.

Zampini, M., Sanabria, D., Phillips, N., & Spence, C. (2007). The multisensory perception of flavor: Assessing the influence of color cues on flavor discrimination responses. *Food Quality & Preference, 18*, 975–984.

Zampini, M., Wantling, E., Phillips, N., & Spence, C. (2008). Multisensory flavor perception: Assessing the influence of fruit acids and color cues on the perception of fruit-flavored beverages. *Food Quality & Preference, 19*, 335–343.

Zellner, D. A. (2013). Color-odor interactions: A review and model. *Chemosensory Perception, 6*, 155–169.

Zellner, D. A., Bartoli, A. M., & Eckard, R. (1991). Influence of color on odor identification and liking ratings. *American Journal of Psychology, 104*, 547–561.

Zellner, D. A., & Durlach, P. (2002). What is refreshing? An investigation of the color and other sensory attributes of refreshing foods and beverages. *Appetite, 39*, 185–186.

Zellner, D. A., & Durlach, P. (2003). Effect of color on expected and experienced refreshment, intensity, and liking of beverages. *American Journal of Psychology, 116*, 633–647.

Zellner, D. A., & Kautz, M. A. (1990). Color affects perceived odor intensity. *Journal of Experimental Psychology: Human Perception and Performance, 16*, 391–397.

Zellner, D., Strickhouser, D., & Tornow, C. (2004). Disconfirmed hedonic expectations produce perceptual contrast, not assimilation. *American Journal of Psychology, 117*, 363–387.

Zellner, D. A., & Whitten, L. A. (1999). The effect of color intensity and appropriateness on color-induced odor enhancement. *American Journal of Psychology, 112*, 585–604.

Zhou, X., Wan, X., Mu, B., Du, D., & Spence, C. (2015). Examining colour-receptacle-flavour interactions for Asian noodles. *Food Quality & Preference, 41*, 141–150.

Multisensory Flavor Priming

Garmt Dijksterhuis[1,2]
[1]University College Roosevelt, University of Utrecht, Middelburg, The Netherlands,
[2]University of Copenhagen, Sensory and Consumer Science Section, Copenhagen, Denmark

1 Introduction

All three words in the title of this chapter ("Multisensory flavor priming") are rather technical terms, jargon if you will. They may be interpreted differently by different readers, even those who are specialized in the field of perception. As they signify difficult concepts, especially when they are combined, as they are here, it is probably good to introduce them in some detail before embarking on a discussion of multisensory flavor priming proper. In the earlier chapter "Attention and Flavor Binding," several matters concerning the complexity of the area are introduced and explained. In this chapter, I will focus on flavor too, but from the point of view of priming.

One special difficulty arises with the concept of flavor. There is no unitary sensory input that results in the perception of flavor. Rather, "flavor" is the result of several rather different sensory systems working together. In particular, this makes it complicated to talk about "flavor priming," because there are many different components of flavor that can presumably be primed, or that can themselves act as primes.

The topic of priming has received a great deal of attention in the social psychological literature lately (cf. Bargh, 2006; Kahneman, 2012). Other recent publications have approached priming from different viewpoints (Loersch & Payne, 2011; Smeets & Dijksterhuis, 2014). Priming is typically defined as an implicit memory process, "a nonconscious influence of past experience on current performance or behavior" (Schacter & Buckner, 1998, p. 185). It occurs in tasks that do not require participants to access their memory explicitly, or even with participants who fail to recollect having learned anything with respect to the task. As priming always entails a learning part, and a task part where an effect of priming can occur, different situations arise with respect to participants being, or not being, aware of these parts. In this chapter, a taxonomy of different resulting priming paradigms is presented.

The literature on flavor expectations (see Piqueras-Fiszman & Spence, 2015, for a review) can be interpreted in the context of priming too. Expectations are top-down effects that result from certain perceptions of (aspects of) food-related stimuli, including information about the food. Upon subsequent ingestion of the food they can influence the intensity and type of (bottom-up) sensory perception (as, eg, shown by Woods et al., 2011). This can be seen as a case of (conceptual) priming, where the information has an effect on subsequent perception of (or judgment about, or behavior toward) food stimuli.

Taking into account the fact that flavor, and in fact all naturalistic perception, always occurs in a certain context, a situated approach along the lines laid out in Papies and Barsalou (2015) is introduced. This approach enables an interpretation of flavor priming, the effect of expectations, and related subjects in a very broad sense.

1.1 Multisensory

The term "multisensory" in the title of this chapter is probably the easier term to explain. However, matters can grow pretty complex rather quickly. "Multisensory" refers to the fact that our perceptual system receives a multitude of different inputs. Traditionally, these inputs have been separated on the basis of the type of stimulation, anatomical structures receiving the stimulation, and the phenomenology associated with experiencing the stimulation. The different sensory systems have been studied in isolation for many years, starting with the "higher" systems (vision, audition), and only relatively recently have the "lower" systems (smell, touch, taste) started to receive their due scientific attention (cf. Köster, 2000).

The traditional view that there are five sensory systems (vision, audition, touch, smell, and taste) breaks down when you study any system in greater detail. Depending upon how you define a "sense," one can easily count 12 or perhaps 40 different sensory systems that somehow transduce changes taking place outside the individual into experiences inside the individual (cf. Dijksterhuis, 2012, 2016; Durie, 2005). Flavor can be seen as one functional sensory system, but also as several distinct anatomical/ neurological systems (cf. Auvray & Spence, 2008; and Stevenson's chapter: Attention and Flavor Binding).

In addition to the multitude of sensory systems, many of these systems interact (cf. Calvert, Spence, & Stein, 2004; and Stevenson's chapter: Attention and Flavor Binding). This is illustrated in Fig. 7.1, where the simplest situation is depicted on the left. Here unisensory stimulation transduced by sensory system A results in an experience of stimulus a. The nature of this experience will not be discussed; since it would take us far outside the realm of this text, into philosophy and phenomenology. Another sensory system (B in the right part of Fig. 7.1) can interfere with the workings of system A. This influence can result in a', an altered experience of a.

There are two different ways in which this experience can be altered: quantitatively and qualitatively (see Driver & Spence, 2000). A quantitative effect occurs when a' is perceived to be weaker or stronger than a, but has not lost its character of a percept in sensory system A. It may also be possible that another property of a could have

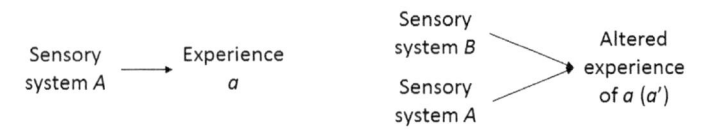

Figure 7.1 Left: Sensory system A, giving rise to a subjective experience, a percept a, of a stimulus transduced by that system. Right: Interaction of sensory systems A with B leads to a change in the percept a. It is altered into a'.

changed, such as its location (as in the ventriloquist effect, where the location of an auditory stimulus changes under the influence of visual cues). Here a' is still a percept in the sensory system A, but it has been altered in some way.

A qualitative change in the experience of a means that it has a changed character, a' cannot be ascribed anymore to have clearly resulted from sensory system A. This will be illustrated shortly. Of course, in practice, combinations of the two types of interaction will also occur.

Olfaction can be seen as a synthetic sensory system in that the participant cannot break down an olfactory experience into its constituents in terms of single receptor exciting molecules (Kubovy, 1981, 1988; Kubovy & Schutz, 2010; Kubovy & Van Valkenburg, 2001). However, in composite odorant mixtures, some of the constituents can be perceived, but the breakdown happens in terms of meaningful and remembered odors (often odor mixtures too) that can be named ("I smell diesel and honey."). Only very experienced and trained professionals (eg, perfumers or flavorists) will be able to name a number of constituting substances. But these substances have become meaningful and remembered odors to them. Flavor can thus be seen as synthetic insofar as it fuses individual odor, taste, and mouthfeel sensations into a single percept, but—after some training—participants can to some extent break the percept down into some of its constitutes (cf. Auvray & Spence, 2008).

It is well known that outside the psychology laboratory, all our senses are almost continuously stimulated. There is thus room for many more interactions between sensory systems than we can keep track of. In the last couple of years, many sensory interactions have been studied in the laboratory, but also in more applied settings such as product experience (cf. Schifferstein & Spence, 2008).

Many of the multisensory interactions are known as illusions, for example, the size–weight illusion, where the size of a lifted object has an effect on its perceived weight (see Koseleff, 1957). This is an instance of a quantitative interaction, whereby the visually perceived *size* of the object affects the *quantity* of the experienced heaviness of the *weight* of the stimulus. An example of a qualitative interaction can be found in a class of visual illusions that can give the observer a clear perception of a rotating image. One of these is shown in Fig. 7.2 (Pinna & Brelstaff, 2000). The *static visual image* appears to rotate when one *moves* toward or away from the figure. This is an example of a qualitative interaction whereby the experience of a *rotating* pattern results from the interaction of the *static visual pattern* and a head *movement* toward and away from the image. A rotating pattern is qualitatively different from the static visual pattern, with (apparent) rotation being introduced into the experience.

"Multisensory" can thus simply be defined as involving more than one sensory system, but it quickly leads to very complex situations, as illustrated above.

1.2 Flavor

Popularly, flavor is defined as the combination of smell and taste. However, many sources in addition include aspects of mouthfeel in their definition of flavor (cf. Lawless & Heymann, 2010) and trigeminal stimulation (ISO, 1992, 2008). Stevenson (in the chapter: Attention and Flavor Binding) explains aspects of flavor

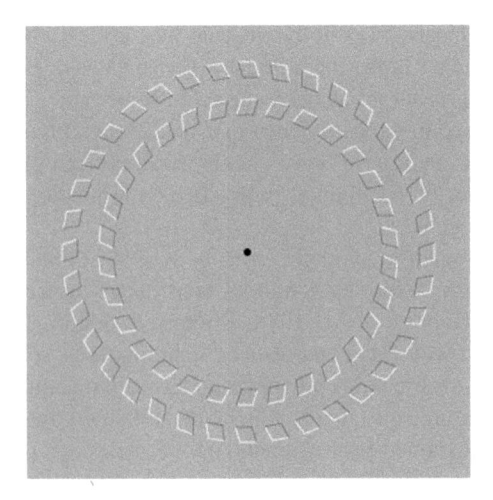

Figure 7.2 An illusion of rotation of a static pattern.
Source: From Pinna and Brelstaff (2000).

Figure 7.3 Illustration of the sensory complexity of the experience of flavor. Mouthfeel is composed of several sensory systems (oral temperature, pain, touch), combined with smell and taste it comprises "flavor."

perception, in particular, how flavor is the bound amalgamate of a number of sensory interactions. Flavor itself is thus clearly multisensory (cf. Spence, Smith, & Auvray, 2015).

The experience of flavor is the result of joint stimulation of the senses of smell and taste, giving rise to a qualitative interaction. The flavor percept is not uniquely connected to either smell or taste anymore. As flavor is always experienced in the mouth, mouthfeel should probably be included in its definition. Oral stimulation will, in naturalistic environments, always accompany the smell and taste experiences that come with the normal perception of food while eating.

"Smell," in this context, always refers to retronasal smell, the olfactory perception of volatiles that travel from the oral cavity to the nasal cavity. Orthonasal smell, where volatiles enter the nose through the nostrils, is not considered part of "flavor." Obviously, orthonasal smell will, by way of sensory interaction, be able to modify flavor perception. It can also give rise to top-down effects by creating an expectation of what is about to be experienced in the mouth, because food on its way to the mouth will often be perceived by orthonasal smell too, just prior to ingestion.

Although the experience of flavor can sometimes be broken down into the separate components of smell, taste, and mouthfeel by trained tasters, it can be seen as a synthetic sense (cf. Auvray & Spence, 2008). There likely are some sensory interactions that even well-trained tasters cannot break down.

The definition of flavor as a unitary sensory system is troublesome, as there is no single external stimulus, nor a distinct anatomical structure that results in an experienced flavor. There is a unitary experience of "flavor" however, which probably motivated it being seen as a unitary sensory system, albeit a special one. In this chapter, "flavor" will be used to be composed of the interaction of several sensory systems, as shown in Fig. 7.3, often summarized as smell, taste, and mouthfeel. Stimulation of any of the sensory systems, smell, taste, mouthfeel (including temperature, pain, and touch) in isolation will result in an experience that is clearly qualitatively different from that of flavor. Flavor constitutes a percept in its own right, and flavor perception can thus be seen as a functionally distinct sensory system. The common fact that a flavor percept is felt to take place in the mouth, while an important part of it takes place in the nose, is an illusion called "oral referral" (Lim & Johnson, 2011; see also chapter: Oral Referral). Even trained tasters experience this illusion, that is, they will not "feel" their olfactory epithelia as the location of the smell part of a flavor percept unless a specific trigeminal (painful) substance excites the pain nerves in their nose.

The mouthfeel component of flavor makes matters even more complicated than they already are. "Mouthfeel" is comprised of a multitude of experiences originating in the oral cavity. These can be those of temperature, pain, and several variations of touch (see Fig. 7.3). That temperature is comprised of separate systems for cold and warmth, pain for stinging and aching, and touch for pressure and vibration (and more), will not be discussed.

Flavor is thus a combination of a number of other sensory systems. It does illustrate the complexity of the flavor system, at the same time as illustrating the possibilities of further exploring the system, for example, in the quest for new flavor experiences or novel applications in commercial food products. It also points at a relatively limited appreciation, and practical exploitation, of the flavor system in its complex entirety.

1.3 Priming

A simple definition of priming is that it is an implicit memory effect (Schacter, 1987) in which exposure to one stimulus (the prime) influences a response to another, subsequently presented, stimulus (schematically presented in Fig. 7.4). The exposure can be of any kind, as can the measured response. Note that "stimulus," can be interpreted broadly. The prime can be a simple controlled stimulus, and it can be a complex amalgam of stimuli, or even a whole situation. A "situation" actually is, of course, a complex mixture of very many stimuli. The second part of a priming study always entails some measurement of the hypothesized effect of the prime. As indicated in Fig. 7.4, this measurement can address a perception or a judgment or a participant's behavior (cf. Schacter & Buckner, 1998).

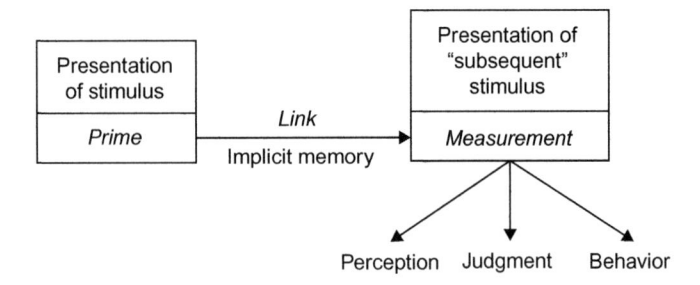

Figure 7.4 Priming schematically: A stimulus presented as a prime, through a link in implicit memory, leads to a change in later perception of, judgment about, or behavior toward, a subsequent presentation of that stimulus.

The simplest form of priming is repetition priming, whereby the second stimulus is identical to the first, and the first stimulus can be shown to facilitate the perception of the second. The facilitation can, for example, be shown in terms of a faster reaction time to the second stimulus or a higher proportion of correct identifications (eg, from among a set of confusable stimuli). Note that the effect can be positive or negative. Although most priming effects are "positive," that is, the effect is a faster response or more correct identifications, negative priming also exists (Mayr & Buchner, 2007).

Loersch and Payne (2011) introduce their Situated Inference Model where they present three steps in a priming situation:

1. Prime exposure
2. Misattribution
3. Afforded questions.

These researchers distinguish between the "true source of the mental content," that is an effect of the prime, and the misattribution the participant makes about its source. This misattribution step seems to presuppose that the participant is aware of a prime being presented. If the prime presentation takes place outside of awareness, the concept of misattribution is not needed. Misattribution may still take place, but it may often rather reside in the experimenter who designed a specific priming situation and a specific measurement, but has to conclude that an unexpected, and uncontrolled, aspect of the situation may have acted as a prime. In the context of food and, in particular, of flavor, this is even likely to occur, as flavor is a multisensory stimulus with many constituents that can act as a prime, or can be acted upon by other primes.

In the measurement situation, the "Misattributed content is used to answer the question afforded by the focal target," according to Loersch and Payne (2011, Fig. 7.1, p. 237). This presupposes that participants are aware that they are in a measurement situation. Introducing the misattribution concept is again not needed when the measurement takes place outside of the awareness of the participant, as will be the case with many (flavor) priming situations. This is the case, in particular, when one takes a situated approach where the context of the stimulation can be as important as the stimulation itself. In food- and flavor-related contexts, there will be many processes and influences taking place outside of the awareness of the participant.

According to Loersch and Payne's (2011) model, both the prime and the measurement can be influenced by a host of external effects. As a result, primes can produce very different responses depending on the measurement situation and on the types of response it may afford. The concept of affordance is important in this respect. The response resulting from the measurement situation will be one from among a set of potential responses afforded by that situation. Priming occurs when the one response actually displayed can be shown to have been made more likely by the prime. However, different participants may show different responses because every (measurement) situation may afford different responses to different individuals, based on their different personal histories and experiences (cf. Barsalou, 2015). In practice (here: in eating situations outside the laboratory) this will make designing and performing a strictly controlled, ecologically valid, priming study very difficult. Both the presentation of the prime and the measurement of a potential effect will have to be carefully designed, and a large number of participants will likely be needed to attain enough statistical power to show a priming effect.

There are different mechanisms that may underlie priming, depending on the type of process hypothesized as responsible for the effect of the original stimulus (the "prime") on the measurement. Typically, one finds reports of perceptual and conceptual priming. In perceptual priming, the sensory stimulation at the measurement situation is often better (faster, more correctly) perceived when the prime was a similar (or identical) stimulus compared to when it was different. Conceptual priming refers to a link between the prime and the effect that involves some cognitive processing, even if taking place outside of awareness. It is based on a similarity in meaning or another meaning-based relationship between concepts (often words). For example, in a measurement task where words need to be identified or completed, the prime "apple," presented as a picture, a word, or by a bite of it, will more likely result in a response "pear" than a response "wall," or perhaps even "steak." Conceptual priming may be difficult in olfaction as odors are not easily decomposed into explicit perceptual features (Koenig, Bourron, & Royet, 2000). That said, such a form of priming does appear to exist (Holland, Hendriks, & Aarts, 2005; Dijksterhuis, Zandstra, De Wijk, & Smeets, 2013). In the study by Holland et al. (2005), participants were seen to perform more cleaning actions (wiping crumbs from a table) when they were primed with the smell of a cleaning product (citrus), compared to a condition without a smell prime. This finding probably results from previously learned associations between a specific odor (citrus) and its use in cleaning products. Dijksterhuis et al. (2013) suggest that it may have been the valence of the odorous stimuli rather than any other "meaning" attributed to the stimuli that gave rise to an effect. In their study, less cleaning behavior was observed in a foul smelling (sulfur) condition, as compared to conditions with more agreeable smells (orange, grass). Both these studies applied orthonasal stimuli ambiently presented in a non-food situation.

A third form of priming is affective priming. Liking is a special aspect, important in particular in the context of food and eating. People are tempted only to seek out pleasant flavor experiences; they will normally ingest what they (expect to) like.

So effects of flavor onto hedonic aspects of the stimuli (be they food products, a meal, or simple stimuli), are important to consider. Veldhuizen, Oosterhoff, and Kroeze (2010) performed a study in which a flavor was presented as a prime for words that participants had to evaluate as positive or negative. The responses to words that were affectively congruent with the prime (a good- or bad-tasting flavor) were faster than to flavor-word pairs with incongruent affective values. This means that the affective value of the prime can work as a prime on subsequent stimuli that have similar affective value.

Flavor, not being a simple sensory system, may need to be regarded a bit differently from other (single) sensory systems. According to Prescott and Stevenson (2015), a flavor can be seen as a "functionally distinct sense that is cognitively 'constructed' from the integration of distinct physiologically defined sensory systems" (p. 1007). Some cognitive processing, albeit not necessarily conscious, appears necessary for flavor perception, as flavor is an integration of several other sensory systems. This observation leads to the conjecture that conceptual priming, which must also include some cognitive processing, may be more likely than perceptual priming in flavor perception.

In purely perceptual flavor priming, any of the sensory inputs (olfaction, taste, and mouthfeel) could be acting as a prime, onto subsequent stimulation at the measurement stage. When the measurement concerns flavor (the composite) and the prime was an odor, only one constituent of the flavor stimulus was presented as a prime. The flavor, being a "cognitive construct," may thus—by that definition—not be likely to be perceptually primed. For this reason, in flavor priming, conceptual priming could be more abundant than perceptual priming, compared to other sensory systems like, for example, olfaction or vision.

2 A taxonomy of primes

As priming is defined here as an implicit effect, this means that the primed participant is not aware of being somehow under the influence of a prime. However, this does not necessarily imply that the participant should never be aware of all aspects of the priming—and measurement—situation. Several distinct cases can be identified based on a participant's awareness of the prime, the measurement, and/or the link between these:

1. The participant did or did not consciously perceive the *prime* (P)
2. The participant is, or is not, aware of the *link* between a prime and its result
3. The participant is, or is not, aware of a *measurement* (M) of the result, taking place.

These three dichotomies can give rise to the eight different situations summarized in Table 7.1. Not all of these may be meaningful in the context of priming. Those studies where participants are aware of the prime, the link, and the measurement (type 1 in Table 7.1) are generally not considered priming studies, as priming is here defined as an implicit process, meaning it acts outside the awareness of the participant. It is possible, and likely, that such a "type 1" study shows an effect of the "prime" on the measurement, for example, in the case where explicit information (the "prime") has

Table 7.1 **A taxonomy of eight possible situations in priming**

Type	Prime (P)	Link	Measurement (M)
1	Perceived	Aware	Aware
2	Perceived	Aware	Not aware
3	Perceived	Not aware	Aware
4	Perceived	Not aware	Not aware
5	Not perceived	Aware	Aware
6	Not perceived	Aware	Not aware
7	Not perceived	Not aware	Aware
8	Not perceived	Not aware	Not aware

an effect on later judgments. An example of such a study in a flavor context can be found in Woods et al. (2011). In this study, participants receive an explicit instruction that they are about to experience a specific taste. The participants are in an MRI-scanner (magnetic resonance imaging, popularly known as "brain scanning"), so they are certainly aware of the measurement situation. The explicit instruction refers to the taste of a sip of a drink they will receive while being scanned, hence a link between the instruction and the measurement is also made known to the participants. What they do not know is whether or not the instruction in reality has any relation to the drink or not, and they may also not be aware that the instruction is there to potentially change their response to the drink.

Not all types of priming shown in Table 7.1 can be easily linked to specific situations in flavor (and food-related) priming situations. Below some cases, in a general context, and, where possible, in a flavor context, will be presented. The reader is invited to add specific instances s/he may know of and that are missed in this chapter.

One can argue that awareness of a link is difficult to uphold when one is not aware of a prime and/or a measurement. This is true; however, in such cases, special circumstances may somehow be assumed to obfuscate the perception of the prime and/or of the measurement leading to it not being consciously perceived.

In the remainder of this chapter, the different types will also be used to illustrate the complexity of flavor priming and flavor perception.

When the participant is aware of the link between the prime and the measurement (cases 2, 5, and 6 in Table 7.1 excluding the "not priming" type 1), it is at first sight difficult to assume a proper, implicit, priming paradigm. However, when the participant does in addition not perceive the prime and/or the measurement (type 6), one could argue that enough implicitness is still present for a proper priming situation to occur. In this case, a participant would be aware of the *existence* of a link between P and M, but not having perceived P and/or M, s/he does not know if and when a priming situation (and measurement) has taken place. So if the measurement shows an effect of the prime, the participant can be assumed to have been in a proper priming paradigm and being under the influence of the prime. An example of such a case could be seen in advertisements. Many advertisements will not be consciously perceived by consumers. Consumers are aware, however, of the fact that advertisements are there, and that it is their aim to make them purchase the goods advertised. When a consumer buys a product s/he is often unaware of the fact that s/he may have been primed to favor the particular product chosen over an alternative. Again, s/he is aware that this is the aim of advertisements for the product, so awareness of the link exists, but both the prime and the measurement will often take place outside awareness.

In another very common situation in priming studies, the participant is certainly aware of the fact that a measurement is taking place (types 3, 5, 7 in Table 7.1). In particular this case occurs in experimental settings in the laboratory. Examples of this are studies where the effect of a prime is measured using, for example, a Lexical Decision Task (LDT) or some form of a word or picture completion task (eg, Gaillet, Sulmont-Rossé, Issanchou, Chabanet, & Chambaron, 2013). There is no doubt that the participant knows s/he is being measured for something, and the participant may presume there is a link between the stimulation received previously and this measurement. However, s/he does not necessarily know they are potentially under the influence of a prime (type 5 in Table 7.1). Proper instruction will distract the participants from finding out what the prime is, and what the link could be.

In other experimental paradigms even the priming stimuli can be consciously perceived (type 3 in Table 7.1), as for example, where letters or symbols are presented, but some occur with a higher frequency than others, so that some form of implicit, statistical, learning takes place, the effect of which can be shown in the measurement. The participant in this case does not know which words are presented with which frequency, so is unaware of the prime. S/he may be aware of the fact that a link likely exists between the presented stimuli and the subsequent stimuli, but due to the uncertainty of the stimulation s/he cannot figure out what the link to the measurement will be.

Type 4 in Table 7.1 refers to the case where a prime is consciously perceived, but there is no awareness of a link, or of a measurement. This is a situation akin to many situations in which a stimulus is perceived which increases its familiarity and thereby its attractiveness. It is a mere-exposure type situation. The participant is not aware that the exposure does lead to it being selected in later circumstances. The classical example is media-exposure, whereby a political candidate's popularity can be shown to correlate highly to the media exposure, irrespective of his/her political message (Zajonc, 1968). Generalizations of the mere exposure effect exist and are not limited to social stimuli (Zajonc, 1980). In a food context, repeated exposure to a food item that is initially not liked will increase its acceptance and can even lead to a liking for it (eg, Wardle, Herrera, Cooke, & Gibson, 2003; Hausner, Hartvig, Reinbach, Wendin, & Bredie, 2012).

A "proper" priming situation (type 8), where prime, link, and measurement are not known to the participant, can be seen in a study performed in the context of nudging (for nudging see Thaler & Sunstein, 2009). The study concerned a situation in which healthy snacks and pieces of fruit, or less healthy snacks, are for sale in a train station food kiosk. The participants are normal commuters routinely using the train station. The measurement consists of the commuters buying the healthy snacks, or, instead, buying the unhealthy snacks. The prime in this case is not a, clearly visible, piece of fruit or healthy snack, but a more abstract concept. The prime here is the *position* of the fruit/healthy snack. In this study, it is positioned close to the cash register, in order for commuters to notice it. Normally this is the place where the unhealthy snack products are displayed. This example also illustrates that it is not always straightforward what constitutes the prime proper in a priming study. The link thus will also not be obvious to the commuter. It is the fact that the position of the healthy product is expected to lead to a higher incidence of it being purchased, as compared with normal purchase, when they are displayed at their regular position, somewhere else in the kiosk. The measurement is done without any commuter knowing, it is just counting how many pieces of healthy snacks are sold from this position, in comparison to when they are at their normal position (Kroese, Marchiori, & De Ridder, 2015). The result of this specific study was that more healthy snacks were sold when repositioned close to the cash register (no fewer unhealthy snacks were sold however).

3 Odor and flavor priming

Odor priming is obviously distinct from flavor priming, for one thing odor priming concerns only olfactory perception. Flavor concerns the multisensory integration of several distinct sensory systems as explained above (Fig. 7.3). Odor, however, is an important part of multisensory flavor perception. In their paper, Smeets and Dijksterhuis (2014) point at several unique properties of olfactory perception in comparison to other sensory systems. This has implications for priming studies in olfaction, about which they conclude that "laboratory environments are not meaningful contexts when trying to establish an appropriate understanding of odorants and their sources" (p. 7). It appears that an odor devoid of its normal context becomes meaningless and often cannot be identified. This is particularly noteworthy when one subscribes to a holistic viewpoint of perception. In naturalistic settings seldom (never) an odor occurs without a context (nor does any stimulus). Very often the context that a retronasal odor is encountered in, is one of food or eating, where the odor is part of a flavor perception. The same can be said of the other constituents, taste and mouthfeel, of the flavor concept, as illustrated in Fig. 7.3. In addition, also an orthonasal odor is very often encountered in the context of a food item or a dinner. Therefore the situation that a flavor is encountered in must be seen as an integral part of the flavor perception, just as was concluded for odors by Smeets and Dijksterhuis (2014).

Furthermore, in contrast to an odor, in normal conditions a flavor cannot be administered outside of awareness, as oral ingestion is necessary. This limits the type of flavor priming studies to types 2, 3, and 4 from Table 7.1. This is a clear difference from (orthonasal) odor stimuli. Odors can be ambiently administered, outside the awareness of the participants; they can be administered peri- (or sub-)threshold.

Smeets and Dijksterhuis (2014) distinguish three awareness circumstances under which odors can be perceived:

1. Attentively, where an odor can be verbally labeled by participants
2. Semiattentively, where a participant may notice, and report, something special in the environment or situation they are in, but cannot identify it to be olfactory
3. Inattentively, where participants show no evidence of being aware of something particular in the environment or situation they are in.

A flavor will always be consciously present. It is impossible to *not* notice a flavor stimulus being delivered to the oral cavity. Unlike ambient odors around their threshold that can remain unnoticed, while they can serve as primes (cf. Degel & Köster, 1999), a flavor cannot remain unnoticed. A particular flavor will thus always be consciously perceived, although a special constituent of the whole of the flavor experience could possibly remain outside of awareness. A special taste, odor, or mouthfeel can remain outside of awareness while it could be perceived under different circumstances. So one particular, aware, flavor might be perceived, and can be reported, while an unconscious perception of another flavor takes place at the same time. In theory, this may happen, although it does sound rather like a theoretical possibility only. An example that may come close to this is the case where, for example, a specific odor is added to the whole of a flavor, such that it may go unnoticed, but can act as a prime of types 5 through 8 from Table 7.1. As an example, it is sometimes reported that a little addition of a "rancid odor" will lead to a "real butter" flavor perception, even though a rancidity percept itself may never reach awareness. This is an example of a type 8 priming from Table 7.1, there is no conscious awareness of a measurement (an increased number of judgments of "real butter"), of a link between the added ("rancid") odor and the butter perception, nor of the added ("rancid") odor itself.

When seen in the context of the complexity of flavor stimuli (and their perception, cf. Fig. 7.3), it is not entirely clear what the consequences of such an addition (of a subtle mouthfeel, odor, or taste component) to the flavor as a whole will be. Detailed psychophysical experimenting with different levels of the addition may be needed to bring to light the relationships between the addition and its influence on the consciously perceived, flavor percept. The circumstances under which such additions can work as successful primes is another research question.

3.1 Flavor in applied contexts

In naturalistic settings a flavor is always encountered in the context of an eating or drinking situation. Flavor perception is even exclusively coupled to a situation where oral ingestion takes place. This means that there will be a number of constants that always occur in conjunction with a flavor percept.

In the home situation one can think of cooking, which is a very complex behavior with many flavor-related constituents and many opportunities for primes, even primes outside the direct constituents of flavor, such as orthonasal odors or the color of the food object or meal components. However, the consumers of the food are in many practical cases not the cook, so they will not be under the influence of the many primes present during cooking. The same is true for many food buying (shopping) contexts, where many eaters of the bought food are not necessarily the buyers. Thus the primes that could act in these situations are not always relevant in applied or commercial contexts (see Fig. 7.5).

In out-of-home consumption situations consumers will receive many impressions of the situation they eat in, including the menu, price, staff, background music, etc. These can act as primes for the flavor they perceive, and their appreciation of it. In both home

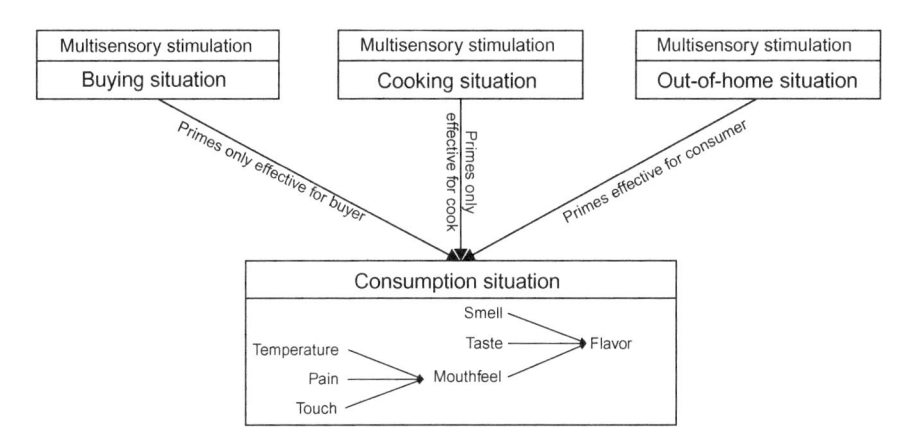

Figure 7.5 Different situations with multisensory stimulation and their relevance for providing primes to eaters, in the context of affecting a flavor experience.

and out-of-home situations the eater will encounter the appearance of the food or drink itself, and typically a glass or plate, utensils, a chair, a table, and last but not least, company. All these constants of the eating situation can, and probably will, provide primes for the flavor of the food/drink (see Fig. 7.5; cf. Spence & Piqueras-Fiszman, 2014).

4 Bottom-up and top-down priming

The fact that the environment (situation or context), of a (flavor) perception should be seen as an integral part of the perception, places the concept of multisensory flavor priming in a broader context. This broader context was also introduced by Smeets and Dijksterhuis (2014) who point at the difference in approach of (odor) priming studies from a perception/psychophysical viewpoint, and from the point of social psychology and embodied/situated cognition. In the former, priming is seen as a bottom-up process where activation by an external stimulus leads to a spreading of neural activation affecting subsequent perception of subsequent stimuli. In social psychological priming studies, the effect of a prime is rather seen as an effect of prior (implicit) learning during earlier exposures to similar situations (eg, Bargh, 2006). This learning will mainly provide top-down effects. In this section, Smeets and Dijksterhuis (2014) are followed in combining both ideas.

There have been several recent publications where a top-down effect on taste or flavor is shown (cf. Woods et al., 2011; Le Berre, Boucon, Knoop, & Dijksterhuis, 2013; Dijksterhuis, Boucon, & Le Berre, 2014). The sources of top-down effects can be multitude. They can result from prior learning, which may take years and would probably often not be seen as priming, although they fall under our definition of priming. They can also result from immediate prior stimulation, where the priming is an almost instantaneous effect of this stimulation.

The typical naturalistic eating situation contains many potential primes. These can be the visual aspects of the food product that later, upon ingestion, will affect its flavor. The visual aspect may create an expectation, which acts as a top-down effect on perception. This would not be unlike conceptual priming. One can also argue that the visual aspect itself is a perceptual prime for the later flavor perception. Analogously, the (orthonasal) smell of the food item may act as such a top-down effect, before the flavor proper is perceived. A mechanism akin to this was used by Le Berre et al. (2013) and Dijksterhuis et al. (2014) who demonstrated that the first bite of a food item can create an expectation for its remainder and showed that there are effects on its taste/flavor. The distinction between bottom-up (multi-)sensory stimulation and top-down effects becomes blurred when one acknowledges the fact that sensory stimulation can lead to the creation of expectations that subsequently act as a top-down effect on the flavor perceived. Fig. 7.6 attempts to summarize a multisensorially perceived eating situation giving rise to both perceptual and conceptual priming processes affecting flavor perception.

4.1 *A situated and embodied approach*

In the broader context provided by a situated viewpoint, all stimulation surrounding a (flavor) perception can be relevant to the flavor perceived. The previous section listed the food or drink items themselves, a plate or a glass, utensils, a chair, a table, and last but not least company. One part of a situated approach was not introduced, namely, the participant him-/herself. A proper embodied viewpoint would certainly count with (the body of) the eater. In particular, during eating this will be important as eating is an activity with obvious bodily (physiological) consequences. Another important part of the situation is the emotional state of the eater. According to some emotion theories this can also be seen as a bodily effect (cf. Damasio, 2001). Here bodily/physiological effects will be mentioned as separate from emotional effects as the latter may also be interpreted by many to rather mean "feelings" (interpreted—bodily felt–emotions), according to an appraisal type of emotion theory.

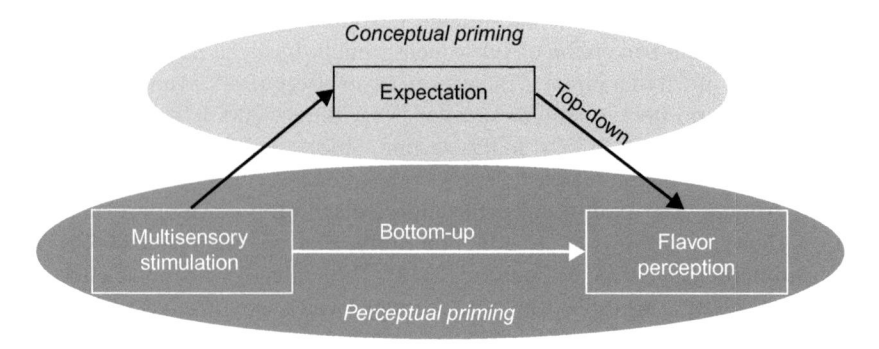

Figure 7.6 A multisensorially perceived eating situation will give rise to both perceptual and conceptual priming processes affecting flavor perception.

So far several sources for flavor priming (five "S"-words) have been identified:

- Sensory: the bottom-up activation of sensory systems
- Surroundings: the physical environment of the food stimulus
- Social: the company in the eating situation
- Somatic: the physiology of the eater
- Sentimental: the emotional state of the eater.

The next section presents a short account of a conceptual framework, with suggestions for a holistic (situated) approach to studying food choice behavior. It is based on ideas from priming, situated/embodied cognition, and (food-related) perception. Priming is seen as a neural-activation (bottom-up)-related effect as well as a top-down effect (see Fig. 7.6). The framework builds on the theory that parts of an event can trigger a simulation-type activity of a previously encountered event (cf. Barsalou, 2015; Papies & Barsalou, 2015). The simulation is thought of as a multisensory mental activation in which specific aspects of an event are somehow mentally re-created. The simulation will typically remain covert to the participant him/herself, but it can lead to measurable effects in especially designed (eg, priming) experiments. Typically the results are faster responses, better recognition, more spontaneous mentioning, and so on, of the aspects under consideration.

The approach is situated to the extent that it recognizes that the whole perception, or rather the total experience, of stimulation is the key concept for an organism. To a tasting organism, including humans (eg, food consumers), it is normally of no value to know the exact nature of the stimulation and of the sensory systems that ultimately lead to the experience of a flavor. This is the ecological approach to perception introduced by Gibson (1966), presented in a functional realm of food perception and flavor by Prescott (2012). In the next section this approach is taken one step further in presenting a situated approach to flavor (and food-related) perception.

5 A suggestion for a conceptual framework for situated (flavor) perception and behavior

Fig. 7.7 presents a schematic account of a conceptual framework in which flavor perception, including flavor priming, can be seen to take place. In the Appendix the framework is presented in a somewhat more formal detail. Here it is presented in terms of concepts and ideas.

In this framework an *event* is an object (eg, in this context a food-related stimulus) *including* the whole of aspects of the situation it occurs in (see box "Current event with aspects" in Fig. 7.7). Aspects of previous encounters of events have been stored in memory (box "Situated aspects previous event" in Fig. 7.7). Aspects of the current event can be the earlier identified sources for flavor priming (sensory, surroundings, social, somatic, sentimental), in addition to any other sources not listed here that may be relevant in specific cases. When an aspect of a current event is similar enough to an aspect of a previous encounter, a simulation of the previous event occurs (box "Simulated event with aspects" in Fig. 7.7). This simulation refers to a mental activity

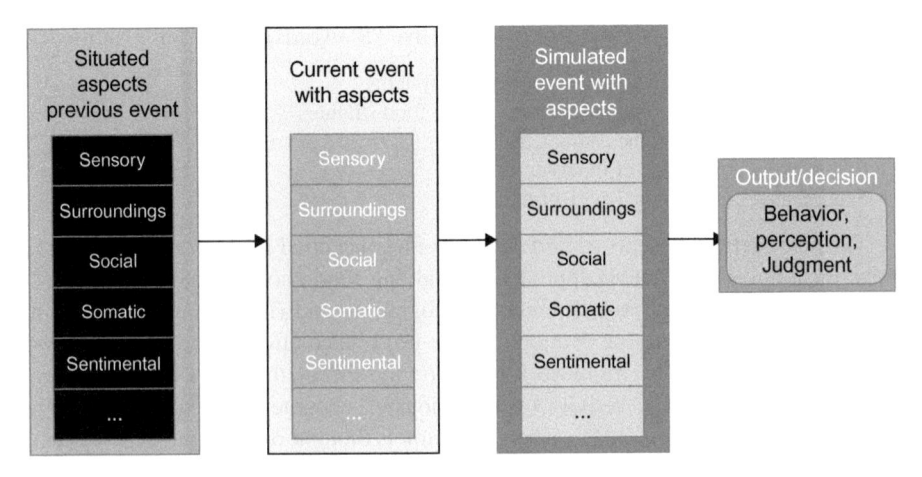

Figure 7.7 Schematic of the proposed framework for situated (flavor) perception and behavior.

akin to the one occurring during external stimulation, but without the external stimulation. When the intensity of the total simulation, that is, of the simulated whole *event* including all aspects, is larger than some threshold, an effect of the simulated event on behavior, perception, or judgment is conjectured to occur (box "Behavior, perception, judgment" in Fig. 7.7). In a priming context this will be a measurement showing an effect of an earlier presented prime.

6 Conclusions

As flavor is such a complex concept, comprising smell, taste, and mouthfeel (at the very least), flavor priming is also not a simple matter. In this chapter, priming is presented in some detail, in particular in a context of situatedness of all stimulation. Because of the special nature of flavor perception it was necessary to regard priming in a wider framework.

Flavor is a "cognitively constructed" sense (Prescott & Stevenson, 2015), which leads to the conjecture that it may be more prone to conceptual priming compared to other sensory inputs.

Flavor is never encountered in isolation (nor is any stimulus outside the laboratory) but flavor always implies a rather specific situation. It always includes oral ingestion and hence a context of food or drink. As a result there are some constants that will always surround flavor perception and will be part of any flavor-related priming situation. Five aspects (five "S"-words) of this situation were listed that are relevant sources occurring in flavor priming: sensory, surroundings, social, somatic, and sentimental.

Particularly in applied contexts, the flavor of a food product, be it packaged consumer goods or out-of-home meals, will be under the influence of its surroundings.

The producers of these products will do their utmost to optimize the flavor experience and will take into account possibilities of providing primes that will positively affect the experience. Some of these primes will not reach the food consumer, while other aspects of the situation can act as uncontrolled unwanted primes.

The complexity of "flavor" as an experience is sketched, and a situational approach was taken, where the environment the flavor is encountered in is recognized to provide many cues that will affect the experience of the flavor. Studying flavor devoid of a context is studying a different entity compared to flavor in the food situation it is normally encountered in. Results from studies where flavor has been isolated will be very hard to generalize to other contexts.

Writing about flavor is automatically writing about food and about applications of flavor. Recognizing the special "constructed" nature of flavor and viewing flavor in the situated framework presented in this chapter may suggest new directions for flavor research and may open new ways of delivering flavors.

Acknowledgments

Professor Larry Barsalou and Dr Esther Papies (Glasgow University, UK) are thanked for discussions on an earlier version of the manuscript.

Appendix

The framework presented in Section 5 (and Fig. 7.7) is presented here in some more formal detail. In this framework, a previous event is assumed to have taken place, including the whole of the situation it occurs in (see box "Current event with aspects X_i" in Fig. 7.A1). Aspects of a previous encounter of an event have been stored in memory, with a "strength" or "intensity" p_i, where i is an index referring to a specific aspect, of which there are N, so $i=1,\ldots,N$ (box "Situated aspects P_i previous event" in Fig. 7.A1). Intensities of aspects of the current event (sensory, surroundings, social, somatic, sentimental, etc.) are named x_i. The intensities x_i and p_i of the aspects are the quantitative results of a chosen psychophysical measurement of the perceived intensity of the aspects. For the moment the type of measurement is not discussed, but it is assumed that corresponding measurements can be done.

When an aspect X_i of a current event is similar enough to an aspect P_i of a previous encounter, a simulation of the previous event occurs (box "Simulated event with aspects S_i" in Fig. 7.A1). The difference between X_i and P_i can be written as $\Delta_i = d(X_i, P_i)$. When this difference is small enough, say smaller than some threshold value T_s, so $\Delta_i < T_s$, a simulation process will occur, leading to a mental (re-)activation, S_i, of aspect P_i from previous encounters. In its simplest form, d is a distance function that measures a psychologically relevant distance between X_i and P_i. The exact form of the function d can be chosen. The value Δ_i can also be seen as a measure for an affective distance. In this case a function d would work on the hedonic value of the

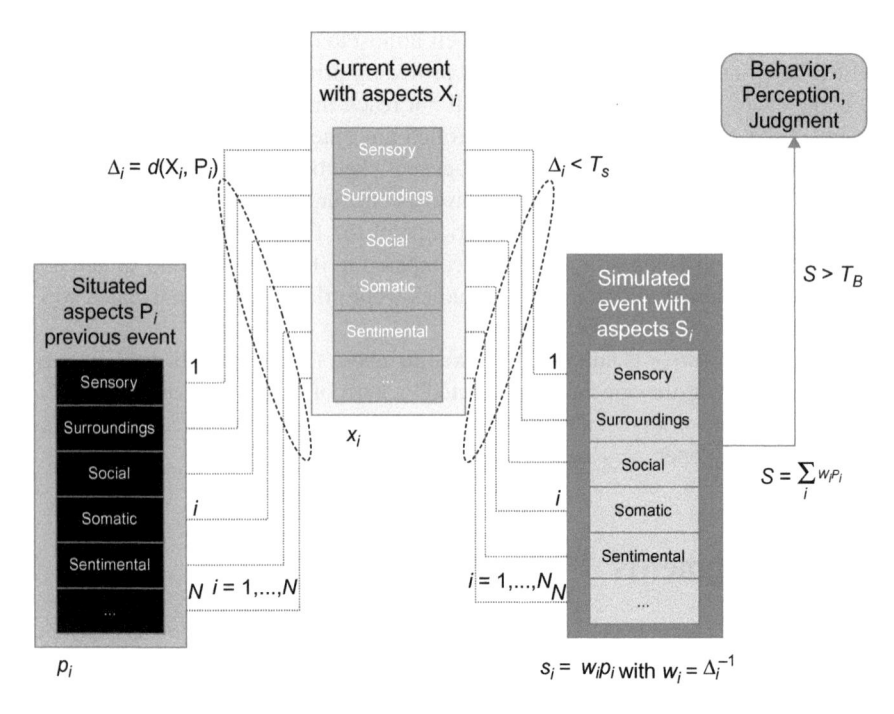

Figure 7.A1 Formalized presentation of the proposed framework.

aspects: $\Delta_i = d(h(X_i), h(P_i))$, where h is a function giving the hedonic value of the aspects X_i and P_i.

S_i, the result of a simulation of a previous aspect P_i, occurs with an intensity s_i, where this intensity depends on the size of the difference Δ_i. The larger this difference, the lower the intensity of the simulation will be. The smaller this difference, the higher the intensity of the simulated aspect will be. A reasonable form for the relationship between the intensities s_i and p_i is one where the intensity of the simulation s_i is weighted by the inverse of the difference Δ_i. This gives $s_i = w_i p_i$ with the weight $w_i = \Delta_i^{-1}$. Obviously, the case $\Delta_i=0$ needs to be prevented, which may be done by choosing an appropriate function for d. It can probably be assumed that in reality a simulated and a previous aspect will never be identical.

The intensity of the total simulation, that is, of the simulated whole *event* including all s_i, can be written as $S = \Sigma_i w_i p_i$. When the total simulation intensity S is larger than a threshold, say $S > T_B$, an effect of the simulated event will occur (box "Behavior, Perception, Judgment" in Fig. 7.A1). This can lead, depending on the chosen operationalization, to a choice for the current event (situated stimulus) from among a set of alternatives (effect on behavior), a faster recognition of an aspect of the current event (effect on perception), or an increased liking of (aspects of) the current event (effect on judgment).

References

Auvray, M., & Spence, C. (2008). The multisensory perception of flavour. *Consciousness and Cognition, 17*, 1016–1031.

Bargh, J. (2006). What have we been priming all those years? On the development, mechanisms, and ecology of nonconscious social behavior. *European Journal of Social Psychology, 36*, 147–168.

Barsalou, L. (2015). Situated conceptualization: Theory and application. In Y. Coello & M. H. Fischer (Eds.), *Foundations of embodied cognition*. East Sussex, UK: Psychology Press.

Calvert, G. A., Spence, C., & Stein, B. E. (Eds.), (2004). *The handbook of multisensory processes*. Cambridge: MIT Press.

Damasio, A. (2001). Fundamental feelings. *Nature, 413*, 781.

Degel, J., & Köster, E. P. (1999). Odors: Implicit memory and performance effects. *Chemical Senses, 24*, 317–325. http://dx.doi.org/10.1093/chemse/24.3.317.

Dijksterhuis, G. B. (2012). The total product experience and the position of the sensory and consumer sciences: More than meets the tongue. *New Food, 15*(1), 38–41.

Dijksterhuis, G.B. (2016). Perceptie en illusie. (Perception and illusion. In Dutch.) In: *Onbeperkt Van Abbe* (Van Abbe unlimited: In Dutch). Eindhoven: Van Abbe Museum.

Dijksterhuis, G. B., Boucon, C., & Le Berre, E. (2014). Increasing saltiness perception through perceptual constancy created by expectation. *Food Quality and Preference, 34*, 24–28.

Dijksterhuis, G.B., Zandstra, E., De Wijk, R.A., & Smeets, M.A.M. (2013). *Smelly and dirty: Valence, not semantics, of odours prompt cleaning behaviour*. Poster at 10th Pangborn Sensory Science Symposium. August 2013. Rio de Janeiro, Brazil.

Driver, J., & Spence, C. (2000). Multisensory perception: Beyond modularity and convergence. *Current Biology, 10*, R731–R735.

Durie, B. (2005). Doors of perception. *New Scientist, 2484*, 33–36.

Gaillet, M., Sulmont-Rossé, C., Issanchou, S., Chabanet, C., & Chambaron, S. (2013). Priming effects of an olfactory food cue on subsequent food-related behaviour. *Food Quality and Preference, 30*, 274–281.

Gibson, J. J. (1966). *The senses considered as perceptual systems*. Boston: Houghton Mifflin Company.

Hausner, H., Hartvig, D. L., Reinbach, H. C., Wendin, K., & Bredie, W. L. P. (2012). Effects of repeated exposure on acceptance of initially disliked and liked Nordic snack bars in 9–11 year-old children. *Clinical Nutrition, 31*(1), 137–143.

Holland, R., Hendriks, M., & Aarts, H. (2005). Smells like clean spirit. nonconscious effects of scent on cognition and behavior. *Psychological Science, 16*(9), 689–693.

ISO, (1992). *Standard 5492: Terms relating to sensory analysis. International Organization for Standardization*. Vienna: Austrian Standards Institute.

ISO, (2008). *Standard 5492: Terms relating to sensory analysis. International Organization for Standardization*. Vienna: Austrian Standards Institute.

Kahneman, D. (2012). A proposal to deal with questions about priming effects. *Nature*. http://www.nature.com/polopoly_fs/7.6716.1349271308!/suppinfoFile/Kahneman%20Letter.pdf Retrieved October 12, 2012.

Koenig, O., Bourron, G., & Royet, J. P. (2000). Evidence for separate perceptive and semantic memories for odours: A priming experiment. *Chemical Senses, 25*(6), 703–708.

Koseleff, P. (1957). Studies in the perception of heaviness. *Acta Psychologica, 13*, 242–252.

Köster, E. P. (2000). Psychophysics and sensory analysis in the "lower" senses. In C. Bonnet (Ed.), *Fechner day 2000: Proceedings of the 16th annual meeting of the international society for psychophysics* (pp. 79–84)). Strasbourg, France: the International Society for Psychophysics.

Kroese, F. M., Marchiori, D. R., & De Ridder, D. T. D. (2015). Nudging healthy food choices: A field experiment at the train station. *Journal of Public Health, 37*, 1–5.

Kubovy, M. (1981). Concurrent-pitch segregation and the theory of indispensable attributes. In M. Kubovy & J. R. Pomerantz (Eds.), *Perceptual organization* (pp. 55–98). Hillsdale, NJ: Erlbaum.

Kubovy, M. (1988). Should we resist the seductiveness of the space:time:vision:audition analogy? *Journal of Experimental Psychology: Human Perception and Performance, 14*, 318–320.

Kubovy, M., & Schutz, M. (2010). Audio-visual objects. *Review of Philosophy & Psychology, 1*, 41–61.

Kubovy, M., & Van Valkenburg, D. (2001). Auditory and visual objects. *Cognition, 80*, 97–126.

Lawless, H., & Heymann, H. (2010). *Sensory evaluation of food: Principles and practices.* New York: Springer.

Le Berre, E., Boucon, C., Knoop, M., & Dijksterhuis, G. B. (2013). Reducing bitter taste through perceptual constancy created by an expectation. *Food Quality and Preference, 28*, 370–374.

Lim, J., & Johnson, M. B. (2011). Potential mechanisms of retronasal odor referral to the mouth. *Chemical Senses, 36*, 283–289.

Loersch, C., & Payne, B. K. (2011). The situated inference model: An integrative account of the effects of primes on perception, behavior, and motivation. *Perspectives on Psychological Science, 6*, 234–252.

Mayr, S., & Buchner, A. (2007). Negative priming as a memory phenomenon. A review of 20 years of negative priming research. *Zeitschrift für Psychologie/Journal of Psychology, 215*, 35–51.

Papies, E., & Barsalou, L. (2015). Grounding desire and motivated behavior. A theoretical framework and review of empirical evidence. In W. Hofmann & L. F. Nordgren (Eds.), *The psychology of desire* (pp. 36–60). New York: Guilford Press.

Pinna, B., & Brelstaff, G. J. (2000). A new visual illusion of relative motion. *Vision Research, 40*, 2091–2096.

Piqueras-Fiszman, B., & Spence, C. (2015). Sensory expectations based on product-extrinsic food cues: An interdisciplinary review of the empirical evidence and theoretical accounts. *Food Quality & Preference, 40*, 165–179.

Prescott, J. (2012). Multimodal chemosensory interactions and perception of flavor. In M. M. Murray & M. Wallace (Eds.), *The neural bases of multisensory processes* (pp. 691–704). Boca Raton, FL: CRC Press.

Prescott, J., & Stevenson, R. J. (2015). Chemosensory integration and the perception of flavor. In R. L. Doty (Ed.), *Handbook of olfaction and gustation* (pp. 1008–1028). Wiley & Sons. Chapter 45.

Schacter, D. L. (1987). Implicit memory: History and current status. *Journal of Experimental Pschology: Learning, Memory, and Cognition, 13*, 501–518.

Schacter, D. L., & Buckner, R. L. (1998). Priming and the brain. *Neuron, 20*, 185–195.

Schifferstein, H. N. J., & Spence, C. (2008). Multisensory product experience. In H. N. J. Schifferstein & P. Hekkert (Eds.), *Product experience* (pp. 133–161). Amsterdam: Elsevier.

Smeets, M. A. M., & Dijksterhuis, G. B. (2014). Smelly primes—when olfactory primes do or do not work. *Frontiers in Psychology: Cognitive Science, 5*, 96. http://dx.doi.org/10.3389/fpsyg.2014.00096.

Spence, C., & Piqueras-Fiszman, B. (2014). *The perfect meal. The multisensory science of food and dining*. Chichester: Wiley Blackwell.
Spence, C., Smith, B., & Auvray, M. (2015). Confusing tastes and flavours. In D. Stokes, M. Matthen, & S. Biggs (Eds.), *Perception and its modalities* (pp. 247–274). Oxford, UK: Oxford University Press.
Thaler, R. H., & Sunstein, C. R. (2009). *Nudge. Improving decisions about health, wealth and happiness*. New York: Penguin Books.
Veldhuizen, M. G., Oosterhoff, A. F., & Kroeze, J. H. A. (2010). Flavors prime processing of affectively congruent food words and non-food words. *Appetite, 54*, 71–76.
Wardle, J., Herrera, M. -L., Cooke, L., & Gibson, E. L. (2003). Modifying children's food preferences: The effects of exposure and reward on acceptance of an unfamiliar vegetable. *European Journal of Clinical Nutrition, 57*, 341–348.
Woods, A. T., Lloyd, D. M., Künzel, J., Poliakoff, E., Dijksterhuis, G. B., & Thomas, A. (2011). Expected taste intensity affects response to sweet drinks in primary taste cortex. *Neuroreport, 22*(8), 365–369.
Zajonc, R. (1968). Attitudinal effects of mere exposure. *Journal of Personality and Social Psychology, 9*, 1–27.
Zajonc, R. (1980). Feeling and thinking: Preferences need no inferences. *American Psychologist, 35*, 151–175.

Flavor Liking

John Prescott
TasteMatters Research & Consulting, Sydney, Australia

8

Our responses to the hedonic quality of flavors appear immediate, almost reflexive in nature. We never need to think one way or another about whether our "taste buds" are asking us to accept or reject a food. One possible exception to this is for flavors that we have not encountered before where, if we get over our initial wariness regarding novelty (Pliner & Pelchat, 1991), we tend to actively search for some point of reference—that is, for similar, familiar flavors—prior to deciding on issues of acceptability. This search is, in itself, telling, as will become evident later. However, the apparently instantaneous nature of flavor hedonics hides a wide range of learned responses that rely on the integration of the odor component of flavors with other stimuli—tastes—that are already hedonically valenced. Importantly, since odors form the characteristic quality that identifies flavors, there is evidence that the underlying dimension of olfaction is hedonic in nature (Khan, Luk, Flinker, Aggarwal, & Lapid, 2007). That is, although odors differ from one another in chemical structure, it is their degree of pleasantness that distinguishes different odors for the consumer.

Flavors are defined here as always encompassing both tastes and odors, although recognizing that somatosensory qualities (textures, for example), and visual and auditory qualities can also contribute to flavor perceptions and preferences in different contexts. Flavors need not be, but almost always are, also accompanied by intake and metabolic utilization of energy, nutrients, or pharmaco-active agents present in foods and beverages. Flavors are also experienced within particular external contexts which include places and occasions of eating, interactions with others, and one's own emotional state. All of these factors contribute to flavor hedonics via association and integration.

1 Development of liking

For changes in liking to occur, repeated exposure to an otherwise neutral, or perhaps even disliked, flavor or odor is necessary and sometimes sufficient (Methven, Langreney, & Prescott, 2012). The fact that exposure to the odor compounds of flavors in utero leads to increased liking for those flavors post-partum (Mennella, Jagnow, & Beauchamp, 2001) appears to be a fundamental example of this process. Since liking can develop without any obvious pairing with another positive stimulus (Zajonc, 1968), there remains a basic question of the processes underlying such *mere* exposure (ME) effects. Even so, explanations for ME effects typically involve associative links. The first of these, developed largely from studies that have used rapidly presented, brief visual stimuli, is that the repeated stimulus becomes increasingly more easily processed by the brain. This processing fluency hypothesis suggests that since processing a stimulus is effortful, any reduction in such effort will be rewarding, resulting in positive evaluation of the stimulus (Reber, Winkielman, & Schwarz, 1998). It has also been suggested that novel stimuli initially produce unpleasant levels of arousal, and that repeated exposure is rewarding because of the reduction in arousal that comes with increased familiarity (Berlyne, 1970). This may explain why rapidly repeated stimuli are associated with positive mood states in research participants, with these mood states consequently becoming associated with the stimulus (Monahan, Murphy, & Zajonc, 2000).

Whether such processes are relevant to the development of flavor preferences is unclear. But what is more consistent with what we do know is the idea that repetition of a stimulus that is novel is associated with a reduction in fear of aversive consequences—in other words, a sort of "safety signal" (Zajonc, 2001). Both the concept of the "Omnivore's Paradox"—in which novel foods could be either edible or toxic—and the phenomenon of food neophobia are underpinned by fear of novel foods (defined as those foods to which we have not been exposed). Whether through the reduction of the fear of illness or of an unpleasant taste, the flavors of new foods become accepted through repeated tasting (Birch and Marlin, 1982; Pliner et al., 1993).

In the case of brief visual stimuli, active attention appears to have no impact on the development of liking for the repeatedly exposed stimulus. However, priming studies of longer visual stimuli have shown that attended stimuli are evaluated more positively than those that are ignored (Raymond, Fenske, & Tavassoli, 2003). Development of liking for odors has also been shown to be facilitated by explicit attention to a repeated odor in contrast to a similarly exposed, but ignored, odor (Prescott, Kim, & Kim, 2008). This may be another example of liking being derived from the pairing of the stimulus with a positive emotional state that arises from the strong reciprocal interactions between attention and emotion (Vuilleumier & Huang, 2009). Yet another example of this may be the fact that choice itself—that is, choosing between two equally liked alternatives—acts to increase liking for the chosen option (Coppin, Delplanque, Cayeux, Porcherot, & Sander, 2010). One rationale for this effect is that choices are an indication of exerting control over our environment and it is this perception of having such control that gives rise to positive affect (Leotti & Delgado, 2011).

Mere exposure effects can also be seen as a form of incidental transfer of hedonics from a variety of contextual properties that accompany any repeated stimulus exposure. Thus, a new food flavor could be accompanied by a pleasant environment, enjoyable company, a general state of relaxation, and so on. It has been suggested, for example, that comfort foods, which appear to be a widespread phenomenon across cultural boundaries, are liked because they are foods that have most often been eaten in the company of those with whom we have important relationships, including parents, siblings, and partners. One possibility is that the feelings of psychological comfort that were experienced at the time of eating such foods become "encoded" with the flavor of those foods (Troisi & Gabriel, 2011), thus providing a positive emotional experience while eating.

2 Evaluative conditioning

This type of explanation for ME effects is clearly evoking what has been termed "evaluative conditioning," which is the process by which a given stimulus—be it a picture or a car or an odor—changes in liking due to proximity to something that already has valence, that is, a positive or negative value (Levey & Martin, 1975). In the first demonstration of evaluative conditioning with flavors, participants repeatedly sampled one tea flavor with sucrose and another tea flavor in water (Zellner, Rozin, Aron, & Kulish, 1983). The sweet-paired tea came to be liked more than the tea experienced in water, even when subsequently tasted without sucrose. Similar changes in odor liking based upon pairing with pleasant or unpleasant tastes have subsequently been demonstrated (Baeyens, Eelen, Van den Bergh, & Crombez, 1990; Yeomans, Mobini, Elliman, Walker, & Stevenson, 2006).

The type of learning involving a positive (sweet) or negative (bitter) taste paired in solution with the odor component of flavor is typically referred to as flavor–flavor learning. The outcome—a change in liking for the odor or flavor—occurs independently of any consequence that the ingestion of the tastant may have—for example, provision of calories. Thus, pairing an odor in solution with a sweet taste results in increased liking for the odor, irrespective of whether the sweetness comes from a nutritive sweetener such as sucrose or a nonnutritive sweetener like aspartame (Mobini, Chambers, & Yeomans, 2007). Such changes appear to be permanent and, once established, no longer require further association with the taste (Baeyens, Crombez, Van den Bergh, & Eelen, 1988).

Flavor–flavor learning appears to be one form of more general evaluative learning. Hence, pairing odors with a wide variety of positive or negative stimuli produces, respectively, increases or decreases in liking. This is strongly evident in the co-exposure of odors with activities that varied across participants in terms of their pleasantness. For example, the same activity (eg, a massage) produced increased liking for the odor to the extent it was a positive experience for that individual (Baeyens, Wrzesniewski, De Houwer, & Eelen, 1996). Hence, those who found the massage unpleasantly painful failed to show increased liking for the odor, in contrast to those who enjoyed the massage. It has also been shown experimentally that, at least in children, observing the behavior of an adult consuming a beverage can produce an increase in liking for the flavor of that drink (Baeyens, Vansteenwegen, & De Houwer, 1996). Such observational learning forms the basis of much advertising in which liked or admired individuals—actors, sports stars, and so on—are paired with products. The assumption is that the positive emotions evoked by the person "rub off" on the product.

When the post-ingestive consequences of food consumption are considered, additional flavor learning beyond the impact of the affective value of the associated taste (flavor–consequence learning) can be demonstrated. A considerable number of animal (Myers and Sclafani, 2001) and human (Prescott, 2004; Yeomans et al., 2008; Kern et al., 1993) studies have shown that odor–taste pairings can produce learned preferences when the tastant provides valued nutrients. So, while the *ingestion* of sucrose increases liking over and above that produced when the sucrose is only tasted but not consumed, no additional increase is seen with the ingestion of aspartame (Mobini et al., 2007). It is assumed that the body's utilization of the energy provided by sucrose, but not aspartame, has positive consequences that become associated with the flavor. A similar effect is seen with those flavors that have been paired with glutamate (Prescott, 2004; Yeomans, Gould, Mobini, & Prescott, 2008), with fat (Johnson, McPhee, & Birch, 1991; Capaldi & Privitera, 2007), and also with caffeine (Chambers, Mobini, & Yeomans, 2007). In the latter case, the effect appears to relate to the positive state that occurs after caffeine relieves the effects of caffeine abstinence.

In either flavor–flavor or flavor–consequence learning, the value of the tastant is crucial. Hence, while sweetness is generally a positive taste, there are those who find medium to strong sweetness relatively unpalatable (so-called "sweet dislikers") (Kim, Prescott, & Kim, 2014) and, for such individuals, evaluative conditioning is less evident following odor/sweetness pairing. Similarly, those less sensitive to bitterness show decreased dislike for bitter-paired odors (Yeomans, Prescott, & Gould, 2009). In flavor–consequence learning, hedonic changes rely on the metabolic value of the associated nutrient/energy, with hunger being a prerequisite for learning to occur (Yeomans & Mobini, 2006). One ingenious study demonstrated flavor–consequence learning with fat producing increased flavor liking in the presence of a bitter tastant that would be expected to otherwise produce decreases in flavor liking due to flavor–flavor learning (Capaldi & Privitera, 2007).

The distinction between flavor–flavor and flavor–consequence learning can be seen in terms of the distinction between the development of food "liking" versus food "wanting," the latter a construct that has been explored in terms of both distinct neural and motivational substrates (Castro & Berridge, 2014; Garcia-Burgos & Zamora, 2015; Winkielman & Berridge, 2003). Wanting reflects a drive to consume, the effects of which can be observed in eating that is independent of energy needs.

Since wanting reflects a motivational state, a factor such as relative hunger can modulate whether flavor–consequence learning is effective in influencing subsequent intake. In particular, wanting can be triggered by sensory cues—odors, visual or auditory cues—that have been associated with nutrient learning. It has been demonstrated that a novel soup flavor paired with ingested monosodium glutamate (MSG), relative to a non-MSG control, not only increased in rated liking, even when tested without added MSG, but also produced behavioral changes including increases in ad libitum food intake and rated hunger following an initial tasting of the flavor (Yeomans et al., 2008). In other words, the soup flavor became a cue that motivated the desire or want for consumption of the soup. Such effects correspond to the common experience of being stimulated to eat by the sights or smells of food.

The effects of such stimulation can be found in research showing that consumption of a food in response to cues can occur even after consuming the same food to satiation (Ferriday & Brunstrom, 2008). In addition, since wanting and liking can be quite separate, this also helps explain why overeating can take place even if the food being eaten becomes less and less pleasant during eating due to allesthesia (Rolls & Rolls, 1997). Wanting and liking can be differentiated in other important ways too. In the former, the effects can be measured in appetitive terms using measures of intake and hunger. Flavor–consequence learning is maximized when conditioning and evaluation of learning take place under relative hunger (Yeomans & Mobini, 2006). A recent study also dissociated wanting and liking by conditioning using an unpalatably salty level of MSG added to a soup provided for lunch over several days (Dermiki et al., 2015). Following conditioning, the elderly participants ate more of the unpalatable MSG-paired soup, but yet failed to show any increase in liking for this same soup flavor.

Understanding the associative learning basis of flavor preferences has important practical implications, including sheding light on the origin of the triggers for unrestrained eating that represents a risk for obesity. Flavors not only become liked by pairing with energy but also become cues that engage wanting for foods, in particular palatable (often high-sugar, high-fat) foods, since these are both common and effective sources of energy. Regardless of our weight, we tend to eat more of a palatable food than of one that is not palatable. However, those who are obese respond to palatable foods by eating even more than do normal-weight individuals (Nisbett, 1968). Is it just that the obese find palatable foods more pleasurable? This is a reasonable enough conclusion, but may be confusing palatability and energy content. Since wanting—increased appetite and hunger—can be triggered by sensory cues that have come to signal energy, therefore cues such as the odor and flavor of food are not only appealing but they also motivate us to consume. For the obese, the sensory cues for the presence of energy may be especially powerful in triggering the desire to eat (Hofmann, van Koningsbruggen, Stroebe, Ramanathan, & Aarts, 2010).

Recognizing the role of experience and learning in flavor preferences also provides insight into the origins of variations in such preferences at both the level of the individual and of the culture. Culturally based "flavor principles," for example, arise from combinations of odors, tastes, and textures (and cooking techniques) that have been paired and become liked through repeated exposure, flavor–flavor and flavor–consequence learning. This can explain relatively superficial differences in similar cuisines: pumpkin as a sweet (USA) rather than a savoury (UK) vegetable. But flavor principles also have deeper implications. Combinations of different ingredients provide a characteristic flavor that unites foods within a culture, and identifies them as originating from that culture (Rozin & Rozin, 1981). Such a characteristic and preferred combination of flavorings may also assist in ensuring that a culture can be flexible in the range of foods that it consumes, thus maintaining dietary variety and an adequate intake of nutrients. A problem with incorporating new foods into diets lies in our tendency to reject those foods that are unfamiliar (food neophobia). Since food neophobia is largely based on fear of a food tasting unpleasant, using a familiar combination of flavors is a way to address this problem. This explains how whole cultures can successfully incorporate new foods, if for example a staple becomes unavailable. The addition of already acceptable and familiar flavor principles has also been proposed as a mechanism through which children (and perhaps adults as well) can develop preferences for new foods that might otherwise be rejected (Stallberg-White & Pliner, 1999; Pliner & Stallberg-White, 2000).

A special case of flavor–consequence learning is the so-called "taste aversion." Experiencing a flavor followed by nausea commonly results in a long-lasting and intense dislike (aversion) to that flavor. Survey research suggests that many North Americans have one or more aversions, many of these being formed on the basis of a single flavor–nausea pairing (Logue et al., 1981). Humans also experience taste aversions that are essentially cognitive in nature (Batsell & Brown, 1998). That is, they involve associations between foods and either feelings of disgust or negative information about the food. One other difference between cognitive and nausea-based aversions is that, for most of us, nausea will reliably induce an aversion to a recently experienced flavor. In contrast, a characteristic of cognitive aversions is that the same experience of a food in a particular context can be disgusting to one person and not to another. Most recently, a form of flavor aversion arising from pairing a flavor with exercise, but without nausea, has also been reported in humans (Havermans, Salvy, & Jansen, 2009). The most plausible explanation for this so far is that, since exercise results in a net loss of energy (calories), the body may be seeking to protect itself by avoiding any signals (flavors) that might predict such energy loss.

Not all foods are equally vulnerable to the development of taste aversions (Midkiff & Bernstein, 1985) and this variation appears to relate to the distinctiveness of the flavor, since without being able to experience the flavor of the food, the likelihood of a taste aversion is the same regardless of the food's composition (Bernstein, 1999). Equally, taste aversions are much more likely to occur when illness follows the experience of a relatively novel flavor (Bernstein, 1999). Both of these findings point to the importance of the flavor as a unique predictor of the likelihood of illness.

3 Models of evaluative learning

There is still debate centered on how evaluative learning takes place (Hofmann, De Houwer, Perugini, Baeyens, & Crombez, 2010). Most accounts place emphasis on the formation of associative links in memory between the novel stimulus (sometimes referred to as the CS, as in Pavlovian conditioning) and the already liked/disliked stimulus (the Pavlovian US). Once these links have been established, experiencing the CS alone then elicits the affective valence of the US, resulting in reports of changed liking. One such model of evaluative learning explains that the two stimuli (the CS and US) are actually encoded as a compound (also known as a *configural*) stimulus. That is, the CS and US are viewed as simply different components subsumed within the same stimulus. This explanation has been used to account for the sensory integration that occurs when a novel odor is repeatedly paired with a taste, resulting in the transfer of the taste property to the odor (eg, the sweet smell of vanilla) (Stevenson, Boakes, & Prescott, 1998). In this case, each experience of an odor invokes a search of memory for prior encounters with that odor. If, in the initial experience of the odor, it was paired with a taste, a crossmodal (taste plus smell) configural stimulus—that is, a flavor—is encoded in memory. Subsequently sniffing the odor alone will evoke the most similar odor memory—the flavor—that will include both the odor and the taste component. Thus, for example, sniffing caramel odor activates memorial representations of caramel flavors, which includes a sweet taste component. This results either in perceptions of smelled taste properties such as sweetness or, in the case of a mixture, a perceptual combination of the memorial odor representation with the physically present taste in solution—thus explaining the ability of sweet-smelling or salty-smelling odors to enhance their respective tastes (Stevenson et al., 1999; Nasri et al., 2011).

This model can be equally well applied to hedonic changes because the valence is an intrinsic property (like sweetness or bitterness) of the taste US. Even if the associative pairings that lead to hedonic changes are more or less identical to those leading to perceptual changes, these two consequences of pairing can be shown to be independent from one another (Yeomans & Mobini, 2006). The crucial difference here lies in the responsiveness of evaluative odor–taste learning to factors that alter the hedonic value of the US. Hence, while sweetness is generally a positive taste, there are those who find medium to strong sweetness as relatively unpalatable and for such individuals, evaluative conditioning is less evident. Similarly, those who are less sensitive to bitterness show decreased dislike for bitter-paired odors (Yeomans et al., 2009). Neither factor, however, influences the fact that the sweet- or bitter-paired odors become sweet or bitter smelling, respectively, after the same pairing.

The extent to which holistic/configural models of evaluative conditioning can explain the processes underlying flavor–consequence learning is unclear. In these cases, rather than being encoded as a configural stimulus, the flavor and its metabolic consequences may be linked as a typical Pavlovian CS–US pair in which the CS acts to predict the occurrence of the US. Thus, the change in valence of the CS may be more analogous to the fear or appetitive responses seen in animal studies when a CS predicts a shock or food presentation, respectively. Another model of evaluative conditioning—the referential model—suggests that the CS leads to either expectancy of US or promotes awareness of US properties. The valence of the CS in this model reflects the average valence of stimuli with which it has been previously paired (Baeyens & De Houwer, 1995). One major issue with this explanation, at least in the case of taste aversions, is that they can be shown to occur independently of conscious awareness of any CS–US association.

4 Mechanisms of binding of odors and tastes

The development of flavor perception has been explained in terms of the sensory qualities (tastes, retronasal odors, tactile qualities) occurring together in the mouth being bound into a single perception, essentially a perceived property of the food (Small & Prescott, 2005). In neuronal terms, these different sensory inputs are part of one sensory receptive field (Small, 2015). The idea of flavors as the outcome of sensory integration has received support from a variety of sources, including cellular recordings in animals (Rolls & Bayliss, 1994), fMRI studies of neural activation in humans (Small et al., 2004), as well as psychophysical studies of odor–taste interactions following repeated co-exposure mentioned above. In each case, these studies support the idea of a discrete flavor sense in which the flavor is processed as a unique entity.

The binding of different sensory input into a flavor relies on association in time and place (the mouth). The temporal association of food qualities is not problematic. However, while both gustatory and tactile receptors are spatially located in the mouth, olfactory receptors are not. The question then arises of how odors become bound to taste and touch.

Central to flavor binding is the *olfactory location illusion*, in which the odor components of a food appear to originate in the mouth (Rozin, 1982). The extent to which either concurrent taste or touch, or both, is chiefly responsible for the capture and referral of olfactory information to the oral cavity is not known. However, the somatosensory system is more strongly implicated since it provides more detailed spatial information than does taste (Lim & Green, 2008). Thus, tactile stimulation is able to capture taste, presumably by providing superior spatial information and enhancing localization (Lim & Green, 2008; Todrank & Bartoshuk, 1991). Tactile information may therefore have an important role in binding tastes, perhaps together with odors, both to one another and to a physical stimulus such as a food. Moreover, in neuroimaging studies, odors presented retronasally (via the mouth) have been shown to activate the mouth area of primary somatosensory cortex (where oral touch information is processed), whereas the same odors presented via the nose do not (Small, Gerber, Mak, & Hummel, 2005). This distinction, which occurs even when participants are unaware of the route of stimulation (achieved via cannulae inserted to deliver odors either at the external nares or the nasopharynx), suggests a likely neural correlate of the binding process, and supports the idea that somatosensory input is the underlying mechanism.

The perceptual outcome of this process has been described in terms of the development of congruency between the odor and the taste (Schifferstein & Verlegh, 1996; Frank, van der Klaauw, & Schifferstein, 1993). That is, the binding of the odor and taste makes them more alike, a result of the transfer of properties—both hedonic and perceptual—from the taste to the odor. This is the reason why congruency is essential for an odor such as caramel to enhance a sweet taste—they share a common property, sweetness of both taste and smell. Recent studies suggest that, once established, odor–taste congruency may significantly enhance the referral of odors to the mouth. Hence, sweetness was the only taste effective in inducing a sense of vanilla odor being localized to the tongue; similarly, saltiness was the only taste to increase referral to the tongue of soy sauce odor (Lim & Johnson, 2011). Subsequent research has suggested that the taste-congruent odors can be referred to the mouth even in the absence of an accompanying taste (Lim & Johnson, 2012).

In the same way that our hedonic (and sometimes perceptual) responses to foods are influenced by expectations, sensory binding in flavor is subject to "top-down" influences. It can be shown, for example, that directing attention toward the discrete sensory elements of a flavor—the sweet taste and the vanilla odor—rather than the overall compound (vanilla) flavor during learning inhibits integration of the various sensory elements and does not produce the hedonic (or perceptual) changes in the odor (Prescott, Johnstone, & Francis, 2004; Prescott & Murphy, 2009). Similarly, highlighting the different taste, odor, and tactile properties of a flavor during an evaluation of how much the flavor is liked reduces the magnitude of those liking ratings (Prescott, Lee, & Kim, 2011). This is exactly what trained sensory panels are asked to do when analytically evaluating foods or other consumer goods, and these data constitute a good reason why such panels can not reflect the hedonic response of consumers, who respond in a much more synthetic manner.

5 Conclusions: The adaptive significance of flavors

If it is the case that odors, tastes, and tactile stimuli are encoded together, the question is what is the adaptive significance of this? Why, for example, do discrete neural circuits across a range of brain structures represent flavors rather than simply odors and tastes separately? One possibility lies in the complementarity of tastes and odors in identifying foods. At birth (or in the case of salt, shortly thereafter), we are hedonically inflexible when it comes to basic tastes—sweet, sour, salty, bitter, and umami. Our likes and dislikes appear pre-set as an adaptive mechanism to ensure the intake of nutrients (the sweetness of carbohydrates; the saltiness of sodium; the umami taste of proteins) and avoid toxins or otherwise harmful substances that are frequently bitter or sour. In contrast, there is little evidence that odor preferences are other than the result of experience and it is this plasticity and flexibility that allows an increased range of food sources which, in turn, promotes survival.

Through learning, the integration of odors with tastes attaches additional meaning to the odor that is primarily hedonic. A flavor that is not bitter, not too sour, and quite sweet provides pleasure. In other words, we are motivated to consume it because of its prior associations with the pleasure of sweet taste and the calories that the sweetness, and subsequently, the odor signals. Moreover, the perceptual qualities that are transferred also have hedonic consequences—sweet-smelling odors are pleasant and this quality may in itself motivate consumption even if we cannot identify the actual odor or its source. There is even evidence suggesting that such odors activate the same reward pathways as tasted sweetness (Prescott & Wilkie, 2007). Conversely, a bitter or sour odor is likely to elicit rejection, especially if we cannot recognize the odor. As such, both hedonic and perceptual changes to odors may help compensate for the fact that odor identification is particularly difficult even for common foods (Lawless & Engen, 1977). A pleasant odor or flavor is, by definition, one that has been previously experienced with positive taste or post-ingestive consequences and hence is more than likely safe to consume.

Research on flavor perception has paralleled to some extent research on multisensory perception more generally. It is increasingly recognized that our brain functions as an organ that integrates sensory signals to optimize the representation of information, particularly about those events and objects important to our survival. Flavor perception is no different, with one important exception. This is that, by binding odors with tastes, which have intrinsic hedonic properties, flavors too become intrinsically hedonic. Emphasizing the importance of individual sensory qualities within flavors therefore risks losing their adaptive significance. Certainly when we consume foods, flavors are far more important than the sum of their parts.

References

Baeyens, F., Wrzesniewski, A., De Houwer, J., & Eelen, P. (1996). Toilet rooms, body massages, and smells: Two field studies on human evaluative odor conditioning. *Current Psychology*, *15*(1), 77–96.

Baeyens, F., Vansteenwegen, D., & De Houwer, J. (1996). Observational conditioning of food valence in humans. *Appetite*, *27*, 235–250.

Baeyens, F., Crombez, G., Van De Bergh, O., & Eelen, P. (1988). Once in contact always in contact: Evaluative conditioning is resistant to extinction. *Advances in Behaviour Research and Therapy, 10*, 179–199.

Baeyens, F., & De Houwer, J. (1995). Evaluative conditioning is a qualitatively distinct form of classical conditoning: A reply to Davey (1994). *Behaviour Research and Therapy, 33*(7), 825–831.

Baeyens, F., Eelen, P., Van den Bergh, O., & Crombez, G. (1990). Flavor-flavor and color-flavor conditioning in humans. *Learning and Motivation, 21*, 434–455.

Batsell, W. R., & Brown, A. S. (1998). Human flavor-aversion learning: A comparison of traditional aversions and cognitive aversions. *Learning and Motivation, 29*, 383–396.

Berlyne, D. E. (1970). Novelty, complexity, and hedonic value. *Perception & Psychophysics, 8*, 279–286.

Bernstein, I. L. (1999). Taste aversion learning: A contemporary perspective. *Nutrition, 15*(3), 229–234.

Birch, L. L., & Marlin, D. W. (1982). I don't like it; I never tried it: Effects of exposure on two-year-old children's food preferences. *Appetite, 3*, 353–360.

Capaldi, E. D., & Privitera, G. J. (2007). Flavor-nutrient learning independent of flavor-taste learning with college students. *Appetite, 49*(3), 712–715.

Castro, D. C., & Berridge, K. C. (2014). Advances in the neurobiological bases for food "liking" versus "wanting.". *Physiology & Behavior, 136*, 22–30.

Chambers, L., Mobini, S., & Yeomans, M. R. (2007). Caffeine deprivation state modulates expression of acquired liking for caffeine-paired flavours. *Quarterly Journal of Experimental Psychology, 60*(10), 1356–1366.

Coppin, G., Delplanque, S., Cayeux, I., Porcherot, C., & Sander, D. (2010). I'm no longer torn after choice: How explicit choices implicitly shape preferences of odors. *Psychological Science, 21*(4), 489–493.

Dermiki, M., Prescott, J., Sargent, L. J., Willway, J., Gosney, M. A., & Methven, L. (2015). Novel flavours paired with glutamate condition increased intake in older adults in the absence of changes in liking. *Appetite, 90*, 108–113.

Ferriday, D., & Brunstrom, J. M. (2008). How does food-cue exposure lead to larger meal sizes? *British Journal of Nutrition, 100*, 1325–1332.

Frank, R. A., van der Klaauw, N. J., & Schifferstein, H. N. J. (1993). Both perceptual and conceptual factors influence taste-odor and taste-taste interactions. *Perception & Psychophysics, 54*(3), 343–354.

Garcia-Burgos, D., & Zamora, M. C. (2015). Exploring the hedonic and incentive properties in preferences for bitter foods via self-reports, facial expressions and instrumental behaviours. *Food Quality and Preference, 39*, 73–81.

Havermans, R. C., Salvy, S.-J., & Jansen, A. (2009). Single-trial exercise-induced taste and odor aversion learning in humans. *Appetite, 53*, 442–445.

Hofmann, W., van Koningsbruggen, G. M., Stroebe, W., Ramanathan, S., & Aarts, H. (2010). As pleasure unfolds: Hedonic responses to tempting food. *Psychological Science, 21*(12), 1863–1870.

Hofmann, W., De Houwer, J., Perugini, M., Baeyens, F., & Crombez, G. (2010). Evaluative conditioning in humans: A meta-analysis. *Psychological Bulletin, 136*(3), 390–421.

Johnson, S. L., McPhee, L., & Birch, L. L. (1991). Conditioned preferences: Young children prefer flavors associated with high dietary fat. *Physiology & Behavior, 50*, 1245–1251.

Kern, D. L., McPhee, L., Fisher, J., Johnson, S., & Birch, L. L. (1993). The postingestive consequences of fat condition preferences for flavors associated with high dietary fat. *Physiology & Behavior, 54*, 71–76.

Khan, R. M., Luk, C.-H., Flinker, A., Aggarwal, A., Lapid, H., Haddad, R., et al. (2007). Predicting odor pleasantness from odorant structure: Pleasantness as a reflection of the physical world. *The Journal of Neuroscience, 27*(37), 10015–10023.

Kim, J.-Y., Prescott, J., & Kim, K.-O. (2014). Patterns of sweet liking in sucrose solutions and beverages. *Food Quality and Preference, 36*, 96–103.

Lawless, H., & Engen, T. (1977). Associations to odors: Interference, mnemonics, and verbal labeling. *Journal of Experimental Psychology Human Learning and Memory, 3*(1), 52–59.

Leotti, L. A., & Delgado, M. R. (2011). The inherent reward of choice. *Psychological Science, 22*(10), 1310–1318.

Levey, A. B., & Martin, I. (1975). Classical conditioning of human "evaluative" responses. *Behaviour Research and Therapy, 13*, 221–226.

Lim, J., & Green, B. G. (2008). Tactile interaction with taste localization: Influence of gustatory quality and intensity. *Chemical Senses, 33*, 137–143.

Lim, J., & Johnson, M. B. (2011). Potential mechanisms of retronasal odor referral to the mouth. *Chemical Senses, 36*, 283–289.

Lim, J., & Johnson, M. B. (2012). The role of congruency in retronasal odor referral to the mouth. *Chemical Senses, 37*, 515–521.

Logue, A. W., Ophir, I., & Strauss, K. E. (1981). The acquisition of taste aversions in humans. *Behaviour Research and Therapy, 19*, 319–333.

Mennella, J. A., Jagnow, C. P., & Beauchamp, G. K. (2001). Prenatal and postnatal flavor learning by human infants. *Pediatrics, 107*(6), E88.

Methven, L., Langreney, E., & Prescott, J. (2012). Changes in liking for a no added salt soup as a function of exposure. *Food Quality and Preference, 26*, 135–140.

Midkiff, E. E., & Bernstein, I. L. (1985). Targets of learned food aversions in humans. *Physiology & Behavior, 34*, 839–841.

Mobini, S., Chambers, L. C., & Yeomans, M. R. (2007). Interactive effects of flavour-flavour and flavour-consequence learning in development of liking for sweet-paired flavours in humans. *Appetite, 48*, 20–28.

Monahan, J. L., Murphy, S. T., & Zajonc, R. B. (2000). Subliminal mere exposure: Specific, general, and diffuse effects. *Psychological Science, 11*, 462–466.

Myers, K. P., & Sclafani, A. (2001). Conditioned enhancement of flavor evaluation reinforced by intragastric glucose. I. Intake acceptance and preference analysis. *Physiology & Behavior, 74*, 481–493.

Nasri, N., Beno, N., Septier, C., Salles, C., & Thomas-Danguin, T. (2011). Cross-modal interactions between taste and smell: Odour-induced saltiness enhancement depends on salt level. *Food Quality and Preference, 22*(7), 678–682.

Nisbett, R. E. (1968). Taste, deprivation, and weight determinants of eating behavior. *Journal of Experimental Social Psychology, 10*(2), 107–116.

Pliner, P., & Stallberg-White, C. (2000). "Pass the ketchup, please": Familiar flavors increase children's willingness to taste novel foods. *Appetite, 34*, 95–103.

Pliner, P., Pelchat, M., & Grabski, M. (1993). Reduction of neophobia in humans by exposure to novel foods. *Appetite, 20*, 111–123.

Pliner, P., & Pelchat, M. L. (1991). Neophobia in humans and the special status of foods of animal origin. *Appetite, 16*(3), 205–218.

Prescott, J. (2004). Effects of added glutamate on liking for novel food flavors. *Appetite, 42*(2), 143–150.

Prescott, J., Kim, H., & Kim, K.-O. (2008). Cognitive mediation of hedonic changes to odors following exposure. *Chemosensory Perception, 1*, 2–8.

Prescott, J., & Wilkie, J. (2007). Pain tolerance selectively increased by a sweet-smelling odor. *Psychological Science, 18*(4), 308–311.

Prescott, J., & Murphy, S. (2009). Inhibition of evaluative and perceptual odour-taste learning by attention to the stimulus elements. *Quarterly Journal of Experimental Psychology, 62,* 2133–2140.

Prescott, J., Lee, S. M., & Kim, K.-O. (2011). Analytic approaches to evaluation modify hedonic responses. *Food Quality and Preference, 22*(4), 391–393.

Prescott, J., Johnstone, V., & Francis, J. (2004). Odor/taste interactions: Effects of different attentional strategies during exposure. *Chemical Senses, 29,* 331–340.

Raymond, J. E., Fenske, M. J., & Tavassoli, N. T. (2003). Selective attention determines emotional responses to novel visual stimuli. *Psychological Science, 14*(6), 537–542.

Reber, R., Winkielman, P., & Schwarz, N. (1998). Effects of perceptual fluency on affective judgments. *Psychological Science, 9*(1), 45–48.

Rolls, E. T., & Rolls, J. H. (1997). Olfactory sensory-specific satiety in humans. *Physiology & Behavior, 61,* 461–473.

Rolls, E. T., & Bayliss, L. L. (1994). Gustatory, olfactory, and visual convergence within the primate orbitofrontal cortex. *The Journal of Neurosciences, 14*(9), 5437–5452.

Rozin, E., & Rozin, P. (1981). Culinary themes and variations. *Natural History, 90*(2), 6–14.

Rozin, P. (1982). "Taste-smell confusions" and the duality of the olfactory sense. *Perception & Psychophysics, 31*(4), 397–401.

Schifferstein, H. N. J., & Verlegh, P. W. J. (1996). The role of congruency and pleasantness in odor-induced taste enhancement. *Acta Psychologica, 94,* 87–105.

Small, D.M. 2015 *Deconstructing flavor,* Paper presented at the 11th Pangborn Symposium: Gothenburg, Sweden.

Small, D. M., & Prescott, J. (2005). Odor/taste integration and the perception of flavor. *Experimental Brain Research, 166,* 345–357.

Small, D. M., Voss, J., Mak, E., Simmons, K. B., Parrish, T., & Gitelman, D. (2004). Experience-dependent neural integration of taste and smell in the human brain. *Journal of Neurophysiology, 92,* 1892–1903.

Small, D. M., Gerber, J. C., Mak, Y. E., & Hummel, T. (2005). Differential neural responses evoked by orthonasal versus retronasal odorant perception in humans. *Neuron, 47,* 593–605.

Stallberg-White, C., & Pliner, P. (1999). The effects of flavor principles on willingness to taste novel foods. *Appetite, 33,* 209–221.

Stevenson, R. J., Prescott, J., & Boakes, R. A. (1999). Confusing tastes and smells: How odors can influence the perception of sweet and sour tastes. *Chemical Senses, 24,* 627–635.

Stevenson, R. J., Boakes, R. A., & Prescott, J. (1998). Changes in odor sweetness resulting from implicit learning of a simultaneous odor-sweetness association: An example of learned synesthesia. *Learning and Motivation, 29,* 113–132.

Todrank, J., & Bartoshuk, L. M. (1991). A taste illusion: Taste sensation localised by touch. *Physiology & Behavior, 50,* 1027–1031.

Troisi, J. D., & Gabriel, S. (2011). Chicken soup really is good for the soul: "Comfort food" fulfills the need to belong. *Psychological Science, 22*(6), 747–753.

Vuilleumier, P., & Huang, Y.-M. (2009). Emotional attention. Uncovering the mechanisms of affective biases in perception. *Current Directions in Psychological Science, 18*(3), 148–152.

Winkielman, P., & Berridge, K. C. (2003). Irrational wanting and subrational liking: How rudimentary motivational and affective processes shape preferences and choices. *Political Psychology, 24*(4), 657–679.

Yeomans, M. R., Prescott, J., & Gould, N. J. (2009). Acquired hedonic and sensory characteristics of odours: Influence of sweet liker and propylthiouracil taster status. *Quarterly Journal of Experimental Psychology, 62*(8), 1648–1664.

Yeomans, M. R., Leitch, M., Gould, N. J., & Mobini, S. (2008). Differential hedonic, sensory and behavioral changes associated with flavor-nutrient and flavor-flavor learning. *Physiology & Behavior, 93*, 798–806.

Yeomans, M. R., Gould, N., Mobini, S., & Prescott, J. (2008). Acquired flavor acceptance and intake facilitated by monosodium glutamate in humans. *Physiology & Behavior, 93*, 958–966.

Yeomans, M. R., & Mobini, S. (2006). Hunger alters the expression of acquired hedonic but not sensory qualities of food-paired odors in humans. *Journal of Experimental Psychology Animal Behavior Processes, 32*(4), 460–466.

Yeomans, M. R., Mobini, S., Elliman, T. D., Walker, H. C., & Stevenson, R. J. (2006). Hedonic and sensory characteristics of odors conditioned by pairing with tastants in humans. *Journal of Experimental Psychology Animal Behavior Processes, 32*(3), 215–228.

Zajonc, R. B. (1968). Attitudinal effects of mere exposure. *Journal of Experimental Social Psychology, 9*(2), 1–27.

Zajonc, R. B. (2001). Mere exposure: A gateway to the subliminal. *Current Directions in Psychological Science, 10*(6), 224–228.

Zellner, D. A., Rozin, P., Aron, M., & Kulish, C. (1983). Conditioned enhancement of human's liking for flavor by pairing with sweetness. *Learning and Motivation, 14*, 338–350.

Flavor Memory

Jos Mojet[1] and Ep Köster[2]
[1]Food & Biobased Research, Wageningen University and Research Centre, Wageningen, The Netherlands, [2]Helmholtz Institute, Psychology, University of Utrecht, Utrecht, The Netherlands

1 Introduction

Defined in a narrow sense, the flavor of a food refers to the combination of olfactory and taste impressions created when eating. In a broader sense, it also includes a number of other sensory experiences such as: (1) pain and astringency as a result of trigeminal stimulation in mouth, larynx, and nose; (2) texture created by both kinesthetic and touch sensations in the mouth and throat; and (3) temperature both in the sense of actual temperature of the food (Moskowitz, 1972; Talavera, Ninomya, Winkel, Voets, & Nilius, 2007) and in the sense of chemesthesis, such as the cooling effects created by menthol (Cliff & Green, 1994). These sensory experiences all interact with olfaction and taste and will most certainly be strongly linked to them in memory. Therefore, we propose to use the broader definition of flavor in this chapter, notwithstanding the fact that the perception of the other sensations is treated in several of the other chapters in this volume. Furthermore, we propose to treat flavor memory in its naturally occurring, incidentally learned, and implicit form (ie, based on mere and often even unconscious feelings that the flavor is familiar, or has been experienced before), even if the verification of the memory is carried out quite explicitly by submitting people to recognition experiments. Thus, we will not devote much attention to those experiments in which participants, during the initial exposure to the food, are made aware of the fact that their memory is going to be tested later. Furthermore, we will also not direct our attention to explicit flavor recognition and identification which, although important to professional flavorists, occur in normal life almost exclusively in gastronomic settings, while during everyday eating we devote little attention to the food as long as it meets our expectations (Köster, Møller, & Mojet, 2014). How explicitly people deal with their food determines their appreciation, their choice, and their intake in several ways (Higgs, 2009; Piqueras-Fiszman & Spence, 2015; see also below on the role of expectations).

Prescott (2012) provides an excellent overview of the role of psychological attitudes (synthetic vs analytic) in the interactions and the integration of olfaction and taste in the development of flavor perception. He discusses the many instances in which it has been shown that odors and tastes when paired together acquire each other's sensory qualities through association (Frank & Byram, 1988; Frank, Ducheny, & Mize, 1989; Stevenson, Boakes, & Prescott, 1998; Stevenson, Prescott, & Boakes, 1995, 1999). Across a series of studies, Prescott and his colleagues demonstrated that analytical attitudes evoked by experimental task demands prevented synthesis of smell and taste (Le Berre et al., 2008; Prescott, Johnstone, & Francis, 2004; Prescott, Lee, & Kim, 2011; Prescott & Murphy, 2009). Thus, in his recent overview, Prescott came to the conclusion that "If it is the case that odours and tastes are automatically co-encoded as a flavour in the absence of task demands that focus attention on the elements, then experimental designs in which the odour and taste elements appear together without such an attentional strategy are likely to predispose towards synthesis" (Prescott, 2012, p. 556).

Since it is our intention in this chapter to describe the workings of flavor memory in everyday life, we will not discuss many of the laboratory experiments in which such explicit task demands are evoked.

On the other hand, we will devote special attention to implicit learning and memory as the ecologically most frequent form of flavor memory. By its strong links with remembered situations and the accompanying emotions it has an important influence on flavor perception through the often unconscious expectations against which those perceptions are matched. Recently, Piqueras-Fiszman and Spence (2015) have extensively reviewed the role of expectations based on product-intrinsic or product-extrinsic food cues or both and discussed the theories that explain the effects that arise when the actual product perception does not match the expectations. Like in their misfit theory of spontaneous conscious odor perception, Köster, Møller et al. (2014) stress the essential role of flavor memory as a warning system against the dangers of poisoning and the fact that discrepancies between expectations and actual perceptions ("misfits") are therefore more often consciously noted than "fits" (ie, those cases in which all is in accordance with expectations).

Much of the traditional research on memory has neglected these aspects, or at least not fully understood their significance for understanding the workings of memory. As we will argue below, this has often led to artificial and false conclusions about the function of flavor memory. Thus, it has been shown that food memory is based on change and novelty detection (Morin-Audebrand, Mojet, Chabanet, Issanchou, Møller, Köster & Sulmont-Rossé, 2012) rather than on precise recollection of the foods that were eaten earlier, as is often assumed in analogy to memory in vision and audition. The consequences of this for the validity of measuring techniques in memory research and data treatments such as forced choice methods and signal detection measures are also critically discussed. Finally, we will devote some attention to the development of food and flavor memory in the different life phases and in relation to disease and loss of sensitivity.

2 Different forms of memory

Memory is involved in all behaviors that are dependent on, or are modified by, previous experiences. Memory may, or may not, be based on awareness of the relationship with such previous experiences on the part of the subject involved in the behavior. If it is not, we speak of implicit learning (no awareness of the learning) and implicit memory (no awareness of the fact that we have memory of it). If it is based on awareness, we speak of explicit learning and memory. In normal everyday life, most of our sensory learning and memory is implicit and functions without being consciously noted (Köster & Mojet, in press). We automatically execute most of our practical acts (opening doors, buttoning shirts, etc.) without any explicit realization of the memory and learning involved, unless something is wrong (the door sticks, a button is missing). In relation to food, the same is true: we eat most foods without much awareness of the way in which we learned to eat and like them or of the memory for their intrinsic flavor, unless something about them is different from what was expected. Thus, it could be shown (see Morin-Audebrand et al., 2012, for an overview), that memory for implicitly experienced foods was not based on the recollection of an internal representation of the earlier-eaten food, but on the detection of novelty and change. Thus, recognition hit rates were never significantly higher than 50% (ie, based on chance guessing). On the other hand, the correct rejection rates of distractor variants that deviated to only a small and hardly detectable degree from the original food were always much higher and were responsible for the significant recognition indices (d'). This proved to be the case for texture (Mojet & Köster, 2002, 2005), for taste (Köster, Prescott, & Köster, 2004), for flavor in the narrow sense (Møller, Mojet, & Köster, 2007; Sulmont-Rossé, Møller, Issanchou, & Köster, 2008), for sweetness, thickness, and flavor (Laureati et al., 2008; Laureati, Pagliarini, Mojet, & Köster, 2011; Morin-Audebrand, Sulmont-Rossé, & Issanchou, 2006; Morin-Audebrand et al., 2009).

These findings contributed to the formulation of the misfit theory of spontaneous conscious odor perception (MITSCOP) (Köster, Møller et al., 2014). According to this theory, in olfaction (and probably also in the other senses involved in flavor perception with the possible exception of pain), memory is based on whether the same pattern of stimulation has occurred previously in a given situation or not. Thus, instead of the precise recollection of an internal representation of the earlier perceived stimuli, it is novelty or change detection that plays the major role in olfactory memory. This is in accordance with the first step in olfactory perception and memory according to the mnemonic theory of Stevenson and Boakes (2003). These authors hypothesize that the incoming pattern of glomerular responses is immediately and simultaneously compared with all previous patterns encountered and classified as new (and therefore possibly dangerous) or old (and therefore harmless, ie, survived before). The pattern is then immediately linked to the situation in which it occurs and this association is stored in the memory. However, according to Köster, Møller et al.'s (2014) misfit theory, it is not stored as a mental representation of the odor itself, or as a so-called odor object that can later be retrieved from memory, as Stevenson and Boakes and most other authors working on olfactory memory suppose. Although such ideational odor objects may indeed be formed by the association of different flavor components from olfaction and taste in explicit odor perception, they do not seem to be used as functional and retrievable representations in implicitly learned odor memory and odor recognition. As it should be in an efficient warning system, which is located at the gateways to vital functions such as breathing and food intake, immediate detection of unusual odors and/or food is much more important than the recognition and identification of the flavor features. Known stimuli, which have been survived before, are not dangerous and can be neglected as harmless. Similar arguments have been put forward by Boesveldt, Frasnelli, Gordon, and Lündstrom (2010) who also stress (p. 316) that "In the olfactory system the classification of odors into the food or nonfood category (by the detection of possibly bad odors) is of eminent importance." Thus, the misfit theory explains why we do not devote spontaneous attention to most known odors, but are immediately on our toes when unexpected novelties or changes take place. This seems a much better explanation of human inattentiveness to smells than the theory proposed by Sela and Sobel (2010), according to which inattentiveness is due to "change blindness." As pointed out by Köster, Møller et al. (2014), the misfit theory can explain many more olfactory phenomena than the theory put forward by Sela and Sobel, such as: the early bump in odor-evoked autographical memory (Chu & Downes, 2000), the absence of early and late memory effects in serial odor learning (Miles & Hodder, 2005), and the resistance to counter-conditioning and to extinction of olfaction–taste combinations in the formation of so-called "odour objects" (Stevenson et al., 1995, 1998, 1999; Stevenson, Boakes, & Wilson, 2000a, 2000b). Note here also that the arguments for the change blindness in the change-blindness theory are based on speculation and supposed analogies to vision, instead of on experimental evidence (Köster, 2002; Köster, Möller, Mojet, & Schaal, in preparation; Møller, Piper, Hartvig, & Köster, 2009).

The fact that novelty and change detection seem to be essential, at least in odor memory, but probably also in the texture and temperature memories involved in flavor perception, seems to be related to the predominant warning function of these senses. This is probably also the reason why the sensations evoked by these types of stimuli are so immediately and strongly associated with the situation in which they are experienced and, as a result, usually evoke this previous situation much more rapidly than the name of the odor or flavor. Retrieval takes about 4 s for the names of rather well-known odors (Cameron, Anderson, & Møller, 2015a; Köster, Van der Stelt et al., 2014). Of course, such a timeframe (of 4 s) would be fatal as a warning system for possibly poisonous stimuli such as odors and flavors that are already in our body when we detect them. Immediate novelty detection is therefore essential for such "near" senses while identification is not. The same may hold for another "near" sense, skin sensitivity, which shares with olfaction the fact that it shows complete adaptation (ie, a complete loss of sensitivity to monotonous stimulation). This demonstrates that continued perception of known impressions is (or quickly becomes) uninteresting and perhaps even unwanted, since it might hamper the alertness for the detection of new and possibly dangerous stimuli. Just as we quickly loose sensitivity to continuous smells, we do not feel the pressure of our garments, but remain sensitive to the slightest unexpected touch. This well-known phenomenon is also closely linked to expectations based on familiarity or habituation and may even be provoked by them in some cases as an interesting experiment showed in monkeys (Graziano, Alisharan, Hu, & Gross, 2002).

In the "far" (or distal) senses of vision and audition, where spatial orientation is important, there is no complete adaptation, and even if there is some reduction of sensitivity under strong light or conditions of loud auditory stimulation, the general contact with the spatial stimulus constellation remains intact. Furthermore, it has been shown (Møller, Köster, Dijkman, De Wijk, & Mojet, 2012) in explicit same–different experiments with odors that "same" judgments take much longer (300 ms) than "different" judgments, whereas in vision the reverse is true ("same" 50 ms faster than "different"; Luce, 1986; Posner, 1986). This also illustrates the importance of quick change detection in olfaction. Thus, it becomes clear that memory serves different functions in "far" and "near" senses and works differently. As pointed out earlier (Köster, 2005) in odor, flavor, or tactile memory, there is also less need for identification, because there is only one type of reaction that can follow the perception of an unexpected and possibly dangerous stimulus (respectively: holding one's breath, spitting the food out, and withdrawing), whereas in seeing and hearing identification plays an important role in the choice of an adequate reaction (hiding, avoiding, fleeing, removing, reacting aggressively, or smiling submissively, etc.). In most cases in vision and audition, there is more time to identify any potentially menacing stimulus (the stimulus often being still at a distance from the body when first perceived), which can help to find the most adequate reaction to it. Probably in relation to this fact, identification is a much faster process in these senses. In fact, even uncommon visual objects are recognized and identified in under a second (Cameron et al., 2015a). On the other hand, and in contrast to odor memory, with its strong situational connections, in vision it often takes rather long to remember exactly where and when one experienced this or a very similar event before.

In odor memory, the immediate and very strong associative situation links are often even the best way to deduct the name and nature of the stimulus itself. Many great novelists (cf. Proust, see also below, Nabokov) have not only described that odors can evoke previous situations very easily and in great detail, but also noted that this effect is not reciprocal and that remembering situations in detail does not evoke the smell memories. Most people (about 70%, Richardson, 1994 and observation in the pre-experiments for the selection of imagers in Møller et al., 2012 and Köster, Van der Stelt et al., 2014) have no odor or flavor imagination and in the persons who have it, it is usually far less vivid than visual or auditory imagination. And since people who do not have it, nevertheless often have very good odor memory (Møller, Wulff, & Köster, 2004) and are no faster on odor naming tasks (Köster, Van der Stelt et al., 2014) than those who do not have it, this may be considered as another indication that odor memory does not depend on the retrieval of internal representations in the form of sensory images.

The situational reference raised by an odor may reassure us or may still raise our level of alertness depending on the memories of the consequences of the stimulus perception and its postingestional effects experienced at the earlier occasion. Furthermore, it is strongly influenced in its hedonic connotations by the surroundings during the earlier encounter with the stimulus and the company with whom the experience was shared (Piqueras-Fiszman & Spence, 2014). In a renewed acquaintance with the flavor, all these implicit memories play a role in the pleasure expected by the perceiver, and the conformity or discrepancy of the renewed experience with the expectations raised by these implicit memories are of great importance for the future liking of the product. Large discrepancies may have deleterious effects on the relationship between the stimulus and the perceiver.

When the memory of the product is much better than the actual experience upon retasting, the resulting disappointment may have a long-term disturbing influence on the appreciation of the flavor. Thus, unannounced changes in the flavor of products that people regularly use, will almost always be noted immediately and may have disastrous influences on the market success of the product and even on the trust in the brand. If, on the other hand, the renewed confrontation with the traditional product comes as a very pleasant surprise, it means that the memory for the experience may have been distorted either by bad reminiscences, such as unpleasant intestinal after-effects or situational memories that are not directly and consciously related to the taste of the previous consumption. This may reduce the chances of a product-repurchase. In the best-case scenario, the confrontation between expectation and the renewed perception leads to a moderately pleasant surprise. Products that show this will have a longer and healthier chance in the market.

The same might well be true for oral texture perception. That is, we immediately know whether a food is too hard or too soft, but we are usually unable to imagine the texture of foods, unless they are so edgy and sharp that pain becomes involved. Nevertheless our memory for texture is highly developed and plays probably the most important role in the recognition of previous experiences after olfactory memory (see also the next section).

Other product-related implicit memory sensations, that we are normally not aware of unless they are changed or lacking, vary from the cooling that plays an important role in food appreciation effects of certain flavors like mint to the pharyngeal warmth that accompanies the umami taste of genuine Parmesan. Only when one is confronted by fake Parmesan that seems to have the right flavor, but misses this warming aftereffect in the throat, one may realize how important a role this sensation plays in the appreciation of Parmesan. In general, the role of pharyngeal sensations is strongly underestimated and even neglected in most sensory research. Nevertheless, it plays a major role in the final appreciation of drinks like wines and beers and may be a reason for deception if it does not meet our expectations. Note that sip-and-spit protocols, as often usual in alcoholic drinks research, prevent attention to these sensations.

Thus, most of the workings of implicit memory remain really hidden in expectations until the moment that there is a mismatch between the expectations based on them and the perception in the actual situation. Sometimes memory may also become explicit when the perception fits the expectation perfectly and thus immediately revives the situation in which the stimulus was encountered earlier, but such cases are rare.

Piqueras-Fiszman and Spence (2015) have given an excellent overview of the sensory expectations that are based on product-extrinsic food cues, such as names and descriptions of product, labels about production- and growing-processes, or about health/ingredients and the ethnicity of the food. They cite a large number of papers dealing with the influence of these on the consumer's perception via expectations. All of these have an influence on the expectations with which people confront the actual tasting of products. In an earlier paper, Piqueras-Fiszman and Spence (2014) discuss the relative scarcity of papers on the influence of the color of the food itself on the variety or monotony of this factor on the pleasantness and intake of the meal (Geier, Wansink, & Rozin, 2012; Kahn & Wansink, 2004; Rolls, Rolls, Rowe, & Sweeney, 1981; Rolls, Rowe, & Rolls, 1982). They also discuss color-based sensory-specific satiety or the progressive boredom that arises with monotonous color stimulation and indicate that more research on the influence of color on the food intake of adults is needed in order to see whether it can play a role in food intake regulation.

Most of the examples of the influence of expectations stress the importance of implicit learning and memory in our everyday appreciation of food. In normal life, it is certainly the most common form, although some people consciously and explicitly involve their ideas about health and environmental considerations. Completely explicit flavor memory is exceptional in normal life and for most people occurs only when they try to remember a flavor in repurchase decisions. At such times, however, it is usually not very successful. On the other hand, explicit flavor memory plays an important role for those of us who are either professionally interested as flavorists and food developers or people who indulge in gastronomic experiences and wine tasting. They try to analyze and identify their sensations while eating or drinking. In gastronomic circles, this is a cultural game. But even gastronomists eat many things without explicit awareness of their expectations and intrinsic product-related memories, like most other people normally do.

3 Flavor learning and memory over lifetime

If flavor memory indeed depends on novelty and change detection, as stated above (Morin-Audebrand et al., 2012), and the fact of having perceived the flavor before or not is a decisive factor in the attention given to it at later occasions (as supposed by the misfit theory of Köster, Møller et al., 2014), it becomes clear that age and experience are very important factors in flavor memory. First encounters with odors and flavors must be very important and indeed they are, as shown by the fact that the memories evoked later in life by odors (and probably also by flavors) go further back than those evoked by visual and auditory stimuli (Chu & Downes, 2000, 2002; Willander & Larsson, 2006, 2007, 2008).

In fact, about 50% of autobiographical memories evoked by odor go back to the first 10 years of life and are probably related to first encounters with the odor, whereas for memories of the other stimuli (visual auditory and verbal) the period at the end of puberty, between 15 and 25 years of age, is the most important source of autobiographical memories.

Of course, certain special odors like the perfumes of our first love, or odors related to other particularly emotional experiences that occur later in life also evoke autobiographical events, but the normal everyday odors are usually not perceived at a later date and thus are not connected to special situations. However, when such odors occur unexpectedly in another situation (a mismatch with the expectations) they may suddenly carry people back to the original situation. Thus, Proust describes how the very specific flavor combination of a madeleine pastry dipped in Linden tea takes him back to the bedroom of his aunt on Sunday mornings before mass during the childhood vacations at her house in Combray. Such an experience is a good example of crossmodal imagery (as described extensively by Spence & Deroy, 2013) in which a stimulus presented in one sensory modality (flavor in this case) evokes imagery in another sensory modality. Olfactory and flavor stimuli seem to be especially able to evoke visual images of earlier emotional situations, provided that both the remembered occasion and the evoking stimulus are uncommon and rather unique. In the case of crossmodal mental imagery evoked by odor it seems particularly important that the odor is uncommon and not identifiable or easily describable. On the other hand, the speciality of the occasion may fix the preference for a special flavor or odor for a very long time and, in some cases, for a lifetime. Thus, the first sip of coffee in your grandmother's house, or the first sip of beer from your father's glass, often create very long-lasting preferences even if the nature of the odor or flavor remains indescribable. Other coffees or beers have a hard time beating these memories.

The importance of first encounters with flavors was also exemplified by Haller, Rummel, Henneberg, Pollmer, and Köster (1999) who showed that the first contact of newborn bottle-fed babies, who immediately after birth were fed with vanilla-flavored milk, had an unconscious, but permanent and positive influence on their preference for vanilla flavor. Furthermore, it has been shown that even prenatal exposure to flavors via the mothers' diet during pregnancy may influence the child's acceptance of them after birth (Hepper, 1995; Marlier, Schaal, & Soussignan, 1998; Mennella, Coren, Jagnow, & Beauchamp, 2001; Schaal & Durand, 2012; Schaal, Marlier, & Soussignan, 2000). Postnatal breast feeding also influences acceptance in relation to maternal food intake (Hausner, Bredie, Mølgaard, Petersen, & Møller, 2008; Hausner, Nicklaus, Issanchou, Mølgaard, & Møller, 2010). Such imprinting does not mean that the flavors are recognized as such, however. They simply lead to less refusal of the foods and possibly to feelings of familiarity. The variety in the mothers' food intake during this period may also lead to greater and more widespread acceptance of food during weaning in their children and may even have long-term effects on the variety in their later food choices (Nicklaus, Boggio, Chabanet, & Issanchou, 2005).

Old people are often carried back to exceptional childhood experiences when consuming traditionally prepared foods and drinks. Often they remember special happy situations like eating at their grandparent's house or at birthday parties of friends. The fact that many older people are not aware of the losses in their sense of smell (Nordin, Monsch, & Murphy, 1995) does not seem to affect these reminiscences very much. Even when they have completely lost their sense of smell, they may remember such occasions provided the texture of the food is right. It is therefore important to cook for them in old-fashioned ways even if that leads to loss of vitamins and is less desirable from a health point of view. Providing food that awakens memories is perhaps the best way to fight old-age anorexia. However, since flavor memory is based on novelty and change detection, deviations from the old flavor and texture are immediately noticed and may lead to severe disappointment. In that case novel and unknown foods are preferable. Such considerations have been included in the training of cooks in some cases and have led to better acceptance of institutional cooking. From a scientific point of view, these examples once again illustrate the importance of implicit flavor memory and its strong linkage to the eating situation. It is the latter that is remembered in full detail and not the flavor itself, as long as it fits our expectations. Flavors can, just like odors (Zucco, Aiello, Turuani, & Köster, 2012), be used to bring back memories of material that was learned while eating the same foods. Thus, Schab (1990) demonstrated that material that was explicitly learned while eating chocolate was better remembered during an examination when the same chocolate was eaten again. This once more illustrates the close relation of odors and flavors to the detailed memory of the situation in which they were encountered. Anderson, Berry, Morse, and Diotte (2005) also showed that shifting from one flavor to another may disrupt the accuracy of memory in a word-learning task by introducing more false positives, while the proportion of correct recognitions of the target words remained unchanged. This is an interesting finding since it might illustrate that even in its effects on a nonfood-related task, such as explicit word learning, flavor acts on the correct rejection rate through false-positive reduction, rather than on the recollection of the earlier perceived target words. This suggests that repeated or nonrepeated flavor experiences automatically may lead to a novelty-detection mode in memory function.

4 Influence of flavor memory on eating behavior

Everybody knows the feeling of hunger that can be elicited by passing a bakery and smelling the odor of freshly baked bread. Recently, Gaillet, Sulmont-Rossé, Issanchou, Chabanet, and Chambaron (2013) verified the specificity of such implicit odor memories. They invited people for a French (four-course) lunch and asked them to wait for a moment when they arrived because of a short delay in the setting of the tables. During this 10-min period the people were seated in a corridor where either an odor of melon (which in France is typically entrée-related) or an odor of pear (typically dessert-related) was diffused at a concentration low enough not to be consciously remarked by the waiting people. When, after the 10 min, the people entered the restaurant they could freely choose from the menu among appetizers varying from green salads to patés and among desserts varying from chocolate and vanilla puddings to fruits. Under the influence of the unnoticed odors, the melon-exposed people chose more salads and less fatty entrées than the pear-exposed and a nonexposed control group and the pear-exposed people chose more fruit desserts than nonexposed or melon-exposed people. Neither of the exposed groups (respectively to melon or pear) differed from the unexposed group in their choice of, respectively, the dessert or the entrée they had not been unconsciously conditioned for. This example once again illustrates the strength and specificity of the memory bond between the (even consciously unnoticed) odor or flavor and the situation it belongs to. At the same time, most of the people could probably not identify or name these odors when they are presented in clearly perceptible concentrations. Unfortunately, this was not verified in the experiment, but it has been shown in other odor experiments that odor identification even disturbs the unconscious functional relationships between the odor and the situation it occurs in (Degel, Piper, & Köster, 2001). Memorized odors and flavors are not "things" in the same way as remembered visual objects are (Köster, 2009; Morin-Audebrand et al., 2012). They are the epiphenomena of the situations that they are linked to and they are not remembered in their own right as detailed objects like tables or chairs. We can usually not describe them in any detail and even have difficulty finding their names. Spontaneously, we usually do not even recognize them specifically as such, but notice them in principle only when they do not fit our expectations. This raises the question of how flavor memory should be studied in order to understand its true function in normal everyday life.

5 How should real-world flavor memory be studied?

Flavors belong to eating occasions and, as pointed out above, they are often the unobtrusive accompaniments of other activities (table conversation, or even watching TV or reading a newspaper). Explicit attention is seldom devoted to them (probably with the exception of the first two or three bites) unless the eating occasion is a special one and new or unusual food is served. This means that in general the memory for the food one has eaten is based on incidental (nonintentional) implicit learning and not on a feature analysis of the flavor as in the case of visual working memory where we automatically register features of objects. It is even questionable whether working memory exists for the senses implicated in flavor. The laboratory experiments trying to prove its existence are highly artificial (Jönsson, Møller, & Olsson, 2011; White, 1998, 2012) and do not provide convincing evidence of feature analysis. On the other hand, there is good evidence that the incidentally learned memory for odors does exist and is even better preserved with age than explicitly learned odor or flavor memory (Møller et al., 2004). In the first session of these laboratory experiments, 20 very uncommon odors were presented in small bottles and two groups of participants (39 young and 41 elderly) were simply asked to rate their liking for them without any allusion to memory. When they came back the next day they were unexpectedly asked to recognize 10 of these odors presented monadically amidst 10 new odors in a random sequence. The elderly participants did at least as well and even slightly better on this test than the younger ones, but when asked to now remember the presented odors during this test for another memory test on the next day, the younger proved to be much better learners than the elderly (who even performed less well than in the implicit test). It could also be shown that the improvement that was seen in the younger group was not due to a better recognition of the previously presented odors, but to a better rejection of the new odors that were used as distractors in the second test. Thus, even intentionally learned odors are not remembered as such and people have to rely on novelty detection of the new odors rather than on recollection of the earlier presented ones. Furthermore, it could be shown that the difference between the elderly and young participants was also not due to the fact that the young provided more or better names for the uncommon odors and remembered them better on the basis of that. In a later experiment (Møller et al., 2007) with the flavors of new soups that had been presented as an intermezzo between two psychological experiments, these results were confirmed and this time the elderly were significantly better than the young in the unintentional learning condition.

Not only does this show that the mechanisms involved in intentional learning are different from those in incidental learning, but it also demonstrates that some of the common practices in recognition memory research may obscure the true nature of the memory processes. Thus, testing the memory with multiple choice methods makes it impossible to decide whether the memory is recollection- or novelty-detection-based. Only monadic presentation of the odors can show whether the odors are recognized as such or not. In alternative forced choice, the decision may be deducted from the fact that the other alternative(s) is (are) rejected as being novel. A similar argument may hold against the use of signal detection measures without separate verification of the hit and correct rejection rates (Morin-Audebrand et al., 2012). If memory depends on novelty detection, novelty should be considered as the signal and correct rejection as the indication of detection, rather than the detection of the original target odor through the hit-rate. In odor or flavor memory research, the traditional hit rates usually do not differ from chance guessing, whereas the novelty-based correct rejection rates deliver the significant contribution to d' as the detection measure. This also means that many of the methods used to measure the odor memory capabilities of the participants using forced choice methods are based on false assumptions when they conclude that people remember odors as such. Furthermore, the old claims about the longevity of odor memory (Engen & Ross, 1973; Herz & Engen, 1996) should be critically reviewed in light of this (Cameron, Houzenga, Köster, & Møller, 2015b).

If all this is true and flavors as such are not remembered with any precision, but just are not remarked as novel, what is the contribution of flavor memory in normal life and how can we study this contribution? If, like odors, flavors immediately seem to link themselves to the situations in which they are encountered and carry these associations with them, autobiographical details are perhaps more important in flavor memory than the sensory characteristics of the product. This may be true at least as long as these sensory characteristics remain unchanged and do not lead to novelty or change detection. Such autobiographical memories and the emotions linked to them color our expectations and may lead to disappointment or pleasant surprise when remembered flavors are actually tasted again. This means that a comparison of the actually perceived flavor with the memory of the same flavor is a good way to observe the development of flavor memory over time.

If the actual perception of the same food in the later test seems much more or less strong, salty, sweet, fruity, etc., than the remembered one, it tells us something about the way flavor memory distorts the remembered qualities of products (Barker & Weaver, 1983; Köster et al., 2004; Laureati et al., 2008; Mojet & Köster, 2002, 2005; Tuorila, Theunissen, & Alhlstrom, 1996; Vanne, Laurinnen, & Tuorila, 1998). If it is now perceived as less or more pleasant than before, it may also tell us something about the feelings and emotions of the situations during the previous encounter with the product. Memory measurements like these have been used with success in confidential research on launching decisions in which several alternative prototypes were compared. The relative memories may be better predictors of product success than the first impression consumer measurements on which many market decisions are based nowadays. More research along these lines would be needed to further investigate these opportunities.

In all this, it remains essential to rely on spontaneous incidental learning without any awareness of taking part in a memory experiment and without drawing special attention to the sensory properties of the tested products. In this respect, some of the mentioned experiments in which people were merely asked to judge the pleasantness of stimuli could perhaps still be criticized, since they drew some special attention and made people react explicitly to the stimuli. So far, just letting people eat normally and later asking them to compare their incidentally collected memories with the now explicitly presented product seems the best method (Köster et al., 2004; Mojet & Köster, 2002, 2005). Another good line of research for the effects of flavor memory is the paradigm used by Gaillet et al. (2013) described above, which does not even introduce any conscious perception of the flavor in the conditioning exposure phase. Even in explicit flavor memory testing, identification and naming should be avoided. As indicated above, identification may disturb the specific function of flavor memory since it "objectifies" the flavor and makes it lose its intimate contact with the situations in which it was perceived before, thus disturbing the implicit expectations that result from these contacts. These implicit expectations play an important role in the product perception (Piqueras-Fiszman & Spence, 2015). They are the main way in which flavor memory influences food choice and food appreciation in our everyday behavior, either by providing intimate and often unconscious feelings of knowing and trustworthiness or by the special attention devoted to the product as a "misfit" in the case of a disconfirmation of the expectation (Cardello, 1994, 1995; Köster, Møller et al., 2014; Schifferstein, Mojet & Kole, 1999).

6 Conclusions

In everyday life, flavor memory is not based on the recollection of the flavor in its multisensory perceptual complexity itself, but rather on the recurrence of the earlier perceived stimulation pattern and its close associations with the situational factors (circumstances, company, emotions, etc.) that were present at the earlier encounter. Just like odor memory, flavor memory does not depend on feature analysis but primarily on novelty and change detection and on these associations. Explicit flavor memory is only used by experts and in gastronomical circles. There is a need for research on implicit real-world flavor memory and its effects on food choice, intake, and satisfaction, which is not met by most of the present laboratory-type investigations.

References

Anderson, M. J., Berry, C., Morse, D., & Diotte, M. (2005). Flavor as context: Altered flavor cues disrupt memory accuracy. *Journal of Behavioural Neuroscience, 3*, 1–5.

Barker, L. M., & Weaver, C. A. (1983). Rapid, permanent loss of memory for absolute intensity of taste and smell. *Bulletin of the Psychonomic Society, 21*, 281–284.

Boesveldt, S., Frasnelli, J., Gordon, A. R., & Lündstrom, J. N. (2010). The fish is bad: Negative food odors elicit faster and more accurate reactions than other odors. *Biological Psychology, 84*, 313–317.

Cameron, E.L., Anderson, M.R., & Møller, P. (2015a). Is verbal labeling of odors useful? Submitted to *Chemosensory Perception*.

Cameron, E.L., Houzenga, C., Köster, E.P., & Møller, P. (2015b). Olfactory and visual memory: Same or different? Submitted to *Chemosensory Perception*.

Cardello, A. V. (1994). Consumer expectations and their role in food acceptance. In H. J. MacFie & D. M. H. Thomson (Eds.), *Measurement of food preferences* (pp. 253–297). London: Blackie Academic.

Cardello, A. V. (1995). Food quality: Relativity, context and consumer expectations. *Food Quality and Preference, 6*, 163–170.

Chu, S., & Downes, J. J. (2000). Long live proust: The odour-cued autobiographical memory bump. *Cognition, 75*, B41–B50.

Chu, S., & Downes, J. J. (2002). Proust nose best: Odors are better cues of autobiographical memory. *Memory and Cognition, 30*, 511–518.

Cliff, M. A., & Green, B. (1994). Sensory irritation and coolness by menthol: Evidence for selective desensitization of irritation. *Physiology and Behavior, 56*, 1026–1029.

Degel, J., Piper, D., & Köster, E. P. (2001). Implicit learning and implicit memory for odors: The influence of odor identification and retention time. *Chemical Senses, 26*, 267–280.

Engen, T., & Ross, B. M. (1973). Long-term memory of odors with and without verbal descriptions. *Journal of Experimental Psychology, 100*, 221–227.

Frank, R. A., & Byram, J. (1988). Taste–smell interactions are tastant and odorant dependent. *Chemical Senses, 13*, 445–455.

Frank, R. A., Ducheny, K., & Mize, S. J. S. (1989). Strawberry odor, but not red color, enhances the sweetness of sucrose solutions. *Chemical Senses, 14*, 371–377.

Gaillet, M., Sulmont-Rossé, C., Issanchou, S., Chabanet, C., & Chambaron, S. (2013). Priming effects of an olfactory food cue on subsequent food-related behaviour. *Food Quality and Preference, 30*, 274–281.

Geier, A., Wansink, B., & Rozin, P. (2012). Red potato chips. Segmentation cues can substantially decrease food intake. *Health Psychology, 31*, 398–401.

Graziano, M. S. A., Alisharan, S. E., Hu, X., & Gross, C. G. (2002). The clothing effect: Tactile neurons in the precentral gyrus do not respond to the touch of the familiar primate chair. *Proceedings of the National Academy of Sciences USA, 99*, 11930–11933.

Haller, R., Rummel, C., Henneberg, S., Pollmer, U., & Köster, E. P. (1999). The influence of early experience with vanillin on food preference later in life. *Chemical Senses, 24*, 465–467.

Hausner, H., Bredie, W. L. P., Mølgaard, C., Petersen, M. A., & Møller, P. (2008). Differential transfer of dietary flavour compounds into human breast milk. *Physiology and Behaviour, 95*, 118–124.

Hausner, H., Nicklaus, S., Issanchou, S., Mølgaard, C., & Møller, P. (2010). Breastfeeding facilitates acceptance of a novel dietary flavour compound. *Clinical Nutrition, 29*, 141–148.

Hepper, P. G. (1995). Human fetal "olfactory" learning. *International Journal of Prenatal and Perinatal Psychological Medicine, 2*, 147–151.

Herz, R. S., & Engen, T. (1996). Odor memory: Review and analysis. *Psychonomic Bulletin & Review, 3*, 300–313.

Higgs, S. (2009). Cognitive influences on food intake: The effects of manipulating memory for recent eating. *Physiology and Behavior, 94*, 734–739.

Jönsson, F. U., Møller, P., & Olsson, M. J. (2011). Olfactory working memory: Effects of verbalization on the 2-back task. *Memory and Cognition, 39*, 1023–1032.

Kahn, B. E., & Wansink, B. (2004). The influence of assortment structure on perceived variety and consumption quantities. *Journal of Consumer Research, 30*, 519–533.

Köster, E. P. (2002). The specific characteristics of the sense of smell. In C. Rouby, B. Schaal, D. Dubois, R. Gervais, & A. Holley (Eds.), *Olfaction, taste and cognition* (pp. 27–44). New York, NY: Cambridge University Press.

Köster, E. P. (2005). Does olfactory memory depend on remembering odors? *Chemical Senses, 30*(Suppl.1), i236–i237.

Köster, E. P. (2009). Diversity in the determinants of food choice: A psychological perspective. *Food Quality and Preference, 20,* 70–82.

Köster, E. P., & Mojet, J. (in press). Sensory memory. In S. A. Kemp, J. Hort, & T. Hollowood (Eds.), *Time–dependent measures of perception in sensory evaluation.* Sussex: Wiley-Blackwell.

Köster, E. P., Møller, P., & Mojet, J. (2014). A "misfit" theory of spontaneous conscious odor perception (MITSCOP): Reflections on the role and function of odor memory in everyday life. *Frontiers in Psychology, 5,* 64.

Köster, E. P., Möller, P., Mojet, J., & Schaal, B. (2016, in preparation). Ecological sense and nonsense in olfactory research.

Köster, E. P., Van der Stelt, O., Nixdorf, R. R., Linschoten, M. R. I., De Wijk, R. A., & Mojet, J. (2014). Olfactory imagination and odor processing: Three same-different experiments. *Chemosensory Perception, 7,* 68–84.

Köster, M. A., Prescott, J., & Köster, E. P. (2004). Incidental learning and memory for three basic tastes in food. *Chemical Senses, 29,* 441–453.

Laureati, M., Morin-Audebrand, L., Pagliarini, E., Sulmont-Rossé, C., Köster, E. P., & Mojet, J. (2008). Influence of age and liking on food memory: An incidental learning experiment with children, young and elderly people. *Appetite, 51,* 273–282.

Laureati, M., Pagliarini, E., Mojet, J., & Köster, E. P. (2011). Incidental learning and memory for food varied in sweet taste in children. *Food Quality and Preference, 22,* 264–270.

Le Berre, E., Thomas-Danguin, T., Béno, N., Coureaud, G., Etiévant, P., & Prescott, J. (2008). Perceptual processing strategy and exposure influence the perception of odor mixtures. *Chemical Senses, 33*(2), 193–199.

Luce, R. D. (1986). *Response times.* New York, NY: Oxford University Press.

Marlier, L., Schaal, B., & Soussignan, R. (1998). Neonatal responsiveness to the odor of amniotic and lacteal fluids: A test of perinatal chemosensory continuity. *Child Development, 69,* 611–623.

Mennella, J. A., Coren, P., Jagnow, M. S., & Beauchamp, G. K. (2001). Prenatal and postnatal flavor learning by human infants. *Pediatrics, 107,* E88.

Miles, C., & Hodder, K. (2005). Serial position effects in recognition memory for odors: A reexamination. *Memory and Cognition, 33,* 1303–1314.

Mojet, J., & Köster, E. P. (2002). Texture and flavour memory in foods: An incidental learning experiment. *Appetite, 38,* 110–117.

Mojet, J., & Köster, E. P. (2005). Sensory memory and food texture. *Food Quality and Preference, 16,* 251–266.

Møller, P., Köster, E. P., Dijkman, N., De Wijk, R. A., & Mojet, J. (2012). Same-different reaction times to odors: Some unexpected findings. *Chemosensory Perception, 5,* 158–171.

Møller, P., Mojet, J., & Köster, E. P. (2007). Incidental and intentional flavour memory in young and older subjects. *Chemical Senses, 32*(6), 557–567.

Møller, P., Piper, D., Hartvig, D., & Köster, E. P. (2009). Comparison of incidental and intentional learning of olfactory and visual stimuli. *Chemical Senses, 34,* A86.

Møller, P., Wulff, C., & Köster, E. P. (2004). Do age differences in odour memory depend on differences in verbal memory? *Neuroreport, 15,* 915–917.

Morin-Audebrand, L., Laureati, M., Sulmont-Rossé, C., Issanchou, S., Köster, E. P., & Mojet, J. (2009). Different sensory aspects of a food are not remembered with equal acuity. *Food Quality and Preference, 20*, 92–99.

Morin-Audebrand, L., Mojet, J., Chabanet, C., Issanchou, S., Møller, P., Köster, E. P., et al. (2012). The role of novelty detection in food memory. *Acta Psychologica, 139*, 233–238.

Morin-Audebrand, L., Sulmont-Rossé, C., & Issanchou, S. (2006). Impact of hedonic appreciation on memory for custard dessert. In *XVII ECRO Congress*, Granada, Spain, September, 4–8.

Moskowitz, H. R. (1972). Effects of solution temperature on taste intensity in humans. *Physiology and Behaviuor, 10*, 289–292.

Nicklaus, S., Boggio, V., Chabanet, C., & Issanchou, S. (2005). A prospective study of food variety seeking in childhood, adolescence and early adult life. *Appetite, 44*, 289–297.

Nordin, S., Monsch, A. U., & Murphy, C. (1995). Unawareness of smell loss in normal aging and Alzheimer's disease: Discrepancy between self-reported and diagnosed smell sensitivity. *Journal of Gerontology B: Psychological Science and Social Sciences, 50B*, 187–192.

Piqueras-Fiszman, B., & Spence, C. (2014). Colour, pleasantness, and consumption behaviour within a meal. *Appetite, 75*, 165–172.

Piqueras-Fiszman, B., & Spence, C. (2015). Sensory expectations based on product-extrinsic food cues: An interdisciplinary review of the empirical evidence and theoretical accounts. *Food Quality and Preference, 40*, 165–179.

Posner, M. I. (1986). *Chronometric explorations of mind* (2nd Ed.). Oxford: Oxford University Press. Revised.

Prescott, J. (2012). Chemosensory learning and flavour: Perception, preference and intake. *Physiology and Behavior, 107*, 553–559.

Prescott, J., Johnstone, V., & Francis, J. (2004). Odor–taste interactions: Effects of attentional strategies during exposure. *Chemical Senses, 29*(4), 331–340.

Prescott, J., Lee, S. M., & Kim, K. D. (2011). Analytic approaches modify hedonic responses. *Food Quality and Preference, 22*(4), 391–393.

Prescott, J., & Murphy, S. (2009). Inhibition of evaluative and perceptual odour-taste learning by attention to the stimulus elements. *Quarterly Journal of Experimental Psychology, 62*(11), 2133–2140.

Richardson, A. (1994). *Individual differences in imaging: Their measurement, origins and consequences*. Amityville, NY: Baywood Publishing Company.

Rolls, B. J., Rolls, E. A., Rowe, E. A., & Sweeney, K. (1981). Sensory specific satiety in man. *Physiology & Behavior, 27*, 137–142.

Rolls, B. J., Rowe, E. A., & Rolls, E. T. (1982). How sensory properties of foods affect human feeding behaviour. *Physiology & Behavior, 29*, 409–417.

Schaal, B., & Durand, K. (2012). The role of olfaction in human multisensory development. In A. J. Bremner, D. Lewkowicz, & C. Spence (Eds.), *Multisensory development* (pp. 29–62). Oxford: Oxford University Press.

Schaal, B., Marlier, L., & Soussignan, R. (2000). Human foetuses learn odours from their pregnant mother's diet. *Chemical Senses, 25*, 729–737.

Schab, F. R. (1990). Odors and the remembrance of things past. *Journal of Experimental Psychology: Learning Memory and Cognition, 16*, 648–655.

Schifferstein, H. W. J., Kole, A. P. W., & Mojet, J. (1999). Asymmetry in the disconfirmation of expectations for natural yoghurt. *Appetite, 32*, 307–329.

Sela, L., & Sobel, N. (2010). Human olfaction: A constant state of change blindness. *Experimental Brain Research, 205*, 13–29.

Spence, C., & Deroy, O. (2013). Crossmodal mental imagery. In S. Lacey & R. Lawson (Eds.), *Multisensory imagery: Theory and applications* (pp. 157–183). New York, NY: Springer.

Stevenson, R. J., & Boakes, R. A. (2003). A mnemonic theory of odor perception. *Psychological Review, 110*, 340–364.

Stevenson, R. J., Boakes, R. A., & Prescott, J. (1998). Changes in odor sweetness resulting from implicit learning of a simultaneous odor-sweetness association: An example of learned synesthesia. *Learning and Motivation, 29*, 113–132.

Stevenson, R. J., Boakes, R. A., & Wilson, J. P. (2000a). Counter-conditioning following human odor–taste and color–taste learning. *Learning and Motivation, 30*, 114–127.

Stevenson, R. J., Boakes, R. A., & Wilson, J. P. (2000b). Resistance to extinction of conditioned odor perceptions: Evaluative conditioning is not unique. *Journal of Experimental Psychology: Learning, Memory, and Cognition, 26*, 423–440.

Stevenson, R. J., Prescott, J., & Boakes, R. A. (1995). The acquisition of taste properties by odors. *Learning and Motivation, 26*, 1–23.

Stevenson, R. J., Prescott, J., & Boakes, R. A. (1999). Confusing tastes and smells: How odours can influence the perception of sweet and sour tastes. *Chemical Senses, 24*, 627–635.

Sulmont-Rossé, C., Møller, P., Issanchou, S., & Köster, E. P. (2008). Effect of age and food novelty on food memory. *Chemosensory Perception, 1*, 199–209.

Talavera, K., Ninomya, Y., Winkel, C., Voets, T., & Nilius, B. (2007). Influence of temperature on taste perception. *Cellular and Molecular Life Sciences, 64*, 377–381.

Tuorila, H., Theunissen, M. J. M., & Ahlstrom, R. (1996). Recalling taste intensities in sweetened and salted liquids. *Chemical Senses, 21*, 29–34.

Vanne, M., Laurinen, P., & Tuorila, H. (1998). Ad libitum mixing in a taste memory task: Methodological issues. *Chemical Senses, 23*, 379–384.

White, T. L. (1998). Olfactory memory: The long and short of it. *Chemical Senses, 23*, 433–441.

White, T. L. (2012). Attending to olfactory short-term memory. In G. M. Zucco, R. S. Herz, & B. Schaal (Eds.), *Olfactory cognition* (pp. 137–152). Philadelphia, PA: John Benjamins Publishing Company.

Willander, J., & Larsson, M. (2006). Smell your way back to childhood: Autobiographical odor memory. *Psychonomic Bulletin and Review, 13*, 240–244.

Willander, J., & Larsson, M. (2007). Olfaction and emotion: The case of autobiographical memory. *Memory and Cognition, 35*, 1659–1663.

Willander, J., & Larsson, M. (2008). The mind's nose and autobiographical odor memory. *Chemosensory Perception, 1*, 210–215.

Zucco, G. M., Aiello, L., Turuani, L., & Köster, E. P. (2012). Odor-evoked autobiographical memories: Age and gender differences along the life span. *Chemical Senses, 37*, 179–189.

Individual Differences in Multisensory Flavor Perception

10

Cordelia A. Running[1] and John E. Hayes[2]
[1]Sensory Evaluation Center, College of Agricultural Sciences, The Pennsylvania State University, University Park, PA, United States, [2]Department of Food Science, College of Agricultural Sciences, The Pennsylvania State University, University Park, PA, United States

1 Introduction

Flavor is not an immutable phenomenon, uniformly experienced identically across individuals. Variation in perception arises through both fundamental properties of biology as well as through experience and expectation. Some of this variation is malleable, through learned association or simple exposure to foods and flavors. Although biological variation based on genomic factors is theoretically immutable within a single person over the course of his or her lifespan, other biological factors, such as receptor expression or the amount of salivary proteins, may vary over time. While sensory scientists are able to successfully use humans as instruments in order to evaluate various aspects of flavor quantitatively, fundamental differences rooted in biology cannot be completely trained away, and will contribute natural variability in measures of flavor perception.

This natural variability is not simply "noise" in sensory measurements, however. These differences among individuals can be used to understand differences in perception, experience, and behavior in the way that humans interact with flavor and food. This chapter will review the current knowledge of biological differences in flavor perception, focusing on the sensory modalities of texture, chemesthesis, taste, and odor.

2 Overview of genetics and molecular biology

2.1 Chromosomes to proteins

Humans have 23 pairs of chromosomes, which contain our genetic code in the form of DNA (deoxyribonucleic acid). DNA takes the form of a helix, made of two strands of sugar-phosphate backbones that have side chains: the nucleotides (adenine [A], thymine [T], guanine [G], cytosine [C]). In its resting state, DNA can be highly condensed, as the strands are wrapped around other proteins that can interfere with the ability to transcribe the genetic code. The field of epigenetics studies the process of the unwinding and accessibility of DNA, and virtually nothing is known about the relationship between epigenetics and flavor perception, although this is likely to change in the coming decade. When condensed, DNA must be unwound and made accessible, then the strands split apart for the bases to be read and transcribed.

Nucleotides are read as codons (groups of three nucleotides) that designate specific amino acids to be used as building blocks. Genes to be transcribed contain promoter regions (which regulate binding of RNA polymerase), a start codon, exons (regions that will be read into the final protein), introns (regions not read into the final protein), and a stop codon.

DNA is transcribed by RNA polymerase into messenger RNA (mRNA). The strand of mRNA is then modified, removing the introns (noncoding regions) through a process called splicing. "Splice variants" occur when the same strand of RNA is spliced in different ways, combining different groups of exons, which results in a different protein. Single genes can therefore encode multiple, distinct proteins from the same original genetic code. After splicing, mRNA complexes with transfer RNA (tRNA). This occurs outside the nucleus of the cell at the ribosomes. Here, tRNA reads the mRNA codons to create a chain of amino acids, which, when assembled, form a protein or peptide.

The primary structure of a protein is a specific sequence of amino acids. These amino acids typically do not simply remain as a long, relaxed string. Instead, the environment around the protein (pH, aqueous/lipid, etc.) and the structure of the amino acids themselves cause the protein to create both a secondary structure (alpha helices and beta sheets) and a tertiary structure (folding, condensing of the whole molecule). In some cases, even small changes in the primary structure of a protein (amino acid sequence, read from the codons in the mRNA) can result in large changes in the secondary or tertiary structure of the protein, changing its overall shape. Changing the final shape of the protein can then change how that protein reacts with its surroundings, including other proteins, small molecules, potential ligands, etc. This process is summarized in Fig. 10.1.

2.2 Terminology

When discussing genetic variability, several terms and concepts are vital to understanding and interpreting study outcomes. Single nucleotide polymorphisms (SNPs, pronounced "snips") occur when a single nucleotide is changed in the genetic code. Sometimes this results in no change in the amino acid that is transcribed (synonymous SNP), and, at other times, the amino acid is altered (nonsynonymous SNP). In some instances, such nonsynonymous SNPs can change the secondary or tertiary structure of the protein, or alter the mechanics of a binding pocket for ligands. This can result in an altered function of the protein. Additionally, SNPs in noncoding regions can also play a role in altering protein expression. For example, SNPs in the promoter region of a gene can alter the regulation of that gene (ie, when/how frequently/in what conditions the gene is turned "on" or "off").

A genetic variant at a specific location is called an allele, and the less common form of an allele is called the minor allele. Accordingly, the minor allele frequency (MAF) is defined as how often the minor allele is found in a given population; the MAF for a specific allele may vary across different ethnic groups. Sometimes, small groups of alleles with SNPs tend to be inherited together, resulting in SNPs that are not statistically independent from each other (ie, if you have one SNP, you likely have

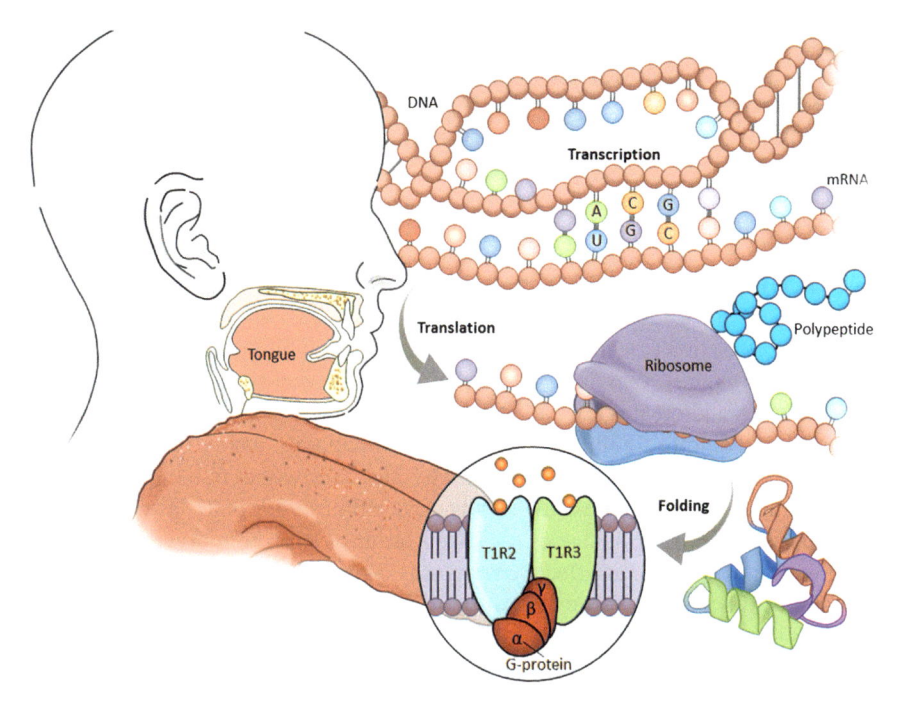

Figure 10.1 DNA is transcribed into RNA, which is translated into proteins. These proteins are then folded and translocated to specific locations within cells. For the chemoreceptors involved in taste, smell, and chemesthesis, these locations typically occur at cell surfaces that interact with the outside environment, where they may interact with ligands from foods or the environment. Small changes in DNA can alter chemosensory function by influencing the shape of these specialized receptors, or by influencing the numbers of these receptors that are expressed.

the other). This is called linkage disequilibrium (LD), and helps explain why sometimes when SNPs are first identified, the protein they encode later turns out to lack a mechanistic relationship to the biological phenomenon of interest (this is discussed in more detail below).

A separate source of genetic variability is copy number variation (CNV). CNV occurs when DNA sections of 1 kb (kilobase, or 1000 nucleotides) or longer are found multiple times in one genome compared to a reference genome. These replicates are then inherited by the next generation, and in the case of chemoreception, higher CNV can result in greater expression of that gene, leading to greater expression of a chemoreceptor.

Phenotypes are observable traits of an individual, and can be any trait that is reliably and stably measurable (sweet preference, food intake, height, weight, etc.). Conversely, genotypes are an individual's collection of genes. When examining SNPs, the diplotype refers to the nucleotides contained on both DNA strands, one from each parent. Phenotypes, therefore, can be products of the genotype (the diplotype), but also include environmental factors beyond the genetic code. Some phenotypes are influenced more by genotype than others. The distinction of phenotype and genotype is critical when discussing individual variability. Especially in flavor perception, observation of a particular phenotype is not a guarantee of a particular underlying genotype, as these associations are often imperfect.

3 Tactile and chemesthetic percepts

Oral somatosensation (touch, pain, movement, temperature) potentially differs across individuals through several different mechanisms. Relevant sensations include astringency, texture, and creaminess, as well as the collection of chemesthetic sensations (eg, burning, cooling, tingling, and buzzing). Importantly, many of the terms used to describe these sensations are more strictly and narrowly defined in the scientific literature than in colloquial usage. This is especially true for both "astringency" and "creaminess." If an individual does not have a clear (or any) definition for these terms, they are often used more as affective responses (astringency for negative and creaminess for positive affect). However, both of these sensations may also have considerable individual variability due to biology. Discerning the biological variability from the semantic variability is challenging, especially when comparing across studies. These caveats should be kept in mind while examining the differences in tactile sensations across individuals, which are discussed below.

3.1 Texture

Differences in saliva can contribute to variation in tactile sensations with respect to astringency. The word "astringent" comes from the Latin *adstringere*, meaning "to bind fast." Thus, from a chemical perspective, astringency is thought to arise from the sensation of salivary proteins binding food components, or food components binding oral surfaces. From the sensory perspective, the phenomenon is likely comprised of multiple sensations of drying, roughing, and puckering of the oral epithelium (Lee & Lawless, 1991), all of which could result from the saliva–food component– epithelium interactions.

Astringency is a common attribute of wines, tea, various fruit juices, beer, coffee, and chocolate, which contain polyphenolic compounds that bind salivary proteins. Humans widely vary in their salivary protein profiles and concentrations (Bradley, 1991), and these differences contribute to variation in astringency perception (Dinnella, Recchia, Fia, Bertuccioli, & Monteleone, 2009; Dinnella, Recchia, Vincenzi, Tuorila, & Monteleone, 2010; Horne, Hayes, & Lawless, 2002). Evidence suggests some of this variation in salivary protein expression is likely due to genetic factors (Törnwall et al., 2011), although specific genes have not yet been identified. The salivary flow rate may also be an important factor in astringency perception, and flow rates, like protein concentrations, also vary widely across individuals (Bradley, 1991). Lower flow rates are generally associated with increased intensity and prolonged perception of astringency (Fischer & Noble, 1994; Horne, Hayes, & Lawless, 2002; Ishikawa & Noble, 1995), but this has not been observed in all studies (Guinard, Zoumas-Morse, & Walchak, 1997). However, astringency is a complex percept that may really be a constellation of related but distinct attributes, and without clear guidance or training, participants may not use the term consistently and precisely.

In the case of high-starch foods, there is some evidence that perception may be influenced by different amounts of salivary alpha amylase, the expression of which is influenced by genetics. Specifically, the gene *AMY1* encodes salivary amylase, and this gene exhibits CNV, where humans can show anywhere from 2 to 15 diploid copies (Perry et al., 2007). Individuals with higher copy numbers have higher salivary amylase activity in both stimulated and unstimulated saliva (Yang et al., 2015) as well as higher concentrations of salivary amylase, which causes increased thinning of starch in the oral cavity over time (Mandel, Peyrot Des Gachons, Plank, Alarcon, & Breslin, 2010).

Numerous reports have also attempted to parse out individual differences in creaminess perception. However, "creaminess" is a broadly defined term (Elmore, Heymann, Johnson, & Hewett, 1999; Frøst & Janhøj, 2007; Schiffman, Graham, Sattely-Miller, & Warwick, 1998). For example, cream soda, cream from cow's milk, thick beverages, cheese products, and even whiskeys can have "creamy" flavor components. Yet whether this term refers to sweetness, vanilla odor, dairy odor, thickness, or some other concept can vary not only from one product to another, but also from one individual to another. Thus, while different ratings for creaminess can certainly be obtained, it is critical to keep in mind that studies looking at "creaminess" may not actually be measuring the same percept.

Some research indicates individuals with greater density of fungiform papillae may perceive greater creaminess and be better at discriminating fat content of foods (Kirkmeyer & Tepper, 2003; Tepper & Nurse, 1998), and individuals who taste 6-*n*-propylthiouracil (PROP) as bitter may also experience greater creaminess from cream (Hayes & Duffy, 2007), potentially through the association of PROP tasting with fungiform papillae density (Hayes & Duffy, 2007). Others have not found an association of PROP tasting with creaminess sensation (Lim, Urban, & Green, 2008). This is not particularly surprising as the mechanism for perceiving increased creaminess by PROP tasters or those with dense fungiform papillae is disputed. Not all studies find a correlation between fungiform papillae density and PROP tasting (for recent examples, see Barbarossa et al., 2015; Garneau et al., 2014; Webb, Bolhuis, Cicerale, Hayes, & Keast, 2015). Some have proposed that SNPs in the gustin gene (carbonic anhydrase VI) are actually responsible for variation in fungiform papillae density (Melis et al., 2013), yet this is also disputed (Feeney & Hayes, 2014). Consequently, the link between heightened perception of creaminess and biological variation is still unclear.

3.2 Chemesthesis

Individual differences that may contribute to variation in chemesthesis (pungency, irritation, chemical heat/cooling) perception and hedonic response include exposure frequency (including culturally determined exposure) and even personality (Byrnes & Hayes, 2013; Törnwall, Silventoinen, Kaprio, & Tuorila, 2012). Individuals who are more sensation-seeking or more sensitive to reward tend to like spicy food more (Byrnes & Hayes, 2013, 2015), and gender differences may potentially modify these relationships. Data on genetic differences are more limited, but some recent work has begun to explore this topic. Twin studies indicate that between 18% and 58% of the variability in pleasantness of spicy or pungent sensations may be genetic (Törnwall, Silventoinen, Kaprio et al., 2012). One potential source of variation is the *TRPA1* gene, which responds to cold and also to chemical irritants, including irritating components of mustard, wasabi, garlic, and ginger among others (Guimaraes & Jordt, 2007). One study has shown that a SNP in *TRPA1* results in differing intensity of hydrogen sulfide, which is both an odorant and an irritant (Schutz et al., 2014).

More recently, data indicate that SNPs in *TRPV1* (transient receptor potential vanilloid receptor 1, which is responsible for burning sensations from heat, capsaicin, and ethanol) may explain some of the variance in individual perception of burning or stinging from ethanol (Allen, Mcgeary, & Hayes, 2014). Notably, the burning sensation only correlated with these SNPs when obtaining time-intensity ratings after 50% ethanol was applied to a circumvallate papilla; correlations were not noted when individuals sipped, swished, and expected a 16% ethanol sample (single time point, whole mouth exposure). More work and replication of these results are required to verify these results, and determine why differences were observed for whole mouth versus circumvallate ratings.

4 Odor

Olfaction is particular interesting from a genetic standpoint as each olfactory receptor cell expresses a single type of receptor, read from a single allele (Chess, Simon, Cedar, & Axel, 1994). However, odorants activate multiple odor receptors, and the receptors recognize multiple odorants; the brain interprets the patterns of activated neurons in order to recognize and identify odors (Buck & Axel, 1991; Hasin-Brumshtein, Lancet, & Olender, 2009). Currently, about 400 functional odor receptor genes have been identified in humans, among a much larger gene family that also includes many pseudogenes (Mainland et al., 2014; Zhang & Firestein, 2009). These pseudogenes contain sequences such as stop codons and missing promoter regions that render the genes nonfunctional, and the number of pseudogenes varies widely across individuals (Menashe, Man, Lancet, & Gilad, 2003; Zhang & Firestein, 2009). Furthermore, of the 400 known odor receptors, only about 40 have identified ligands (Mainland et al., 2014), with the rest currently remaining as orphans (ie, receptors without known ligands). Additionally, CNV accounts for a great deal of genomic variation among individuals in odor receptor genes (Hasin et al., 2008; Olender et al., 2012; Waszak et al., 2010). Whether these CNVs meaningfully contribute to phenotypic differences remains to be determined.

For a half century, it has been well documented that some individuals are odor blind to certain odorants. This inability to smell a particular chemical is called a "specific anosmia," and this lack of sensation was assumed to be due to mutations in odor receptor genes (Amoore, 1977), a supposition that has been at least partially confirmed by modern molecular genetics. These mutations are recessive characteristics (Patterson & Lauder, 1948), which is logical due to the monoallelic nature of odor receptor expression. If an individual has a functional odor receptor on even just one allele, then that individual would express a functional receptor in at least some of the neurons in the olfactory epithelium. Some known specific anosmias relative to foods are discussed below.

4.1 Androstenone and boar taint

Androstenone is a steroid hormone produced in the testes of boars (male pigs) and can be found in the skin and adipose tissue of these animals. The ability to smell the steroid androstenone varies among individuals, with some experiencing no sensation, others a sweet, floral-like sensation, and others an unpleasant sweaty, urinous sensation. When residual androstenone causes pork products to have an unpleasant odor (for individuals who can smell it), it results in a meat defect known as "boar taint." This off-flavor leads to lower acceptability ratings in boar meat, at least for genetically vulnerable individuals (Lunde et al., 2012). Historically, this problem has been largely eliminated via castration of the animals; however this practice is currently under scrutiny for animal welfare reasons. The European Union has recommended phasing out surgical castration by 2018 (European Commission, 2015), which could potentially lead to increased boar taint in pork products from these countries unless alternative solutions are found.

Gene association studies in humans indicate that SNPs in the *OR7D4* olfactory receptor gene are at least partially responsible for differences in the odor of androstenone (Keller, Zhuang, Chi, Vosshall, & Matsunami, 2007). However, twin studies suggest environmental factors may play a larger role than genetic factors (Knaapila et al., 2012). Part of the differences in results observed in studies on androstenone may be because for some individuals, the ability to smell androstenone seems to be inducible, though mechanisms for this remain unclear (Morlein, Meier-Dinkel, Moritz, Sharifi, & Knorr, 2013; Wysocki, Dorries, & Beauchamp, 1989).

4.2 Cilantro

Common descriptions for cilantro odor range from "citrusy" to "soapy" or "bug-like" (Mauer & El-Sohemy, 2012), reflecting a polarized affective response. Some of the variability, particularly for the "soapiness," appears to be genetic (Eriksson et al., 2012; Mauer, 2011). A SNP in *OR4N5*, as well as one near the bitter taste receptor gene *TAS2R1*, may contribute the unpleasant odors in cilantro, as individuals homozygous for both minor alleles at these locations disliked cilantro (Mauer, 2011). Additionally, *OR6A2* variation has been implicated in cilantro soapiness and disliking

(Eriksson et al., 2012). However, assessing the genetic variability is difficult to isolate from cultural exposure to the herb, as this varies widely among ethnic groups.

4.3 Other food odors

Several other food odors have been associated with genetic variation in odorant receptor genes. Sensitivity to odors associated with SNPs in specific genes includes isovaleric acid (cheesy, sweaty), β-ionone (floral), *cis*-3-hexen-1-ol (green, grassy), and guaiacol (smoky) (Jaeger et al., 2012, 2013; Mainland et al., 2014; Menashe et al., 2007). Functional assays have confirmed that SNPs result in a nonfunctional or less functional odorant receptor for isovaleric acid, guaiacol, and β-ionone (Jaeger et al., 2013; Mainland et al., 2014; Menashe et al., 2007). For the odor *cis*-3-hexen-1-ol, the in vitro assays do not reflect the phenotype observed with the genetic variation, indicating that the actual source of the variation may be a gene in LD with these SNPs (Jaeger et al., 2012). For other odors relevant to food, sensitivity has only been associated with regions known to encode odor receptors, such as isobutyraldehyde (malty), β-damascenone (floral) and 2-heptanone (banana) (McRae et al., 2013).

Numerous other food odorants have reported qualitative differences amongst individuals or specific anosmias, including isobutyric acid (Amoore, Venstrom, & Davis, 1968), l-carvone and cineole (Lawless, Thomas, & Johnston, 1995), and trimethylamine (Amoore & Forrester, 1976), among others. There are also anecdotal reports suggesting some individuals may be anosmic to trichloroanisole (TCA), the compound associated with the moldy, musty wine defect known as cork taint. Such reports would be consistent with the observation some individuals fail to reject wine spiked with high levels of TCA (Prescott, Norris, Kunst, & Kim, 2005), although this should be interpreted cautiously, as rejection and detection are different tasks. More broadly, data are currently unavailable on whether there are genetic underpinnings for the variability in odor detection of these compounds.

5 Taste

5.1 Sweetness

Multiple lines of evidence indicate sweetness is an innately pleasurable sensation, and this pleasantness is generally evolutionarily conserved across species. While liking for other tastes like saltiness appears to be learned primarily through exposure, evidence suggests sweetness is liked even prior to birth. When amniotic fluid is made sweeter via injection of the nonnutritive sweetener saccharin, fetuses increase their swallowing rate (DeSnoo, 1937; Liley, 1972); likewise, newborn infants increase suck rate and intensity when sucrose is added to solutions (Maone, Mattes, Bernbaum, & Beauchamp, 1990; Tatzer, Schubert, Timischl, & Simbruner, 1985). In children, sweetness of sucrose relaxes the tension of having a stranger present (Maller & Desor, 1973) and sweetness may work as an analgesic in newborns (Bueno et al., 2013; Harrison et al., 2015). The analgesic effect appears to diminish in older children and adults, for reasons that have yet to be experimentally demonstrated (Harrison et al., 2015).

The sweet taste receptor is a heterodimer of T1R2 and T1R3 proteins, encoded from the genes *TAS1R2* and *TAS1R3*. Different sweeteners appear to bind different regions of this heterodimer, with 3–5 binding sites proposed in current models of receptor–ligand binding (DuBois, 2011; Temussi, 2007). Numerous binding sites would help explain not only the heterogeneity in structures that stimulate a sensation of sweetness, but also the synergism observed when using certain sweeteners in tandem (DuBois, 2011; Hayes, 2008). Differences in both the actual *TAS1R2* and *TAS1R3* genes as well as their promoter regions may alter the perception of sweetness. Considering the numerous proposed binding sites in the protein products of these genes, altered sweetness perception may not be fully quantified by using a limited set of sweet stimuli. Consequently, study outcomes of one sweetener may not predict overall sweetness perception and should be interpreted with caution.

For sucrose, SNPs in the promoter region of *TAS1R3* have been associated with altered sweetness perception. In a study using a ranking task of graded concentrations of sucrose solutions, better performance was observed for individuals with C alleles at two loci in the promoter region of this gene compared to those with T alleles (Fushan, Simons, Slack, Manichaikul, & Drayna, 2009). However, of the 144 participants tested, only 5 were T/T homozygotes at the first loci tested, and only 9 were T/T homozygotes at the second loci. The small sample size for the T allele homozygotes may explain why others have failed to replicate the observed effect (Hayes, Mcgeary, Grenga, & Swift, 2010). However, other research indicates that T allele homozygotes experience less bitter masking from sucrose with quinine and caffeine, but not for urea, 6-*n*-propylthiouracil, or denatonium benzoate (Mennella, Reed, Mathew, Roberts, & Mansfield, 2015). Consequently, this region of *TAS1R3* merits further study for sweetness perception.

The same cohort used by Fushan and colleagues (2009) for the study on *TAS1R3* promotor region SNPs was also studied for differences in *GNAT3*, which encodes a subunit of the gustducin protein. Gustducin is one of the G-proteins coupled to the cell surface receptors for sweet, bitter, and umami taste; activation of the receptors would thus release gustducin intracellularly (Chandrashekar, Hoon, Ryba, & Zuker, 2006). Eleven SNPs in *GNAT3* have been identified that correlated with differences in sorting ability for sucrose solutions (Fushan, Simons, Slack, & Drayna, 2010).

Moving beyond the perception of intensity, hedonic response to sweetness also varies across individuals (Hayes & Duffy, 2008; Lundgren et al., 1978). Whether this variability has a genetic basis is unknown, although some studies suggest differential liking for sweetness may be heritable (Bachmanov et al., 2011; Hwang et al., 2015). A mechanism for this has yet to be uncovered.

5.2 Umami

Umami, the "savory" taste, is exemplified by the amino acid glutamate, which has been extensively used in the food supply as a sodium salt (monosodium glutamate; MSG) or via glutamate-rich ingredients (eg, hydrolyzed yeast extract) to enhance flavor of savory products (soups, meats, cheese, etc.). Theoretically, the taste would have evolved to indicate sources of amino acids, vital for human growth and development. However, why the taste is stimulated predominantly by a single amino acid, glutamate, is unclear. Further, whether this taste is actually "pleasant" is also inconsistent, and may depend on the medium in which it is delivered. Certainly, affective tests indicate that savory foods with added MSG are preferred (despite consumer backlash against the chemical), yet in isolation the taste may not be particularly pleasant (Weiland, Ellgring, & Macht, 2010).

The umami receptor is a heterodimer. The dimer consists of T1R1/T1R3 subunits, so the T1R3 subunit is expressed in both the sweet and umami receptors. Large differences have been found for sensitivity to umami taste, and some individuals may even be unable to taste umami sensations (Lugaz, Pillias, & Faurion, 2002). The reason for this taste blindness is unknown. In our professional experience, many consumers are unfamiliar with the sensation of umami in isolation, so some caution should be used in interpreting results from participants who appear taste-blind to umami.

Numerous SNPs have been investigated in relation to umami sensation. Two relatively common SNPs (one in *TAS1R1* and one in *TAS1R3*) have been identified that covary with recognition thresholds for umami (Shigemura, Shirosaki, Sanematsu, Yoshida, & Ninomiya, 2009), though these results were not consistent for both MSG and inosine monophosphate (IMP, another umami tastant). Others have confirmed the differential activation of the umami receptor with the SNP from *TAS1R3* (Raliou et al., 2011).

5.3 The fatty acid taste—oleogustus

Recent data by multiple research teams suggest that the unique taste of fatty acids belongs within the canonical list of prototypical tastes. In isolation, this percept (termed oleogustus) is clearly unpleasant (Running, Craig, & Mattes, 2015). Of course, bitterness and sourness are also unpleasant in isolation, yet they remain fundamental and enjoyable components of certain foods in certain contexts. Thus, it remains to be determined whether oleogustus contributes positively to affective responses in real foods. Returning to this percept in isolation, very broad differences in both detection thresholds (Running, Mattes, & Tucker, 2013; Stewart et al., 2010) and suprathreshold intensity (Running et al., 2015; Tucker, Nuessle, Garneau, Smutzer, & Mattes, 2015) have been reported. Some individuals may never detect this sensation at all, as detection thresholds were unattainable for some individuals (Running & Mattes, 2014a, 2014b; Tucker & Mattes, 2013). An inability to detect oleogustus could be due to actual lack of functional receptors (as the receptors are still not fully confirmed), a simple lack of understanding of what the sensation is (as there is no lexicon to describe the sensation), or due to the lingering sensation from the lipophilic compounds, which are difficult to clear from the oral cavity and make isolation of this taste sensation challenging.

Prime candidates for the oleogustus receptor are GPR120, a G-protein coupled receptor that responds to long-chain fatty acids, and CD36 (cluster of differentiation 36, also called Fatty Acid Translocase), which is a scavenger receptor that responds to numerous lipophilic compounds (Gilbertson & Khan, 2014; Tucker, Mattes, & Running, 2014). Cellular models demonstrate that CD36 responds to lower concentrations of fatty acids, while GPR120 responds to higher concentrations (Ozdener et al., 2014). Knockout rodent models demonstrate that CD36 may be the primary sensory for fatty acids (Ancel et al., 2015). Strikingly, while in human studies the taste has been described as unpleasant (which is logical as high levels of free fatty acids are an indicator of rancidity), rodents are attracted to the taste and will freely consume fatty acids.

Candidate gene association studies on oleogustus and fat detection have focused primarily on CD36. Two SNPs in CD36 have been associated with fat perception, with individuals homozygous for the A allele displaying higher perception of creaminess, lower ability to discriminate concentrations of added fat in salad dressings, and higher detection thresholds (lower sensitivity) for oleic acid compared to those homozygous for the G allele (Keller et al., 2012; Melis, Sollai, Muroni, Crnjar, & Barbarossa, 2015; Mrizak et al., 2015; Pepino, Love-Gregory, Klein, & Abumrad, 2012). These studies exemplify the difficulty in interpreting data on studies of fat perception, as the higher creaminess but lower sensitivity to oleogustus are seemingly contradictory. However, as mentioned previously, the word "creamy" is notorious for its broad and vague definition, so any results where this word is not strictly defined for participants should be interpreted with caution. *CD36* also has known polymorphisms that cause a deficiency in the protein leading to cardiovascular problems (Curtis & Aster, 1996; Rac, Safranow, & Poncyljusz, 2007; Take et al., 1993; Xu et al., 2013; Yamamoto, Akamatsu, Sakuraba, Yamazaki, & Tanoue, 1994); however, these polymorphisms have not been investigated for their functional ability to detect oleogustus.

Habitual diet may also be important for oleogustus sensitivity, more so than any other taste as studied to date. CD36 expression is decreased in rodents after exposure to triglycerides or fatty acids (Martin et al., 2011; Ozdener et al., 2014), and sensitivity to oleogustus taste is decreased in humans who switch from a low-fat to a high-fat diet (Stewart & Keast, 2012). The time course of this phenomenon has not been established, but certainly recent exposure to fat may influence oleogustus perception. This has clear implications for taste experiments, as experiments of prototypical tastes generally do not control for the content of a previous meal.

5.4 Bitterness

Variability in bitter taste sensation is well documented, and has been reviewed elsewhere (Beckett et al., 2014; Behrens & Meyerhof, 2009; Feeney, O'brien, Scannell, Markey, & Gibney, 2011; Reed, Tanaka, & Mcdaniel, 2006), though this has not necessarily led to consensus regarding the functional importance of this variability. That is, genotypic and phenotypic variation observed in controlled laboratory settings may still fail to influence the diet of free-living humans. The sheer quantity of work conducted on bitter taste genetics dates back to the discovery of dichotomous responses to the compounds phenylthiocarbamide (PTC) and 6-*n*-propylthiouracil (PROP). In the 1930s, chemist Arthur Fox observed that some individuals tasted PTC as bitter while others were "taste-blind" (Fox, 1932). Working with Blakeslee, a famous geneticist of the era, this was quickly shown to be a heritable trait (Blakeslee, 1932). A second thiourea compound, PROP, also displays the same pattern of "tasters" and "non-tasters," and by the 1990s it was clear that the intensity of these compounds was not experienced equally among even just the "tasters," leading to the description of individuals as "non-tasters," "tasters," and "supertasters" (Bartoshuk, 1991; Bartoshuk, Duffy, & Miller, 1994). However, these words, while ubiquitous in the literature and even pop culture (LaZebnik, 2009), should not be interpreted as meaning that "supertasters" of PROP/PTC are extremely sensitive to all oral sensations (Hayes, Bartoshuk, Kidd, & Duffy, 2008; Hayes & Keast, 2011; Roura et al., 2015).

5.4.1 TAS2R38

The difference between tasters and non-tasters of PROP/PTC is driven predominantly by variation in the *TAS2R38* gene, which contains three SNPs that result in two common haplotypes: Proline-Alanine-Valine (PAV) and Alanine-Valine-Isoleucine (AVI). The PAV haplotype leads to the functional ability to detect the PTC and PROP compounds, presumably through binding by the thiourea moiety, though not all thioureas generate the same binomial phenotypical response seen for PTC and PROP. The AVI haplotype leads to lesser functionality of the receptor in response to these chemicals. Notably, both PTC and PROP are synthetic compounds, and relatively few food-based compounds have been discovered (L-vinyl-2-thio-oxazolidone and goitrin) that also stimulate the T2R38 bitter receptor based on this PAV/AVI haplotype (Astwood, Greer, & Ettlinger, 1949; Boyd, 1950; Hofmann, 2009; Wooding et al., 2010). Notably, variability in perceived bitterness of goitrin is not fully accounted for by variation in *TAS2R38*, and very little research has focused on L-vinyl-2-thio-oxazolidone since the 1950s. The response of the T2R38 bitter receptor to goitrin is weak with the PAV form of the protein and negligible for the AVI form (Wooding et al., 2010).

5.4.2 TAS2R16

TAS2R16 encodes a bitter receptor that responds to the beta-glucopyranosides found in some plants (Bufe, Hofmann, Krautwurst, Raguse, & Meyerhof, 2002; Meyerhof et al., 2010). Critically, enzymatic cleavage of these compounds can release cyanide, presumably providing the selective pressure to avoid their ingestion. At least one polymorphism in this gene leads to a protein that has higher activation in vitro to beta-glucopyranosides, and this variant is more common among people with European and Asian heritage compared to African heritage (Soranzo et al., 2005). Additionally, this variant in *TAS2R16* associates with alcohol abuse, with those experiencing greater bitterness showing lower prevalence of alcohol dependence (Hinrichs et al., 2006; Wang et al., 2007). This is presumably mediated through increased bitterness of alcoholic beverages, but this has yet to be directly tested.

Another likely candidate that contributes to individual variability in bitterness perception is *TAS2R31*. Formerly called *TAS2R44*, this gene encodes a bitter receptor that, in addition to sensing a variety of other bitter compounds, appears to be responsible for the bitterness from some nonnutritive sweeteners (Kuhn et al., 2004). An amino acid substitution in the T2R31 receptor results in greater activation in response to saccharin in vitro, as well as lower recognition thresholds in vivo (Pronin et al., 2007). Another SNP in *TAS2R31* corresponds to differential bitterness of both acesulfame potassium (AceK) and saccharin (Roudnitzky et al., 2011). Further work confirms that SNPs in *TAS2R31* are responsible not only for perceptual differences near threshold (ie, at very low concentrations), but also in suprathreshold intensity differences at higher concentrations (Allen, Mcgeary, Knopik, & Hayes, 2013). Notably, these SNPs appear to associate only with the bitterness of nonnutritive sweeteners, but not other bitter tastants that activate the TAS2R31 receptor. Additionally, whether or not willingness to use these sweeteners associates with the genetically derived differences in bitter side tastes has not been tested. Regarding other foods, SNPs in *TAS2R31* also correlate with differences in bitterness from quinine as well as remembered liking of grapefruit (Hayes, Feeney, Nolden, & Mcgeary, 2015).

5.4.3 TAS2R3, TAS2R4, TAS2R5

Other polymorphisms that may influence bitterness perception include SNPs in chromosome 7 in the region of *TAS2R3, TAS2R4,* and *TAS2R5*. These genes, which display three common haplotypes from six SNPs in the region, encode receptors that contribute to bitterness in coffee (Calvo, Pagliarini, & Mootha, 2009; Hayes et al., 2011; Nolden, McGeary, & Hayes, 2016). Individuals with the TTGGAG haplotype may perceive up to twice as much bitterness in coffee compared to the CCCAGT haplotype (Hayes et al., 2011). However, in those individuals, this did not lead to differences in liking or intake of coffee, which could be due to a flavor-conditioning effect (coffee contains caffeine, and people learn to associate the positive psychoactive effect of caffeine with the coffee flavor). Notably, in another study, the TTGGAG haplotype perceived less bitterness from ethanol and capsaicin applied to circumvallate papillae, which is in the opposite direction from the effect observed for coffee (Nolden et al., 2016). The reason for this discrepancy is unknown and warrants further study.

5.4.4 TAS2R19 (formerly TAS2R48)

Variation in *TAS2R19* correlates to bitter perception of quinine (found in tonic water) and grapefruit (Duffy, Hayes, Sullivan, & Faghri, 2009; Reed et al., 2010). However, in vitro assays have not been able to confirm that any of the bitter compounds in grapefruit (limonin, naringin) or quinine actually activate the T2R19 receptor (Meyerhof et al., 2010). More recently, perceived whole-mouth bitterness of quinine and remembered liking of grapefruit were shown to correlate with SNPs in both *TAS2R19* and *TAS2R31*, and these SNPs are in strong LD with each other (Hayes et al., 2015). Consequently, the observed differences in perception of bitterness from quinine or grapefruit are more likely due to genetic variation in *TAS2R31*, whose encoded protein T2R31 does in fact respond to quinine in functional assays (Meyerhof et al., 2010).

5.5 Saltiness

The mechanisms of salty and sour tastes are still disputed. Certainly, receptor candidates have been identified for these tastes (PKD2L1 for sour and the epithelial sodium channel, ENaC, for salty). However, both of these percepts occur through a distinct mechanism from bitterness, savory/umami, and sweetness, as the stimuli perceived as sour and salty are ions. Ionic balance is crucial for a cell to maintain, and flux of ions is a basic mechanism by which cells transmit signals. Thus, trying to isolate these ions as taste stimuli tastants versus fundamental cell maintenance is challenging. That is, complete knockout rodent models for ENaC or its subunits are not viable, as the animals die because the transporter is vital for other cell and organ functions involving sodium balance (Hummler & Vallon, 2005).

Regarding variability in salty taste perception, considerable evidence points to strong environmental and exposure effects. Twin studies would indicate these effects are more sizable than potential genetic factors (Greene, Desor, & Maller, 1975; Wise, Hansen, Reed, & Breslin, 2007). Additionally, habitual dietary sodium levels, hormonal cycle in women, and physiological state all contribute to salt preference and liking (Bertino, Beauchamp, & Engelman, 1982; Hayes, Sullivan, & Duffy, 2010; Wald & Leshem, 2003). To date, evidence for a genetic component to saltiness perception is limited to variations in *TRPV1* and *SCNN1B* that correlate with different suprathreshold intensity of salt solutions, but this has only been explored in a single study (Dias et al., 2013). Another potential source of variation could be the interaction of genetic variation in αENaC and zinc intake, where only women with low zinc intake showed differences in salt taste thresholds that correlated with genotype (Noh, Paik, Kim, & Chung, 2013).

5.6 Sourness

Unlike saltiness, twin studies on sourness perception indicate a large amount of variability may be attributable to genetic factors (Wise et al., 2007). However, precise genes or mechanisms have yet to be identified. Furthermore, sourness intensity does not correlate strongly with liking of foods (Liem, Westerbeek, Wolterink, Kok, & De Graaf, 2004; Törnwall, Silventoinen, Keskitalo-Vuokko et al., 2012). Additional research into mechanisms of sourness will hopefully assist with future work identifying sources of variability.

6 Functional outcome: food intake

The assumed causal chain from individual differences in perception to food intake as mediated by differential liking has recently been reviewed elsewhere, and will not be reiterated in depth here. Interested readers should see Duffy (2007) and Hayes (2014). That said, a discussion of differences in flavor perception would not be complete without at least mentioning the influences such variation putatively has on food choice.

By and large the greatest amount of work connecting flavor perception and/or food intake to genetic variation in flavor perception has been conducted on the bitter taste phenotype. Hundreds of studies of this variation have been conducted on phenotypic response to PROP/PTC, with several dozen looking specifically at food preference, intake or diet, whereas only a handful have looked at the underlying genetic variation. This is understandable, both historically, as the molecular basis for this trait has only been known for little over a decade, and logistically, because it is much easier, less invasive, and less expensive to measure phenotype than genotype. However, the phenotypic measurement of PROP taster status requires accurately measuring the suprathreshold intensity of the sensation, particularly to separate supertasters from medium tasters, and to date there is still no single, generally accepted best practice method for assigning "taster" status. Moreover, any categorical classification system potentially suffers from misclassification bias. For example, if three individuals rate the bitterness of PROP as 25, 48, and 55, and a categorical system divides the group at 50, then the categorical system would group the individual with the 48 rating with the 25 rating, rather than the 55 rating. Thus, this method is clearly flawed (see Hayes and Duffy (2007) for a discussion). The variety of classification schemes, along with disagreement over scaling methods, likely contributes to the diversity of outcomes regarding PROP taster status, food preference, and food intake.

Logically, if some individuals experience more extreme bitter sensations from certain foods, then they may avoid those foods. As stated in Section 5.4.1, the only natural food components that have been shown to activate the T2R38 receptor in a similar pattern to PROP/PTC are goitrin and L-5-vinyl-2-thio-oxazolidone (Astwood et al., 1949; Harris & Kalmus, 1949; Hofmann, 2009; Wooding et al., 2010). Foods, such as cruciferous vegetables, containing these compounds may be expected to exhibit differences in perceived bitterness due to *TAS2R38* genotype, but other explanations (unidentified bitter compounds, other genetic variation that may associate with *TAS2R38*, general hyperguesia, etc.) may be merited for foods that do not contain these compounds (for greater discussion of this topic, see Hofmann (2009)). Nonetheless, numerous studies have investigated the role of PROP/PTC tasting in perception and intake of a wide variety of products, with a large emphasis on items such as bitter cruciferous vegetables and alcohol.

Vegetable bitterness is associated with lower intake of vegetables (Dinehart, Hayes, Bartoshuk, Lanier, & Duffy, 2006), and many studies have also found decreased intake (Colares-Bento et al., 2012; Duffy et al., 2010; Sacerdote et al., 2007) or increased bitterness (Sandell & Breslin, 2006) of vegetables in PAV homozygotes and heterozygotes compared to AVI homozygotes (Colares-Bento et al., 2012; Duffy et al., 2010; Sacerdote et al., 2007). Yet, other studies found no association of *TAS2R38* variation with vegetable intake (Gorovic et al., 2011; Timpson et al., 2005). Considering that culture and exposure also play a large role in food liking and intake, this diversity of findings is not surprising. An interesting facet of this research that remains unexplored is whether different *TAS2R38* alleles, or other genetic variation, associate with condiment use (salt, cheese sauce, etc.) on vegetables, which might be a commonly and widely used means to mitigate bitterness to enable consumption of the food.

As ethanol induces bitter, burning, and sweet sensations, altered experience of any of these sensations could lead to different initial affective response to alcoholic beverages. *TAS2R38* and *TAS2R16* variants are known to correlate with increased alcohol intake (Allen et al., 2014; Duffy, Peterson, & Bartoshuk, 2004; Dotson, Wallace, Bartoshuk, & Logan, 2012; Hayes et al., 2011; Hinrichs et al., 2006; Wang et al., 2007) and reduced bitterness (Allen et al., 2014; Dotson et al., 2012; Duffy et al., 2004; Hayes et al., 2011; Hinrichs et al., 2006; Wang et al., 2007), a relationship that is presumably mediated through differential liking (Nolden & Hayes, 2015). Likewise, recent data suggest *TRPV1* variants may associate with differences in burning sensations (Allen et al., 2014), though this finding still needs to be confirmed. Others have demonstrated that the effects of individual differences in perception on alcohol intake may be most influential during initial exposure to alcohol consumption (Intranuovo & Powers, 1998). This is yet another example, similar to the coffee example above, of how post-ingestive feedback might modulate intake despite genotypes that may make a food less pleasant.

Alternatively, individuals with genotypes that make the food item less pleasant may simply alter their food choices and/or behavior if the other benefits from consumption are sufficiently reinforcing. For example, individuals do not (generally) drink neat ethanol—rather, they can chose from a wide variety of different alcoholic beverages that differ not only in the amount of alcohol they contain, but also in the presence or absence of other taste and odor active substances. Generally, sweetness reduces bitterness via mixture suppression; logically then, some types of alcoholic beverages may be more attractive to those individuals who typically experience more bitterness from ethanol due to genetics. Indeed, when we asked individuals aged 18–45 to rate the remembered liking of 20 different types of alcoholic beverages in a pilot study, the overall liking ratings differed by *TAS2R38* genotype (Nolden and Hayes, unpublished data). Across all beverages, the AV/AV group gave significantly higher liking ratings [$F(1,125) = 4.07$; $P = 0.046$] than the PA/* group (18.2 ± 3.79 vs 9.5 ± 2.08 SEM), consistent with repeated evidence than PA_ carriers tend to consume less alcohol (Allen et al., 2014; Dotson et al., 2012; Duffy et al., 2004; Hayes et al., 2011; Hinrichs et al., 2006; Wang et al., 2007). Moreover, when we split the beverages into sweetened alcoholic beverages (cocktails, sweet wines, etc.) and unsweetened alcoholic beverages (pale ales, hard liquor, etc.), we found a significant interaction between beverage group (sweet vs non-sweet) and *TAS2R38* genotype [$F(1,119) = 6.54$; $P = 0.012$]. As shown in Fig. 10.2, the liking for sweetened beverages does not differ as a function of genetics, whereas individuals who are less genetically responsive to bitterness due to their T2R38 receptor variant (ie, AV_/AV_ homozygotes) liked unsweetened alcoholic beverages substantially more than individuals who carry one or two copies of the PA_ allele. Presumably, such differences could strongly influence beverage choices in an alcohol-rich environment, although this remains to be shown experimentally.

Thus, when there is a strong motivation to increase consumption due to strong post-ingestive consequences (coffee, alcohol, etc.), individuals may simply choose to consume the food in a different manner to mask innately aversive sensations. Indeed, sugars and fats are used to modulate the flavor of these foods and make them more palatable, despite any genetic propensity to sense a food as more or less bitter. For example, alcohol researchers get rats to learn to drink ethanol by using a "sucrose fade," a procedure that is strikingly similar to how many teenagers learn to consume alcoholic beverages. Likewise, the propensity of young coffee drinkers to add large amounts of cream and sugar is well known, if only anecdotally. However, in spite of this apparent propensity, specific data on how common this modulation is by

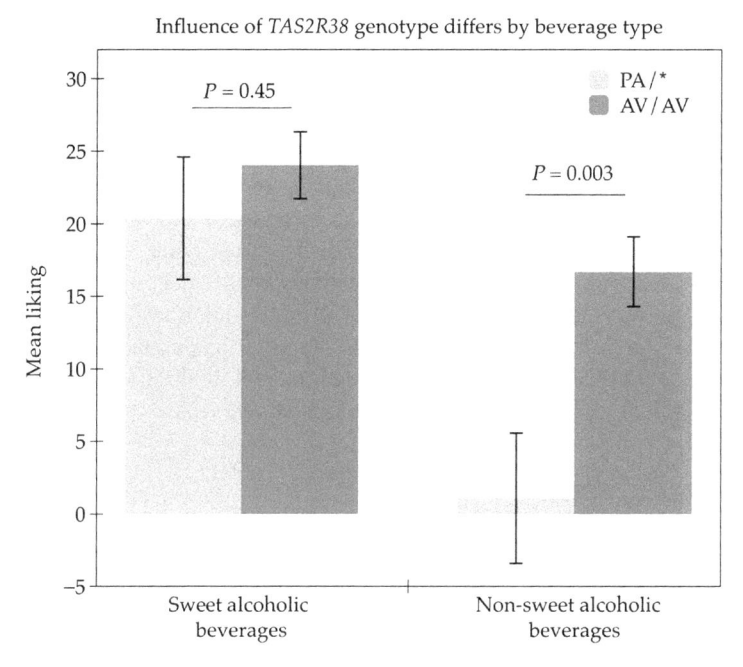

Figure 10.2 Differential remembered liking for various alcoholic beverages as a function of bitter receptor genetics (Nolden and Hayes, unpublished data).

genotype has been minimally explored to date, and is an ideal area for future work linking dietary behavior to genetics.

7 Conclusions and future directions

Humans perceive flavor differently from each other. Considering the numerous anatomical systems that contribute to flavor perception through odor, taste, texture, saliva, chemesthesis, the individual receptor arrays we inherit are quite diverse. Indeed, it is somewhat surprising that we can even find any common basis to discuss these individual experiences. Moreover, other yet unidentified modifiers beyond receptor genetics likely also play a major role—for example, virtually nothing is known about the link between epigenetics and flavor perception. Nonetheless, new insights enabled by molecular genetics and improved mechanistic understanding of chemosensation have the strong potential to influence the fields of food science, sensory evaluation, and psychophysics. Older approaches to assess quality control or product development often had a single individual, a "golden tongue" or a "golden nose," who would evaluate the worthiness of a product. In contemporary practice, most mid-sized and larger food and consumer products companies have now moved to more scientifically rigorous evaluation of flavor and sensation. By using many individuals rather than a handful, the final data catch more of the natural variability in perception, and product development and quality control benefit from this broader knowledge of human experiences.

However, in spite of the variability in our perception of flavor, some fundamentals remain. Sweetness is accepted and bitterness rejected, at least early in life. Young children reject all new foods to some extent. Experience, however, clearly modulates innate predispositions and leads to new affective responses to foods. As children become adults, pressures from society and simple repeated exposures to foods modulate the acceptance of food. Our inherent variability contributes to perception of and willingness to consume certain foods that we may perceive as more bitter, or more pungent, or more stinky than other individuals. In the end, however, we eat not only the foods we like, but also the foods that are familiar and convenient. Thus, while genetics may enable predictions at the population level, biology is not destiny at the individual level, as our food choices are exactly that, choices.

Acknowledgments

The authors wish to thank scientific illustrator Joselyn N. Allen (The Pennsylvania State University) for creating Figure 1 in this chapter.

References

Allen, A. L., Mcgeary, J. E., & Hayes, J. E. (2014). Polymorphisms in TRPV1 and TAS2Rs associate with sensations from sampled ethanol. *Alcoholism, Clinical and Experimental Research, 38*, 2550–2560.

Allen, A. L., Mcgeary, J. E., Knopik, V. S., & Hayes, J. E. (2013). Bitterness of the nonnutritive sweetener acesulfame potassium varies with polymorphisms in TAS2R9 and TAS2R31. *Chemical Senses, 38*, 379–389.

Amoore, J., & Forrester, L. J. (1976). Specific anosmia to trimethylamine: The fishy primary odor. *Journal of Chemical Ecology, 2*, 49–56.

Amoore, J. E. (1977). Specific anosmia and concept of primary odors. *Chemical Senses Flavour, 2*, 267–281.

Amoore, J. E., Venstrom, D., & Davis, A. R. (1968). Measurement of specific anosmia. *Perceptual and Motor Skills, 26*, 143–164.

Ancel, D., Bernard, A., Subramaniam, S., Hirasawa, A., Tsujimoto, G., Hashimoto, T., et al. (2015). The oral lipid sensor GPR120 is not indispensable for the orosensory detection of dietary lipids in mice. *Journal of Lipid Research, 56*, 369–378.

Astwood, E. B., Greer, M. A., & Ettlinger, M. G. (1949). 1-5-Vinyl-2-thiooxazolidone, an antithyroid compound from yellow turnip and from Brassica seeds. *The Journal of Biological Chemistry, 181*, 121–130.

Bachmanov, A. A., Bosak, N. P., Floriano, W. B., Inoue, M., Li, X., Lin, C., et al. (2011). Genetics of sweet taste preferences. *Flavour and Fragrance Journal, 26*, 286–294.

Barbarossa, I. T., Melis, M., Mattes, M. Z., Calo, C., Muroni, P., Crnjar, R., et al. (2015). The gustin (CA6) gene polymorphism, rs2274333 (A/G), is associated with fungiform papilla density, whereas PROP bitterness is mostly due to TAS2R38 in an ethnically-mixed population. *Physiological and Behavioral, 138*, 6–12.

Bartoshuk, L. M. (1991). Sweetness: History, preference, and genetic variability. *Food Technology, 45*, 108–113.

Bartoshuk, L. M., Duffy, V. B., & Miller, I. J. (1994). PTC/PROP tasting—Anatomy, psychophysics, and sex effects. *Physiological and Behavioral, 56*, 1165–1171.

Beckett, E. L., Martin, C., Yates, Z., Veysey, M., Duesing, K., & Lucock, M. (2014). Bitter taste genetics—the relationship to tasting, liking, consumption and health. *Food and Function, 5*, 3040–3054.

Behrens, M., & Meyerhof, W. (2009). Mammalian bitter taste perception. *Results and Problems in Cell Differentiation, 47*, 203–220.

Bertino, M., Beauchamp, G. K., & Engelman, K. (1982). Long-term reduction in dietary-sodium alters the taste of salt. *The American Journal of Clinical Nutrition, 36*, 1134–1144.

Blakeslee, A. F. (1932). Genetics of sensory thresholds: Taste for phenyl thio carbamide. *Proceedings of the National Academy of Sciences, 18*, 120–130.

Boyd, W. C. (1950). Taste reactions to antithyroid substances. *Science, 112*, 153.

Bradley, R. M. (1991). Salivary secretion. In T. V. Getchell, R. L. Doty, L. M. Bartoshuk, & J. B. Snow (Eds.), *Smell and taste in health and disease*. New York, NY: Raven Press.

Buck, L., & Axel, R. (1991). A novel multigene family may encode odorant receptors: A molecular basis for odor recognition. *Cell, 65*, 175–187.

Bueno, M., Yamada, J., Harrison, D., Khan, S., Ohlsson, A., Adams-Webber, T., et al. (2013). A systematic review and meta-analyses of nonsucrose sweet solutions for pain relief in neonates. *Pain Research & Management, 18*, 153–161.

Bufe, B., Hofmann, T., Krautwurst, D., Raguse, J. D., & Meyerhof, W. (2002). The human TAS2R16 receptor mediates bitter taste in response to beta-glucopyranosides. *Nature Genetics, 32*, 397–401.

Byrnes, N. K., & Hayes, J. E. (2013). Personality factors predict spicy food liking and intake. *Food Quality and Preference, 28*, 213–221.

Byrnes, N. K., & Hayes, J. E. (2015). Gender differences in the influence of personality traits on spicy food liking and intake. *Food Quality and Preference, 42*, 12–19.

Calvo, S. E., Pagliarini, D. J., & Mootha, V. K. (2009). Upstream open reading frames cause widespread reduction of protein expression and are polymorphic among humans. *Proceedings of the National Academy of Sciences of the United States of America, 106*, 7507–7512.

Chandrashekar, J., Hoon, M. A., Ryba, N. J. P., & Zuker, C. S. (2006). The receptors and cells for mammalian taste. *Nature, 444*, 288–294.

Chess, A., Simon, I., Cedar, H., & Axel, R. (1994). Allelic inactivation regulates olfactory receptor gene expression. *Cell, 78*, 823–834.

Colares-Bento, F. C., Souza, V. C., Toledo, J. O., Moraes, C. F., Alho, C. S., Lima, R. M., et al. (2012). Implication of the G145C polymorphism (rs713598) of the TAS2r38 gene on food consumption by Brazilian older women. *Archives of Gerontology and Geriatrics, 54*, e13–e18.

Curtis, B. R., & Aster, R. H. (1996). Incidence of the Naka-negative platelet phenotype in African Americans is similar to that of Asians. *Transfusion, 36*, 331–334.

DeSnoo, K. (1937). Das trinkende Kind in Uterus [The drinking baby in the uterus]. *Monatsschr fur Geburtshilfe Gynaekol, 105*, 88–97.

Dias, A. G., Rousseau, D., Duizer, L., Cockburn, M., Chiu, W., Nielsen, D., et al. (2013). Genetic variation in putative salt taste receptors and salt taste perception in humans. *Chemical Senses, 38*, 137–145.

Dinehart, M. E., Hayes, J. E., Bartoshuk, L. M., Lanier, S. L., & Duffy, V. B. (2006). Bitter taste markers explain variability in vegetable sweetness, bitterness, and intake. *Physiological and Behavioral*, *87*, 304–313.

Dinnella, C., Recchia, A., Fia, G., Bertuccioli, M., & Monteleone, E. (2009). Saliva characteristics and individual sensitivity to phenolic astringent stimuli. *Chemical Senses*, *34*, 295–304.

Dinnella, C., Recchia, A., Vincenzi, S., Tuorila, H., & Monteleone, E. (2010). Temporary modification of salivary protein profile and individual responses to repeated phenolic astringent stimuli. *Chemical Senses*, *35*, 75–85.

Dotson, C. D., Wallace, M. R., Bartoshuk, L. M., & Logan, H. L. (2012). Variation in the gene TAS2R13 is associated with differences in alcohol consumption in patients with head and neck cancer. *Chemical Senses*, *37*, 737–744.

Dubois, G. E. (2011). Validity of early indirect models of taste active sites and advances in new taste technologies enabled by improved models. *Flavour and Fragrance Journal*, *26*, 239–253.

Duffy, V. B. (2007). Variation in oral sensation: Implications for diet and health. *Current Opinion in Gastroenterology*, *23*, 171–177.

Duffy, V. B., Hayes, J. E., Davidson, A. C., Kidd, J. R., Kidd, K. K., & Bartoshuk, L. M. (2010). Vegetable intake in college-aged adults is explained by oral sensory phenotypes and TAS2R38 genotype. *Chemosensory Perception*, *3*, 137–148.

Duffy, V. B., Hayes, J. E., Sullivan, B. S., & Faghri, P. (2009). Surveying food and beverage liking: A tool for epidemiological studies to connect chemosensation with health outcomes. *Annals of the New York Academy of Sciences*, *1170*, 558–568.

Duffy, V. B., Peterson, J. M., & Bartoshuk, L. M. (2004). Associations between taste genetics, oral sensation and alcohol intake. *Physiology & Behavior*, *82*, 435–445.

Elmore, J. R., Heymann, H., Johnson, J., & Hewett, J. E. (1999). Preference mapping: Relating acceptance of "creaminess" to a descriptive sensory map of a semi-solid. *Food Quality and Preference*, *10*, 465–475.

Eriksson, N., Wu, S., Do, C., Kiefer, A., Tung, J., Mountain, J., et al. (2012). A genetic variant near olfactory receptor genes influences cilantro preference. *Flavour*, *1*, 22.

European Commission (2015). *European declaration on alternatives to surgical castration of pigs* [Online]. Available: http://ec.europa.eu/food/animals/welfare/practice/farm/pigs/castration_alternatives/index_en.htm]. Accessed 19.11.15.

Father Knows Worst. (2009), Simpsons episode 438. Directed by LaZebnik, R.

Feeney, E., O'brien, S., Scannell, A., Markey, A., & Gibney, E. R. (2011). Genetic variation in taste perception: Does it have a role in healthy eating? *Proceedings of the Nutrition Society*, *70*, 135–143.

Feeney, E. L., & Hayes, J. E. (2014). Exploring associations between taste perception, oral anatomy and polymorphisms in the carbonic anhydrase (gustin) gene CA6. *Physiology & Behavior*, *128*, 148–154.

Fischer, U., & Noble, A. C. (1994). The effect of ethanol, catechin concentration, and pH on sourness and bitterness of wine. *American Journal of Enology and Viticulture*, *45*, 6–10.

Fox, A. L. (1932). The relationship between chemical constitution and taste. *Proceedings of the National Academy of Sciences of the United States of America*, *18*, 115–120.

Frøst, M. B., & Janhøj, T. (2007). Understanding creaminess. *International Dairy Journal*, *17*, 1298–1311.

Fushan, A. A., Simons, C. T., Slack, J. P., & Drayna, D. (2010). Association between common variation in genes encoding sweet taste signaling components and human sucrose perception. *Chemical Senses*, *35*, 579–592.

Fushan, A. A., Simons, C. T., Slack, J. P., Manichaikul, A., & Drayna, D. (2009). Allelic polymorphism within the TAS1R3 promoter is associated with human taste sensitivity to sucrose. *Current Biology, 19,* 1288–1293.

Garneau, N. L., Nuessle, T. M., Sloan, M. M., Santorico, S. A., Hayes, J. E., & Coughlin, B. C. (2014). Crowdsourcing taste research: Genetic and phenotypic predictors of bitter taste perception as a model. *Frontiers in Integrative Neuroscience, 8,* 33.

Gilbertson, T. A., & Khan, N. A. (2014). Cell signaling mechanisms of oro-gustatory detection of dietary fat: Advances and challenges. *Progress in Lipid Research, 53,* 82–92.

Gorovic, N., Afzal, S., Tjonneland, A., Overvad, K., Vogel, U., Albrechtsen, C., et al. (2011). Genetic variation in the hTAS2R38 taste receptor and brassica vegetable intake. *Scandinavian Journal of Clinical and Laboratory Investigation, 71,* 274–279.

Greene, L. S., Desor, J. A., & Maller, O. (1975). Heredity and experience: Their relative importance in the development of taste preference in man. *Journal of Comparative and Physiological Psychology, 89,* 279–284.

Guimaraes, M. Z. P., & Jordt, S. E. (2007). TRPA1: A sensory channel of many talents. In W. B. Liedtke & S. Heller (Eds.), *TRP ion channel function in sensory transduction and cellular signaling cascades.* Boca Raton, FL: CRC and Taylor & Francis.

Guinard, J. X., Zoumas-Morse, C., & Walchak, C. (1997). Relation between parotid saliva flow and composition and the perception of gustatory and trigeminal stimuli in foods. *Physiology & Behavior, 63,* 109–118.

Harris, H., & Kalmus, H. (1949). The measurement of taste sensitivity to phenylthiourea. *Annals of Eugenics, 15,* 24–31.

Harrison, D., Yamada, J., Adams-Webber, T., Ohlsson, A., Beyene, J., & Stevens, B. (2015). Sweet tasting solutions for reduction of needle-related procedural pain in children aged one to 16 years. *The Cochrane Database of Systematic Reviews, 5* CD008408.

Hasin, Y., Olender, T., Khen, M., Gonzaga-Jauregui, C., Kim, P. M., Urban, A. E., et al. (2008). High-resolution copy-number variation map reflects human olfactory receptor diversity and evolution. *PLoS Genetics, 4,* e1000249.

Hasin-Brumshtein, Y., Lancet, D., & Olender, T. (2009). Human olfaction: From genomic variation to phenotypic diversity. *Trends in Genetics, 25,* 178–184.

Hayes, J. (2008). Transdisciplinary perspectives on sweetness. *Chemosensory Perception, 1,* 48–57.

Hayes, J. E. (2014). Measuring sensory perception in relation to consumer behavior. In J. Delarue, J. B. Lawlor, & M. Rogeaux (Eds.), *Rapid sensory profiling techniques and related methods: Applications in new product development and consumer research.* Waltham, MA: Woodhead Pub.

Hayes, J. E., Bartoshuk, L. M., Kidd, J. R., & Duffy, V. B. (2008). Supertasting and PROP bitterness depends on more than the TAS2R38 gene. *Chemical Senses, 33,* 255–265.

Hayes, J. E., & Duffy, V. B. (2007). Revisiting sugar-fat mixtures: Sweetness and creaminess vary with phenotypic markers of oral sensation. *Chemical Senses, 32,* 225–236.

Hayes, J. E., & Duffy, V. B. (2008). Oral sensory phenotype identifies level of sugar and fat required for maximal liking. *Physiological and Behavioral, 95,* 77–87.

Hayes, J. E., Feeney, E. L., Nolden, A. A., & Mcgeary, J. E. (2015). Quinine bitterness and grapefruit liking associate with allelic variants in TAS2R31. *Chemical Senses, 40,* 437–443.

Hayes, J. E., & Keast, R. S. (2011). Two decades of supertasting: Where do we stand? *Physiological and Behavioral, 104,* 1072–1074.

Hayes, J. E., Mcgeary, J. E., Grenga, A., & Swift, R. M. (2010). Do TAS1R3 promoter region SNP rs35744813 a allele carriers show a reduced response to concentrated sucrose? AChemS XXXII, St. Pete's Beach, FL.

Hayes, J. E., Sullivan, B. S., & Duffy, V. B. (2010). Explaining variability in sodium intake through oral sensory phenotype, salt sensation and liking. *Physiological and Behavioral, 100*, 369–380.

Hayes, J. E., Wallace, M. R., Knopik, V. S., Herbstman, D. M., Bartoshuk, L. M., & Duffy, V. B. (2011). Allelic variation in TAS2R bitter receptor genes associates with variation in sensations from and ingestive behaviors toward common bitter beverages in adults. *Chemical Senses, 36*, 311–319.

Hinrichs, A. L., Wang, J. C., Bufe, B., Kwon, J. M., Budde, J., Allen, R., et al. (2006). Functional variant in a bitter-taste receptor (hTAS2R16) influences risk of alcohol dependence. *American Journal of Human Genetics, 78*, 103–111.

Hofmann, T. (2009). Identification of the key bitter compounds in our daily diet is a prerequisite for the understanding of the hTAS2R gene polymorphisms affecting food choice. *Annals of the New York Academy of Sciences, 1170*, 116–125.

Horne, J., Hayes, J., & Lawless, H. T. (2002). Turbidity as a measure of salivary protein reactions with astringent substances. *Chemical Senses, 27*, 653–659.

Hummler, E., & Vallon, V. (2005). Lessons from mouse mutants of epithelial sodium channel and its regulatory proteins. *Journal of the American Society of Nephrology, 16*, 3160–3166.

Hwang, L.-D., Zhu, G., Breslin, P. A. S., Reed, D. R., Martin, N. G., & Wright, M. J. (2015). A common genetic influence on human intensity ratings of sugars and high-potency sweeteners. *Twin Research and Human Genetics, 18*, 361–367.

Intranuovo, L. R., & Powers, A. S. (1998). The perceived bitterness of beer and 6-n-propylthiouracil (PROP) taste sensitivity. *Annals of the New York Academy of Sciences, 855*, 813–815.

Ishikawa, T., & Noble, A. C. (1995). Temporal perception of astringency and sweetness in red wine. *Food Quality and Preference, 6*, 27–33.

Jaeger, S. R., Mcrae, J. F., Bava, C. M., Beresford, M. K., Hunter, D., Jia, Y., et al. (2013). A Mendelian trait for olfactory sensitivity affects odor experience and food selection. *Current Biology, 23*, 1601–1605.

Jaeger, S. R., Pineau, B., Bava, C. M., Atkinson, K. R., Mcrae, J. F., Axten, L. G., et al. (2012). Investigation of the impact of sensitivity to cis-3-hexen-1-ol (green/grassy) on food acceptability and selection. *Food Quality and Preference, 24*, 230–242.

Keller, A., Zhuang, H., Chi, Q., Vosshall, L. B., & Matsunami, H. (2007). Genetic variation in a human odorant receptor alters odour perception. *Nature, 449*, 468–472.

Keller, K. L., Liang, L. C. H., Sakimura, J., May, D., Van Belle, C., Breen, C., et al. (2012). Common variants in the CD36 gene are associated with oral fat perception, fat preferences, and obesity in african americans. *Obesity, 20*, 1066–1073.

Kirkmeyer, S. V., & Tepper, B. J. (2003). Understanding creaminess perception of dairy products using free-choice profiling and genetic responsivity to 6-n-propylthiouracil. *Chemical Senses, 28*, 527–536.

Knaapila, A., Zhu, G., Medland, S. E., Wysocki, C. J., Montgomery, G. W., Martin, N. G., et al. (2012). A genome-wide study on the perception of the odorants androstenone and galaxolide. *Chemical Senses, 37*, 541–552.

Kuhn, C., Bufe, B., Winnig, M., Hofmann, T., Frank, O., Behrens, M., et al. (2004). Bitter taste receptors for saccharin and acesulfame K. *Journal of Neuroscience, 24*, 10260–10265.

Lawless, H. T., Thomas, C. J., & Johnston, M. (1995). Variation in odor thresholds for l-carvone and cineole and correlations with suprathreshold intensity ratings. *Chemical Senses, 20*, 9–17.

Lee, C. B., & Lawless, H. T. (1991). Time-course of astringent sensations. *Chemical Senses, 16*, 225–238.

Liem, D. G., Westerbeek, A., Wolterink, S., Kok, F. J., & De Graaf, C. (2004). Sour taste preferences of children relate to preference for novel and intense stimuli. *Chemical Senses, 29*, 713–720.

Liley, A. W. (1972). Disorders of amniotic fluid. In N. S. Assali (Ed.), *Pathophysiology of gestation*. New York, NY: Academic Press.

Lim, J., Urban, L., & Green, B. G. (2008). Measures of individual differences in taste and creaminess perception. *Chemical Senses, 33*, 493–501.

Lugaz, O., Pillias, A. M., & Faurion, A. (2002). A new specific ageusia: Some humans cannot taste L-glutamate. *Chemical Senses, 27*, 105–115.

Lunde, K., Egelandsdal, B., Skuterud, E., Mainland, J. D., Lea, T., Hersleth, M., et al. (2012). Genetic variation of an odorant receptor OR7D4 and sensory perception of cooked meat containing androstenone. *PLoS ONE, 7*, e35259.

Lundgren, B., Jonsson, B., Pangborn, R. M., Sontag, A. M., Barylko-Pikielna, N., Pietrzak, E., et al. (1978). Taste discrimination vs hedonic response to sucrose in coffee beverage. An interlaboratory study. *Chemical Senses, 3*, 249–265.

Mainland, J. D., Keller, A., Li, Y. R., Zhou, T., Trimmer, C., Snyder, L. L., et al. (2014). The missense of smell: Functional variability in the human odorant receptor repertoire. *Nature Neuroscience, 17*, 114–120.

Maller, O., & Desor, J. A. (1973). Effect of taste on ingestion by human newborns. *Symposium on Oral Sensation and Perception, 4*, 279–291.

Mandel, A. L., Peyrot Des Gachons, C., Plank, K. L., Alarcon, S., & Breslin, P. A. (2010). Individual differences in AMY1 gene copy number, salivary alpha-amylase levels, and the perception of oral starch. *PLoS One, 5*, e13352.

Maone, T. R., Mattes, R. D., Bernbaum, J. C., & Beauchamp, G. K. (1990). A new method for delivering a taste without fluids to preterm and term infants. *Developmental Psychobiology, 23*, 179–191.

Martin, C., Passilly-Degrace, P., Gaillard, D., Merlin, J. F., Chevrot, M., & Besnard, P. (2011). The lipid-sensor candidates CD36 and GPR120 are differentially regulated by dietary lipids in mouse taste buds: Impact on spontaneous fat preference. *PLoS ONE, 6*, e24014.

Mauer, L., & El-Sohemy, A. (2012). Prevalence of cilantro (*Coriandrum sativum*) disliking among different ethnocultural groups. *Flavour, 1*, 8.

Mauer, L. K. (2011). *Genetic determinants of cilantro preference*. Dissertation. Master of Science, University of Toronto.

Mcrae, J. F., Jaeger, S. R., Bava, C. M., Beresford, M. K., Hunter, D., Jia, Y., et al. (2013). Identification of regions associated with variation in sensitivity to food-related odors in the human genome. *Current Biology, 23*, 1596–1600.

Melis, M., Atzori, E., Cabras, S., Zonza, A., Calo, C., Muroni, P., et al. (2013). The gustin (CA6) gene polymorphism, rs2274333 (A/G), as a mechanistic link between PROP tasting and fungiform taste papilla density and maintenance. *PLoS One, 8*, e74151.

Melis, M., Sollai, G., Muroni, P., Crnjar, R., & Barbarossa, I. T. (2015). Associations between orosensory perception of oleic acid, the common single nucleotide polymorphisms (rs1761667 and rs1527483) in the CD36 gene, and 6-n-propylthiouracil (PROP) tasting. *Nutrients, 7*, 2068–2084.

Menashe, I., Abaffy, T., Hasin, Y., Goshen, S., Yahalom, V., Luetje, C. W., et al. (2007). Genetic elucidation of human hyperosmia to isovaleric acid. *PLoS Biology, 5*, e284.

Menashe, I., Man, O., Lancet, D., & Gilad, Y. (2003). Different noses for different people. *Nature Genetics, 34*, 143–144.

Mennella, J. A., Reed, D. R., Mathew, P. S., Roberts, K. M., & Mansfield, C. J. (2015). "A spoonful of sugar helps the medicine go down": Bitter masking by sucrose among children and adults. *Chemical Senses, 40*, 17–25.

Meyerhof, W., Batram, C., Kuhn, C., Brockhoff, A., Chudoba, E., Bufe, B., et al. (2010). The molecular receptive ranges of human TAS2R bitter taste receptors. *Chemical Senses, 35*, 157–170.

Morlein, D., Meier-Dinkel, L., Moritz, J., Sharifi, A. R., & Knorr, C. (2013). Learning to smell: Repeated exposure increases sensitivity to androstenone, a major component of boar taint. *Meat Science, 94*, 425–431.

Mrizak, I., Sery, O., Plesnik, J., Arfa, A., Fekih, M., Bouslema, A., et al. (2015). The A allele of cluster of differentiation 36 (CD36) SNP 1761667 associates with decreased lipid taste perception in obese Tunisian women. *The British Journal of Nutrition, 113*, 1330–1337.

Noh, H., Paik, H. Y., Kim, J., & Chung, J. (2013). Salty taste acuity is affected by the joint action of alphaENaC A663T gene polymorphism and available zinc intake in young women. *Nutrients, 5*, 4950–4963.

Nolden, A., & Hayes, J. (2015). Perceptual qualities of ethanol depend on concentration, and variation in these percepts associates with drinking frequency. *Chemosensory Perception, 8*, 149–157.

Nolden, A. A., McGeary, J. E., & Hayes, J. E. (2016). Differential bitterness in capsaicin, piperine, and ethanol associates with polymorphisms in multiple bitter taste receptor genes. *Physiology & Behavior, 156*, 117–127.

Olender, T., Waszak, S. M., Viavant, M., Khen, M., Ben-Asher, E., Reyes, A., et al. (2012). Personal receptor repertoires: Olfaction as a model. *BMC Genomics, 13*, 414.

Ozdener, M. H., Subramaniam, S., Sundaresan, S., Sery, O., Hashimoto, T., Asakawa, Y., et al. (2014). CD36- and GPR120-mediated Ca^{2+} signaling in human taste bud cells mediates differential responses to fatty acids and is altered in obese mice. *Gastroenterology, 146*, 995–1005.

Patterson, P. M., & Lauder, B. A. (1948). The incidence and probable inheritance of smell blindness to normal butyl mercaptan. *The Journal of Heredity, 39*, 295–297.

Pepino, M. Y., Love-Gregory, L., Klein, S., & Abumrad, N. A. (2012). The fatty acid translocase gene CD36 and lingual lipase influence oral sensitivity to fat in obese subjects. *Journal of Lipid Research, 53*, 561–566.

Perry, G. H., Dominy, N. J., Claw, K. G., Lee, A. S., Fiegler, H., Redon, R., et al. (2007). Diet and the evolution of human amylase gene copy number variation. *Nature Genetics, 39*, 1256–1260.

Prescott, J., Norris, L., Kunst, M., & Kim, S. (2005). Estimating a "consumer rejection threshold" for cork taint in white wine. *Food Quality and Preference, 16*, 345–349.

Pronin, A. N., Xu, H., Tang, H., Zhang, L., Li, Q., & Li, X. (2007). Specific alleles of bitter receptor genes influence human sensitivity to the bitterness of aloin and saccharin.. *Current Biology, 17*, 1403–1408.

Rac, M. E., Safranow, K., & Poncyljusz, W. (2007). Molecular basis of human CD36 gene mutations. *Molecular Medicine, 13*, 288–296.

Raliou, M., Grauso, M., Hoffmann, B., Schlegel-Le-Poupon, C., Nespoulous, C., Debat, H., et al. (2011). Human genetic polymorphisms in T1R1 and T1R3 taste receptor subunits affect their function. *Chemical Senses, 36*, 527–537.

Reed, D. R., Tanaka, T., & McDaniel, A. H. (2006). Diverse tastes: Genetics of sweet and bitter perception. *Physiology & Behavior, 88*, 215–226.

Reed, D. R., Zhu, G., Breslin, P. A., Duke, F. F., Henders, A. K., Campbell, M. J., et al. (2010). The perception of quinine taste intensity is associated with common genetic variants in a bitter receptor cluster on chromosome 12. *Human Molecular Genetics, 19*, 4278–4285.

Roudnitzky, N., Bufe, B., Thalmann, S., Kuhn, C., Gunn, H. C., Xing, C., et al. (2011). Genomic, genetic and functional dissection of bitter taste responses to artificial sweeteners. *Human Molecular Genetics, 20*, 3437–3449.

Roura, E., Aldayyani, A., Thavaraj, P., Prakash, S., Greenway, D., Thomas, W. G., et al. (2015). Variability in human bitter taste sensitivity to chemically diverse compounds can be accounted for by differential TAS2R activation. *Chemical Senses, 40*, 427–435.

Running, C. A., Craig, B. A., & Mattes, R. D. (2015). Oleogustus: The unique taste of fat. *Chemical Senses, 40*, 507–516.

Running, C. A., & Mattes, R. D. (2014a). Different oral sensitivities to and sensations of short-, medium-, and long-chain fatty acids in humans. *American Journal of Physiology. Gastrointestinal and Liver Physiology, 307*, G381–G389.

Running, C. A., & Mattes, R. D. (2014b). Humans are more sensitive to the taste of linoleic and α-linolenic than oleic acid. *American Journal of Physiology. Gastrointestinal and Liver Physiology, 308*, G442–G449.

Running, C. A., Mattes, R. D., & Tucker, R. M. (2013). Fat taste in humans: Sources of within- and between-subject variability. *Progress in Lipid Research, 52*, 438–445.

Sacerdote, C., Matullo, G., Polidoro, S., Gamberini, S., Piazza, A., Karagas, M. R., et al. (2007). Intake of fruits and vegetables and polymorphisms in DNA repair genes in bladder cancer. *Mutagenesis, 22*, 281–285.

Sandell, M. A., & Breslin, P. A. (2006). Variability in a taste-receptor gene determines whether we taste toxins in food. *Current Biology, 16*, R792–R794.

Schiffman, S. S., Graham, B. G., Sattely-Miller, E. A., & Warwick, Z. S. (1998). Orosensory perception of dietary fat. *Current Directions in Psychological Science, 7*, 137–143.

Schutz, M., Oertel, B. G., Heimann, D., Doehring, A., Walter, C., Dimova, V., et al. (2014). Consequences of a human TRPA1 genetic variant on the perception of nociceptive and olfactory stimuli. *PLoS One, 9*, e95592.

Shigemura, N., Shirosaki, S., Sanematsu, K., Yoshida, R., & Ninomiya, Y. (2009). Genetic and molecular basis of individual differences in human umami taste perception. *PLoS One, 4*, e6717.

Soranzo, N., Bufe, B., Sabeti, P. C., Wilson, J. F., Weale, M. E., Marguerie, R., et al. (2005). Positive selection on a high-sensitivity allele of the human bitter-taste receptor TAS2R16. *Current Biology, 15*, 1257–1265.

Stewart, J. E., Feinle-Bisset, C., Golding, M., Delahunty, C., Clifton, P. M., & Keast, R. S. J. (2010). Oral sensitivity to fatty acids, food consumption and BMI in human subjects. *The British Journal of Nutrition, 104*, 145–152.

Stewart, J. E., & Keast, R. S. J. (2012). Recent fat intake modulates fat taste sensitivity in lean and overweight subjects. *The International Journal of Obesity, 36*, 834–842.

Take, H., Kashiwagi, H., Tomiyama, Y., Honda, S., Honda, Y., Mizutani, H., et al. (1993). Expression of GPIV and N(aka) antigen on monocytes in N(aka)-negative subjects whose platelets lack GPIV. *British Journal of Haematology, 84*, 387–391.

Tatzer, E., Schubert, M. T., Timischl, W., & Simbruner, G. (1985). Discrimination of taste and preference for sweet in premature babies. *Early Human Development, 12*, 23–30.

Temussi, P. (2007). The sweet taste receptor: A single receptor with multiple sites and modes of interaction. *Advances in Food and Nutrition Research, 53*, 199–239.

Tepper, B. J., & Nurse, R. J. (1998). PROP taster status is related to fat perception and preference. *Annals of the New York Academy of Sciences, 855*, 802–804.

Timpson, N. J., Christensen, M., Lawlor, D. A., Gaunt, T. R., Day, I. N., Ebrahim, S., et al. (2005). TAS2R38 (phenylthiocarbamide) haplotypes, coronary heart disease traits, and eating behavior in the British Women's Heart and Health Study. *The American Journal of Clinical Nutrition, 81*, 1005–1011.

Törnwall, O., Dinnella, C., Keskitalo-Vuokko, K., Silventoinen, K., Perola, M., Monteleone, E., et al. (2011). Astringency perception and heritability among young finnish twins. *Chemosensory Perception, 4*, 134–144.

Törnwall, O., Silventoinen, K., Kaprio, J., & Tuorila, H. (2012). Why do some like it hot? Genetic and environmental contributions to the pleasantness of oral pungency. *Physiology & Behavior, 107*, 381–389.

Törnwall, O., Silventoinen, K., Keskitalo-Vuokko, K., Perola, M., Kaprio, J., & Tuorila, H. (2012). Genetic contribution to sour taste preference. *Appetite, 58*, 687–694.

Tucker, R. M., & Mattes, R. D. (2013). Influences of repeated testing on nonesterified fatty acid taste. *Chemical Senses, 38*, 325–332.

Tucker, R. M., Mattes, R. D., & Running, C. A. (2014). Mechanisms and effects of "fat taste" in humans. *Biofactors, 40*, 313–326.

Tucker, R. M., Nuessle, T. M., Garneau, N. L., Smutzer, G., & Mattes, R. D. (2015). No difference in perceived intensity of linoleic acid in the oral cavity between obese and nonobese individuals. *Chemical Senses, 40*, 557–563.

Wald, N., & Leshem, M. (2003). Salt conditions a flavor preference or aversion after exercise depending on NaCl dose and sweat loss. *Appetite, 40*, 277–284.

Wang, J. C., Hinrichs, A. L., Bertelsen, S., Stock, H., Budde, J. P., Dick, D. M., et al. (2007). Functional variants in TAS2R38 and TAS2R16 influence alcohol consumption in high-risk families of African-American origin. *Alcoholism, Clinical and Experimental Research, 31*, 209–215.

Waszak, S. M., Hasin, Y., Zichner, T., Olender, T., Keydar, I., Khen, M., et al. (2010). Systematic inference of copy-number genotypes from personal genome sequencing data reveals extensive olfactory receptor gene content diversity. *PLoS Computational Biology, 6*, e1000988.

Webb, J., Bolhuis, D. P., Cicerale, S., Hayes, J. E., & Keast, R. (2015). The relationships between common measurements of taste function. *Chemosensory Perception, 8*, 11–18.

Weiland, R., Ellgring, H., & Macht, M. (2010). Gustofacial and olfactofacial responses in human adults. *Chemical Senses, 35*, 841–853.

Wise, P. M., Hansen, J. L., Reed, D. R., & Breslin, P. A. (2007). Twin study of the heritability of recognition thresholds for sour and salty taste. *Chemical Senses, 32*, 749–754.

Wooding, S., Gunn, H., Ramos, P., Thalmann, S., Xing, C., & Meyerhof, W. (2010). Genetics and bitter taste responses to goitrin, a plant toxin found in vegetables. *Chemical Senses, 35*, 685–692.

Wysocki, C. J., Dorries, K. M., & Beauchamp, G. K. (1989). Ability to perceive androstenone can be acquired by ostensibly anosmic people. *Proceedings of the National Academy of Sciences of the United States of America, 86*, 7976–7978.

Xu, X., Ye, X., Xia, W., Liu, J., Ding, H., Deng, J., et al. (2013). Studies on CD36 deficiency in South China: Two cases demonstrating the clinical impact of anti-CD36 antibodies. *Thrombosis and Haemostasis, 110*, 1199–1206.

Yamamoto, N., Akamatsu, N., Sakuraba, H., Yamazaki, H., & Tanoue, K. (1994). Platelet glycoprotein IV (CD36) deficiency is associated with the absence (type I) or the presence (type II) of glycoprotein IV on monocytes. *Blood, 83*, 392–397.

Yang, Z. M., Lin, J., Chen, L. H., Zhang, M., Chen, W. W., & Yang, X. R. (2015). The roles of AMY1 copies and protein expression in human salivary alpha-amylase activity. *Physiology & Behavior, 138*, 173–178.

Zhang, X., & Firestein, S. (2009). Genomics of olfactory receptors. *Results and Problems in Cell Differentiation, 47*, 25–36.

Pleasure of Food in the Brain

Alexander Fjaeldstad[1,3], Tim J. van Hartevelt[1,2] and Morten L. Kringelbach[1,2]

[1]Department of Psychiatry, University of Oxford, Oxford, United Kingdom, [2]Center of Functionally Integrative Neuroscience (CFIN), Aarhus University, Aarhus, Denmark, [3]Department of Otorhinolaryngology, Aarhus University Hospital, Aarhus, Denmark

1 Introduction

The perception of sensory stimuli dictates how we perceive the world around us and ultimately how we behave, from the simple somatosensory perception of touch to the complex multisensory perception of social interactions, listening to music, or the processing of food stimuli. In a world of nearly endless stimuli, where relevant ones can be sparse, sensory systems have developed in highly differentiated ways in order to enhance the likelihood of survival. This system prioritizes the prudent use of metabolic resources, from efficiency-optimized pathway splitting in peripheral sensory coding and other sparse coding strategies (Gjorgjieva, Sompolinsky, & Meister, 2014) through to efficient central processing and prioritization by means of attention.

Perhaps evolution's boldest trick to ensure the survival of both the individual and the species has been to guide individual behavior towards seeking pleasure and avoiding threats. Pleasure plays an essential role in the reallocation of resources to ensure survival and procreation by rewarding certain favorable behaviors and stimuli (Kringelbach, 2005). A densely connected mesocorticolimbic network of hedonic areas encodes the pleasure of fundamental rewards (food, sex, and social stimuli) and more abstract rewards (eg, money, music, and art) (Kringelbach & Berridge, 2010). Studies of diverse fundamental and abstract pleasures suggest that the valence of reward is processed in a unitary pleasure circuitry (Berridge & Kringelbach, 2015; Kringelbach, 2015).

The scientific study of hedonia (Greek for pleasure: *hédoné* refers to the sweet taste of honey, *hedus*) has, in parallel with the development of neuroimaging techniques, made significant strides over the past few decades. A major step in this development has been the dissection of the reward circuitry into several psychological subcomponents: wanting (motivational processes of incentive salience), liking (core reactions to hedonic impact), and learning (cognitive representations and Pavlovian or instrumental associations) based on careful animal experimentation (Berridge & Kringelbach, 2013; Berridge & Robinson, 2003; Finlayson, King, & Blundell, 2007) (see Fig. 11.1). Separate neurotransmitters within the distinct neuroanatomical hedonic areas have further characterized this subdivision. The process of "wanting" emerges from a large and distributed brain network (and is primarily associated with the neurotransmitter dopamine), while "liking" emerges from the concerted activity of a group of rather small hedonic hotspots within primarily subcortical areas (and is primarily associated with opioids) (Castro & Berridge, 2014; Smith & Berridge, 2005).

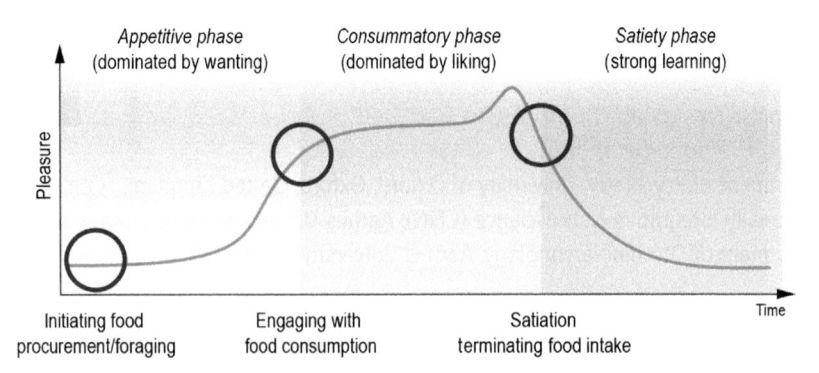

Figure 11.1 The pleasure cycle. The cyclic processing of hedonics is classically described with an appetitive, a consummatory, and a satiety phase (Craig, 1917; Sherrington, 1906). This cyclic processing is supported by multiple brain networks, each with a role to play in the processing of the three different aspects of reward: wanting (motivational processing of incentive salience), liking (the core responses to hedonic impact), and learning (typically instrumental associations or cognitive representations). Although these components can co-occur at any time during the pleasure cycle, wanting tends to dominate the appetitive phase, whereas liking dominates the consummatory phase. Learning is strongest during the satiety phase but can occur throughout the cycle.

Hedonic food processing involves the integration of smell, taste, vision, somatosensation, chemesthesis, and sound with wanting, liking, and learning. While dependent on widespread and highly complex interconnected processing, the multisensory processing of food and drink has been studied and subdivided into unisensory stimuli in behavioral and neuroimaging studies, which both practically and ethically is more feasible than other fundamental and abstract rewards such as sex and art. Although it is highly challenging to account for multisensory awareness (Deroy, Chen, & Spence, 2014), appreciating the contribution of each sensory system has paved the way to a deeper understanding of superadditive and subadditive effects during stimulation of multiple sensory systems (de Araujo, Rolls, Kringelbach, McGlone, & Phillips, 2003) and complex multisensory integration.

In this chapter, the processing of individual sensory stimuli, the integrated multisensory processing related to the fundamental pleasure of food, and the principles of hedonic processing are described.

2 Brain principles of eating

The development of highly specialized olfactory abilities (Rowe & Shepherd, 2016) is likely to have been a driving force for the evolution of the sizeable primate brains (Kringelbach, 2004). The main challenge for the brain is to efficiently balance resource allocation to serve the evolutionary imperatives of survival and procreation (Lou, Joensson, & Kringelbach, 2011). With reward as the primary currency

(Kringelbach & Berridge, 2010), different reward-triggering actions continuously strive for resources to achieve this balance.

This constant balancing of processing consists of a cyclical pattern for all sources of pleasure, which is important to bear in mind when trying to understand the complex brain subcomponents underlying pleasure and reward. The brain principles of eating undergo characteristic cyclical appetitive (before), consummatory (during), and satiety phases (after), which are linked to the different subcomponents of reward: "wanting" (motivational processing dominating the appetitive phase), "liking" (core pleasure component, dominating the consummatory phase), and "learning" (cognitive processing or instrumental association, dominating the satiety phase, but can occur throughout the entire cycle) (see Fig. 11.1) (Berridge & Robinson, 2003; Craig, 1917).

Within this cycle, reward acts as an incentive enticement to initiate, sustain, and switch brain states, where the wanting, liking, and learning components wax and wane throughout the cycle and can co-occur at any time. As shown in Fig. 11.2, the cyclical pleasure model can be extended into an intricate model of eating behavior, where the cyclical changes in hunger levels can be linked to the perceived pleasure, ultimately regulating the initiation of meals in a complex interaction with intrinsic signals, such as blood glucose and metabolite levels, gut–brain signals, feedback from the oral cavity, stomach, intestines, liver, and body mass (Kringelbach, Stein, & Van Hartevelt, 2012). Though these internal cues of hunger or cravings can regulate behavior, top-down food-related triggers, such as visual clues or words, can also activate a cascade of associations that leads to food-seeking behavior (Papies, 2013) and can even initiate pathological processes in eating disorders (García-García et al., 2013). As with the initiation of a meal, the underlying processes terminating a meal are equally complex. Satiety and satiation are two determining factors, which affect the perceived pleasure of continuous eating and are closely linked to the amount of food and energy, but also the type of nutrients gained by the terminated meal (Johansson, Lee, Risérus, Langton, & Landberg, 2015). In case it is a specific food item that is consumed to satiety, something called sensory-specific satiety can occur (de Graaf & Kok, 2010; Kringelbach, O'Doherty, Rolls, & Andrews, 2003). While satiation is defined as the factor terminating the meal, satiety is the feeling of fullness enduring after the meal that suppresses the wanting of further eating. The multilevel processes of terminating the meal are followed by cognitive top-down regulation, chemical signals as well as changes in the blood levels induced by the digested and absorbed meal, which determine the duration of satiety. The abdominal modulatory mechanisms can act directly, or through circulatory metabolites on the hypothalamus and brainstem (Suzuki, Simpson, Minnion, Shillito, & Bloom, 2010). However, the main focus here is to review the brain principles of eating, the multisensory processing, and the underlying hedonic processes.

3 Computational processing related to eating

The computational processing related to eating involves a complex circuitry, which is sustained and modulated by the aforementioned intrinsic signals, sensory stimulation, higher cognitive processes, and the subcomponents of reward. Hunger and attention

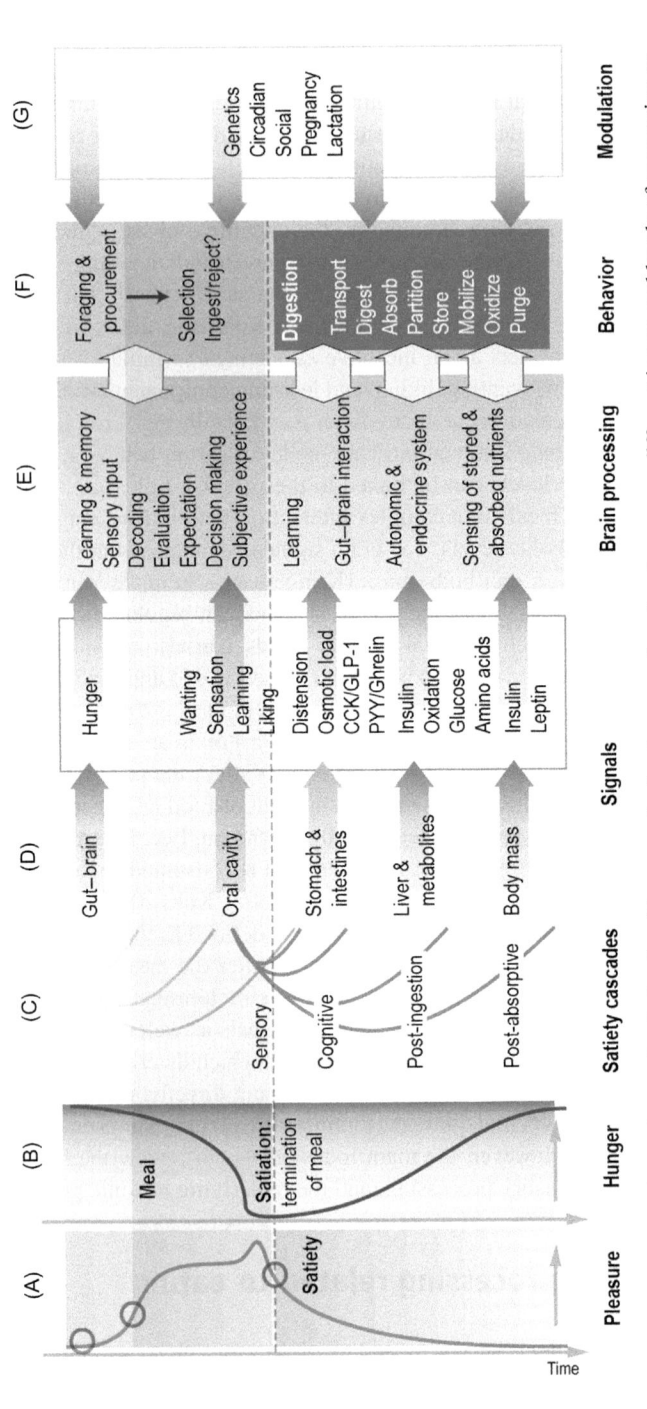

Figure 11.2 Factors influencing eating behavior. The control of eating over time involves many different inputs and levels of processing as illustrated in this model. These intrinsic signals induced by consumption, absorption, and higher cognitive processes can, in unity, illustrate the intake of food over time, which includes the following variables: (A) pleasure/reward, (B) hunger level, (C) satiety and satiation signals, (D) origin of signals, (E) brain processes, (F) behavioral changes and changes in the digestive system, and (G) general modulatory factors (Kringelbach, 2015).

can trigger food-seeking behavior, either by a homeostatic drive arising from biological requirements or by hedonic motivational factors. The multisensory representations of food provide strong triggers of the computational processing related to eating, but learning and cognitive processes can transform these representations. Although sensory stimuli are potent triggers of the hedonic experience of eating, the perception of intensity and identity of sensory stimulation seems to be motivation-independent. The computational processing related to eating is a continuous activity, which can be subdivided into several basic principles and subcomponents described in this section.

3.1 Hunger, homeostasis, and hedonia

A fundamental issue in understanding hunger is the relationship between homeostatic changes in biological needs and the hedonic motivational processes. The initiation of food-seeking behavior, like all other switches of brain states, is driven by changes in either of these processes, signaling that the brain must reallocate resources and a change in behavior is warranted. The homeostatic changes in the internal environment in response to energy and nutrient intake, such as blood glucose and endocrine gut-derived hormones, achieve their effects on the brainstem and the hypothalamus. Within the brainstem, the dorsal vagal complex plays a role in the interpretation and relaying of peripheral signals, while two neural circuits in the arcuate nucleus of the hypothalamus stimulate or inhibit food intake, respectively (Suzuki et al., 2010). The interpretation of gut–brain signals is key in terminating eating behavior, provided that the meal has generated sufficient energy and endocrine response.

Several other factors also determine the time of meal termination and length of satiety phase. Attention, habitual and social aspects during food consumption are highly relevant. Distracting factors such as television usage not only increase the immediate intake, but also shorten the satiety phase (Braude & Stevenson, 2014; de Graaf & Kok, 2010). Social interaction, large portion sizes, or removing visual information about the amount of food eaten during a meal can also increase immediate intake (Bolhuis, Lakemond, de Wijk, Luning, & de Graaf, 2013; Robinson et al., 2013). Although prolonged chewing of the food reduces meal size and later intake, it can also reduce the hedonic experience (Higgs & Jones, 2013). De Graaf and Kok elegantly summarized this general finding of the attention aspects of eating behavior: "foods that can be eaten quickly lead to high food intake and low satiating effects, as these foods only provide brief periods of sensory exposure, which give the human body insufficient cues for satiation" (de Graaf & Kok, 2010).

In environments of plenty, where fast-food options and constant distracting factors during meals are numerous, the multisensory aspects of food intake have drawn increased attention to the hedonic dimension of eating. Through a mismatch between the expected hedonic experience and the actual energy yield, the hedonic system can overthrow the equilibrium of the brain–gut homeostasis and cause over-eating and even contribute to obesity (Zheng, Lenard, Shin, & Berthoud, 2009). The motivational processes of reward are not synchronized with the physiological effects of food consumption and seem to be fundamentally independent of homeostatic regulation of food intake (Finlayson, King, & Blundell, 2008). Analogously, the liking of pleasant food induces reduced sensitivity to several gut–brain satiety signals, as the consumption of pleasant foods is largely driven by an increased activation of the reward, generated by the attractiveness of the flavor (Erlanson-Albertsson, 2005). As such, there always seems to be room for dessert.

3.2 Sensory perception of food-related rewards

The perception of food evolves from an initial distinct sensation, often visual or orthonasal smell cues, from where they merge into a combined multisensory experience during consumption. Food can stimulate all five senses, from the obvious smell and taste to the sight of a well-served dish, the indispensible crunchiness of potato chips, the required warmth when eating a steak, the obligatory fizz in a soda, to the mouth-feel of a tender chateaubriand steak. If any of these qualities differ from what is expected, the consumption will immediately be reevaluated, as any minor alterations can radically change the hedonic experience (Piqueras-Fiszman & Spence, 2015). From highly specialized peripheral receptors, these sensory contributions to the multiple facets of food are processed in localized primary sensory cortices, which share an astonishing homology between human brains (see Fig. 11.3). A comparable pool of receptors and cortical functional correspondence can be found in other vertebrate species, with some evolution-driven functional architectural changes (Mantini et al., 2012).

After reaching the primary cortices, the sensory information is further integrated in multisensory association areas before it is evaluated in the higher association areas such as those included in the hedonic circuitry. As these sensory modalities merge, the food experience extends beyond the simple addition of the attributions of the sensory modalities, and the food is perceived as more object-like, intense, and rewarding (Seubert, Ohla, Yokomukai, Kellermann, & Lundström, 2015). This multisensory merge can be so complete that people fail to appreciate the contributions of other senses (Stevenson, 2014), as the sensory stimuli are described as mere taste. The synaptic routes these sensory stimuli take to reach secondary areas are of utmost importance; while all other sensory modalities undergo modulation in the thalamus, olfactory signals are conveyed directly from the primary olfactory cortex to the secondary olfactory areas, consisting of the amygdala and the orbitofrontal cortex (OFC) (Shepherd, 2005; Van Hartevelt & Kringelbach, 2012), This path, which only involves three neurons (olfactory receptor cells, mitral cells, and olfactory cortical pyramidal neurons), may explain the potent effects of smell on hedonia and memory.

3.2.1 Smell

Chemosensation can be found in all living creatures, even plants (Seo et al., 2001) and bacteria (Nijland & Burgess, 2010) can detect chemicals in their environments and initiate responses accordingly. To be able to sense the environment is an indispensible capability for all life forms in order to survive. However, in the neocortex of humans, olfaction has become a major contributor to the common currency of reward, thus affecting the perception of our environment and subsequently our behavior. The hedonic dimension of olfactory processing is emphasized by the uniform primary

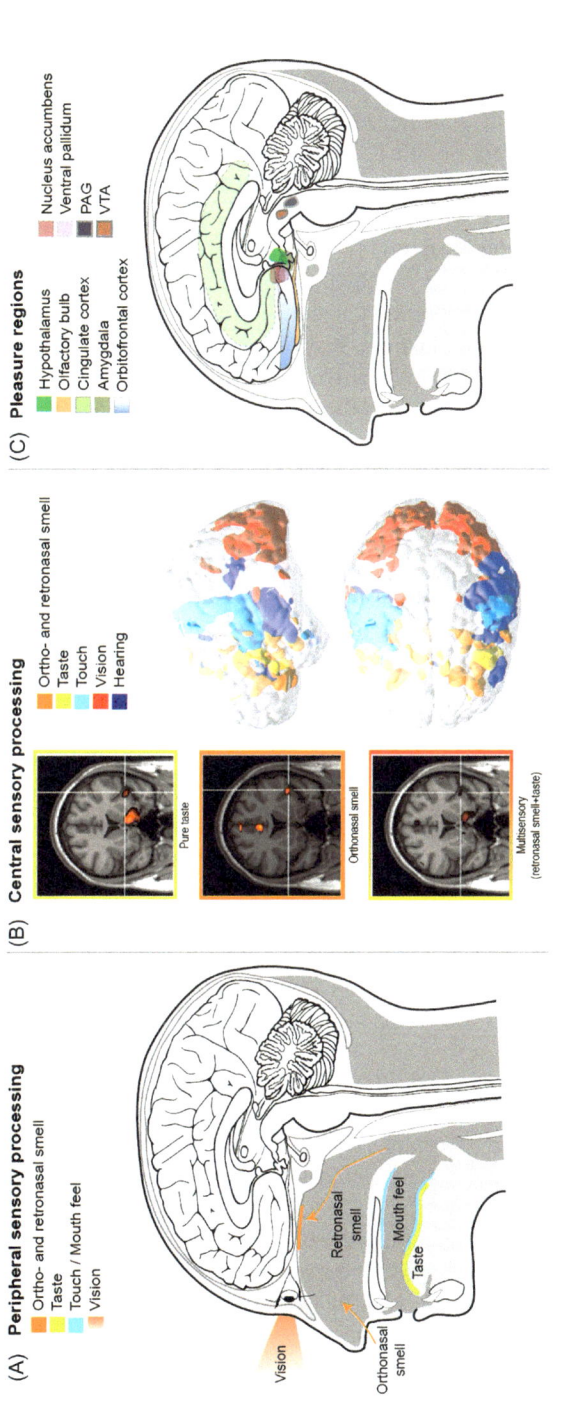

Figure 11.3 Sensory and reward processing. (A) The multisensory contribution to the pleasure of food consumption can involve all of the senses. Their different routes from peripheral receptors in the eyes, ears, nose, and oral cavity reflects the importance of different food qualities: often initiated by the distant processing of the visual appearance, followed by a tactile evaluation of food and the orthonasal smell. If these stimuli support the decision to consume the food, then taste, tactile (mouth-feel), sounds during mastication, and retronasal olfactory processing will provide the brain with cascades of information on the qualities of the food. Smell is the most important contributor to the flavor of food. (B) The topology of central processing is quite uniform between individuals for all sensory modalities: vision (red), hearing (dark blue), touch (light blue), olfaction (orange), and taste (yellow). All senses except olfaction are processed via a thalamic relay (Shepherd, 2005; Tham, Stevenson, & Miller, 2011), which may explain the hedonic potency of olfactory stimuli. (C) The processing of pleasure involves several functionally and topographically distinct regions. The olfactory bulb (orange) is traditionally regarded as a part of the pleasure system, and has an important task in conveying and modulating the olfactory stimuli from the receptors directly to the piriform cortex, entorhinal cortex, and amygdala, among others. The pleasure regions also include the hypothalamus (dark green), cingulate cortex (light green), amygdala (faded green), orbitofrontal cortex (light blue), nucleus accumbens (pink), ventral pallidum (light purple), as well as regions of the brain stem; the periaqueductal gray (PAG, dark purple) is an important region involved in pain, and the ventral tegmental area (VTA, red) is central in dopaminergic regulation.

processing of olfactory signals across species, which is highly correlated to the behavior of approach or withdrawal in animals, and odorant pleasantness in humans (Haddad et al., 2010).

The entry point for olfactory stimuli is in the olfactory epithelium, where odor molecules interact with olfactory receptor proteins. Each sensory neuron expresses a single olfactory receptor (Bargmann, 2006). In most cases, receptors are sensitive to multiple chemical compounds (Hallem & Carlson, 2006) and the specific activation pattern of different receptors leads to the formation of an "odor image" based on the different receptors being activated, where certain individual molecular components in the complex mixture of odorants are more important than others (Lin, Zhang, Block, & Katz, 2005). Although the human sense of smell may have been reduced from approximately 1000 genes to our present 350–390 active genes involved in olfaction (Buck & Axel, 1991; Olender, Lancet, & Nebert, 2008), the strong human cognitive processing capacities compensate greatly for a more rigid sensory receptor array (Shepherd, 2004) compared to our canine friends. Humans are capable of discriminating more than one trillion olfactory stimuli (Bushdid, Magnasco, Vosshall, & Keller, 2014) and can detect minute traces of certain strong-smelling molecules, a skill not reserved for olfactory professionals. Thus, when a perfumer replaces a single one of the innumerable ingredients of a well-known perfume, consumer protests can quickly arise, emphasizing the sensitivity of the untrained nose and the ability to "compare and distinguish a new and incongruous nuance of a familiar scent" (Holley, 1999).

On the receptor level of other senses, it is well-known that the organization of retinal receptors reflects the spatial localization of stimuli and the cochlear receptors reflects the frequencies of auditory stimuli. The grouping of olfactory receptors and the mitral cells in the olfactory bulb is not as clear-cut, however, the organization of the olfactory receptor surface seems to be influenced by the hedonic valence of odorants (Lapid et al., 2011). The extent to which this reflects an innate configuration or reorganization as a result of repeated stimulation remains a topic of debate (Breton-Provencher & Saghatelyan, 2012).

From an individual perspective, it is widely accepted that there is an unusually high genetic diversity in active human olfactory receptor genes, suggesting a highly personalized inventory of functional olfactory receptors (Olender et al., 2012). Much effort has been put into identifying agonists for human odorant receptors, which in time will shed light on the human olfactory capacities at the receptor level (Gonzalez-Kristeller, do Nascimento, Galante, & Malnic, 2015). This will help understand the degree to which "odor fingerprints" are perceived differently in the olfactory bulb and the primary olfactory cortex, and may also give insights to differences in hedonic responses and flavor preferences.

The hedonic valence of olfactory stimuli is registered through several levels of increasingly advanced processing. Identical sensory neurons converge in glomeruli in the olfactory bulb, where the initial neural representations of component coding occur (Lin et al., 2005) before the converged sensory signals reach the primary olfactory cortex. In both the olfactory bulb and the piriform cortex, stable circuits of highly active neurons show activity specifically related to hedonic stimuli (Shakhawat et al., 2014), which underlines the importance of hedonics in the olfactory system.

The direction of olfactory stimuli is an important variable, as the perception of the same odor can differ when perceived through the orthonasal or retronasal route (Small, Gerber, Mak, & Hummel, 2005). The importance of retronasal olfactory stimuli is illustrated by the massive extent to which they interact with higher associative flavor processing compared with orthonasal olfactory stimuli (Shepherd, 2006). As retronasal stimuli naturally occur simultaneously with taste stimuli during consumption, this close connection between the two senses and the subsequent processing would be expected. An illustrative example of this difference is the smell of coffee, where the orthonasal smell of freshly ground coffee beans offers a somewhat distinct experience compared to the flavor of coffee during consumption, where the acidic and bitter taste, temperature, and mouthfeel result in a completely different perception.

3.2.2 Taste

Although taste is popularly described as the primary sensory experience during food consumption, we can only recognize five prototypical taste qualities: sweet (eg, glucose), salty (NaCl), sour (eg, citric acid), bitter (eg, quinine), and most would include umami (eg, monosodium glutamate) (Chaudhari et al., 1996). Additionally, when most foods are masticated they activate other gustatory and somatosensory systems depending on contents: texture and viscosity qualities are conveyed by the somatosensory receptors (Mattes, 2005; Tucker, Mattes, & Running, 2014), while fat is degraded by lingual lipases to free fatty acids that activate taste cells and may become acknowledged as a sixth tastant (Running, Craig, & Mattes, 2015). The five prototypical tastants are detected by chemosensory receptor cells that are arranged in communities of 50–100 cells on taste buds primarily on the tongue, as well as on the soft palate, pharynx, larynx, and on the epiglottis. The suggestion that different regions of the tongue were exclusive for specific tastants has been discarded, and distinct cells in mammalian taste buds, responding to all taste qualities, are distributed on the tongue, the palate, and more sparsely in the pharynx and larynx (Roper, 2013).

In the nucleus of the solitary tract (NST) of the medulla, taste stimuli from the taste buds converge through the facial, glossopharyngeal, and vagal nerves. From the brainstem, sensory taste input is conveyed via the parvicellular part of the ventral posterior medial thalamic nucleus to the primary gustatory cortex, located in the anterior insula/frontal operculum bilaterally (Kringelbach, de Araujo, & Rolls, 2004), where a striking topographic segregation of the microscopic level of the cortex encodes the basic taste qualities in a spatial map (Chen, Gabitto, Peng, Ryba, & Zuker, 2011). Interestingly, the circuitry is slightly different in rodents where there are second-order fibers (ie, NST afferents) that project ipsilaterally to the gustatory parabrachial nuclei in the pons.

The hedonic qualities of tastants are an innate property (Steiner, 1974) in mammalian species, exemplified by the facial responses of newborns: salty taste elicits minor responses; acidic taste causes an aversive crumpling of the lips; bitter taste triggers disgust and distress with an attempt to eject the bitter substance; sugar sparks positive reactions such as licking in rats and something akin to smiling in human babies. In the brainstem and the primary gustatory cortex stereotypical reflexes for each basic taste are based on simple analysis of chemical composition, which are motivation-independent and independent of higher processing areas, as the reflexes prevail even after removal of all neural tissue above the level of the midbrain in animal studies (Grill & Norgren, 1978). These taste neurons are not attenuated by consumption to satiety (Rolls, Scott, Sienkiewicz, & Yaxley, 1988; Yaxley, Rolls, Sienkiewicz, & Scott, 1985), which is an important quality, as the ability to identify food consequently is unaffected by motivational state modulation. This property ultimately enables defensive reflexes to prevail and gives unremitting input to higher-processing brain areas.

In contrast to the brainstem and the primary taste cortex, there is a neural response to satiety in the secondary gustatory cortex (caudolateral OFC). In animal studies the response to glucose stimuli disappears when consumed to satiety alongside a behavioral change from food-seeking behavior to active rejection (Rolls, Sienkiewicz, & Yaxley, 1989). In human neuroimaging studies, satiety changes are also observed in secondary cortices (parahippocampal gyrus, mid-anterior OFC, and prefrontal regions), but not in other areas of the brain and brainstem (Small, Zatorre, Dagher, Evans, & Jones-Gotman, 2001).

3.2.3 Multisensory convergence of smell and taste in flavor

As smell and taste merge, the experience of flavor emerges, which extends beyond the mere addition of the two chemosensory modalities. This multisensory interaction primarily occurs in the OFC and nearby agranular insula (de Araujo, Rolls et al., 2003; Kringelbach et al., 2003).

Recordings from nonhuman primates indicate that sensory neurons in the OFC are either unimodal or multimodal (Rolls & Baylis, 1994). The unimodal areas are frequently found in close proximity to each other, consistent with the hypothesis that the multisensory representations are formed from clustered unimodal inputs (Rolls, 2005). The OFC functions as an integrative circuit in which, for instance, taste-responsive neurons also show sensitivity to somatosensory and olfactory stimulation (de Araujo & Simon, 2009). The sensory inputs enter the OFC through the posterior parts and are integrated in the anterior parts (Kringelbach & Rolls, 2004). Multisensory neurons in the OFC have parallel sensitivities to the input quality from the different modalities: neurons responding best to sweet tastants (glucose) have a stronger response to the visual stimulus of sweet fruit juice or an olfactory stimulus of fruit odors (Rolls & Baylis, 1994). Analogously, a synergistic reinforcement of responses to matched taste and retronasal smell stimuli is found in the mid-anterior OFC region (de Araujo, Rolls et al., 2003). This organization of the OFC is important for its role in the hedonic evaluation of chemosensory stimulation (Small et al., 2001).

Modulations between smell and taste stimuli in the primary cortices have been described, suggesting that this process is not restricted to higher association areas. The gustatory system has a direct influence on the primary olfactory cortex, as the primary olfactory cortex receives both gustatory and olfactory input and convergence of these inputs takes place in single neurons (Maier, Wachowiak, & Katz, 2012). Comparably, olfactory and taste input converge in the far anterior part of the insular cortex in close proximity of the caudal OFC (de Araujo, Rolls et al., 2003). There is a growing body of evidence, indicating that all primary sensory cortices are not limited to unisensory stimulation, but receive other modulatory sensory input as well (Cohen, Rothschild & Mizrahi, 2011; Maier et al., 2012).

However, with the current methods available, the mechanisms responsible for the modulation of activity in primary taste cortex are unclear. Similarly, differences in activity may be based on direct input from different tastants or a subsequent modulation of secondary areas (Kringelbach et al., 2012). This is a crucial point, given that the presumed modulation of activity in the primary cortex by expectation may in fact be misleading (Nitschke et al., 2006). More studies with a higher structural and temporal resolution are needed before these mechanisms are fully understood.

3.2.4 Other sensory alterations of hedonic flavor perception

The superadditive flavor representation is not limited to olfaction and taste, as modulation of flavor perception and hedonic impact can be profoundly altered by input from other senses. Neuroimaging and neurophysiological studies show that the OFC receives input from all five senses, which is integrated in the anterior parts of the OFC and can alter the hedonics of flavor perception (Kringelbach & Rolls, 2004; Schirmer & Kotz, 2006).

Vision is crucial in the prediction of taste associated with consumption, which has a major impact on the selection of food (Rolls, 2005; Spence, 2015b). A simple addition of color to an odorless substance can trick the mind into perceiving an olfactory stimulus (Engen, 1972). By altering the expectation of flavor, color enhances orthonasal olfactory intensity and reduces retronasal intensity (Koza, Cilmi, Dolese, & Zellner, 2005), illustrating the complex interactions between the senses. Visual stimuli can also have a high impact on the hedonic experience from olfactory responses: a strawberry-flavored beverage smells more pleasant when it has a red color compared to a green color (Osterbauer et al., 2005).

A recent comprehensive review shows how sounds can modulate the hedonics of flavor perception (Spence, 2015a), as hedonic valence of sound is integrated in emotional processing (Frey, Kostopoulos, & Petrides, 2000). The hedonic valence of sound modulates odors, irrespective of the hedonic quality of the odor itself. The more the participants liked the preceding sound, the more pleasant the subsequent odor became (Seo & Hummel, 2011). The sound associated with consumption, such as crispness and crunchiness, is an important integrative aspect of eating and is highly correlated with the hedonic experience (Vickers, Peck, Labuza, & Huang, 2014). Modulation of flavor perception by auditory stimuli is a complex process as sound can modulate the perception of odor intensity, pleasantness, and quality (Velasco, Balboa, Marmolejo-Ramos, & Spence, 2014).

The importance of somatosensory perception, or mouthfeel, is underlined by the large areas of the sensory cortex devoted to the lips and the tongue and the equivalently well-developed motor system allocated to control the muscles of the lips and tongue. Food-related decision-making and the sensory perception of food-related rewards depends on multisensory integration of information about texture, viscosity, fat content, temperature, irritation, and pungency, which is mediated by a large number of neural systems (de Araujo & Simon, 2009). The hedonic aspect of touch is illustrated by neuroimaging studies, where it has been found that different areas of the OFC were activated more by pleasant touch and by painful stimuli, respectively, compared to neutral touch. During neutral touch stimuli, the somatosensory cortex was highly active, but during affective stimulation the activation had shifted to the OFC (Rolls et al., 2003). In combination with fat-sensitive chemical receptors, texture is an indicator of fat content, which from a metabolic perspective is an important source of high-energy nutrition and essential fatty acids (Rolls, Critchley, Browning, Hernadi, & Lenard, 1999). In the OFC, the convergence of somatosensory input in single neurons during gustatory stimulation can modulate the response, which explains why food texture is such an important aspect of hedonic perception (Rolls, 2005).

3.3 Perception of quality, intensity, and identity

The perception of a multisensory input is not a mere summation of simple sensory inputs, however it is a complex process of assessing the quality, intensity, and identity of all sensory subcomponents to congregate the unified perception of flavor. It should be noted that some of these subcomponents are so intertwined that it is not possible to assess them separately. In functional neuroimaging studies, it is of the utmost importance to understand the context when designing a study and adjusting for the long list of possible subject- and method-related confounders. In addition to the alteration of flavor perception by stimulation of additional sensory modalities, other factors can have a significant influence, such as hunger state (Small et al., 2001), receptor repertoire (Hallem & Carlson, 2006), comorbidities (Hawkes, 2003; Kayser et al., 2013; Pause et al., 2003), hedonic perception determined by learning (Pazart, Comte, Magnin, Millot, & Moulin, 2014), congruency of multimodal stimuli (Small et al., 2004), and the important aspects of stimulus quality, intensity, and identity.

The qualities of different perceived sensory stimuli can have great interpersonal and intrapersonal variation. In wine studies, the perception of wine is affected by the level of expertise (Sáenz-Navajas, Ballester, Pêcher, Peyron, & Valentin, 2013), where experts show stronger activity (as indicated by the BOLD signal) in the brainstem, hippocampal and parahippocampal formations (Pazart et al., 2014), whereas the activity of amygdala was higher in novice wine tasters (Castriota-Scanderbeg et al., 2005). This difference reflects a modulation of the sensory stimuli with a systematic focus on the gustatory and trigeminal information in experts, which gives indispensable chemosensory information on the qualities of the wine (Pazart et al., 2014). Expertise can also change the hedonic experience, as wine experts seem to prefer wines with less added sugar compared to novice controls (Blackman, Saliba, & Schmidtke, 2010), thus personal preference, conditioned aversions, and expertise can significantly alter the perceived sensory qualities and hedonic valence of a stimulus.

There is a significant correlation between activity of the mid-anterior OFC and a decrease in subjective pleasantness when a liquid food is consumed to satiety, which is specific to that particular food and not to other foods (Kringelbach et al., 2003). This hedonic modulation is known as selective satiety (or sensory-specific satiety) (Rolls & Rolls, 1997), emphasizing the importance of hunger and satiety on perception of hedonic quality. Therefore, when studying behavioral responses or neural activity, it can become difficult to interpret findings when subjects or groups differ in perceived hedonic quality of a stimulus. Any differences could simply reflect hedonic preferences and not an alteration of the intended dependent variable (Croy et al., 2014).

Some features, however, such as identity and intensity, have innate brainstem reflex-mediated responses for tastants that are motivation-independent and unaffected by hunger state (Grill & Norgren, 1978; Kringelbach et al., 2004; Small, Jones-Gotman, Zatorre, Petrides, & Evans, 1997). This independence from hunger and hedonic preferences is important because consumption of all meals has to undergo strict control, as bacteria, toxins, and spoiled foods can be harmful and potentially lethal if ingested. The important principle that the identity and intensity of a taste are represented separately from its pleasantness makes it possible to evaluate what a taste is in more objective terms, and to learn about it, even when we are satiated (Rolls, 2005).

This prompt identification of potentially harmful foods is in sharp contrast to the difficulties humans have when attempting to label olfactory stimuli verbally without cues from other sensory modalities. This well-known problem is called the tip-of-the-nose phenomenon (Lawless, 1977), and represents the strong nonconscious component of olfactory processing, as the ability to verbally identify objects based on smell alone is not as intuitive as a comparable visual stimulus (Yeshurun & Sobel, 2010). The Proustian phenomenon illustrates that this does not reflect a lack of association between olfaction and memory, since a momentary trace of a specific familiar odor can lead to a strong activation of autobiographical memories. These odor-evoked memories are typically experienced as being more emotional, and can elicit more vivid flashbacks compared to verbal or visual clues (Arshamian et al., 2013). The unique connections between the olfactory system and limbic circuitries form the neural basis for this strong "odor-emotional memory" (Sullivan, Wilson, Ravel, & Mouly, 2015), where prior exposure and the learning-dependent mechanisms can have resilient behavioral consequences.

4 Reward processing in the brain

Multisensory perceptions are assessed in the hedonic circuitry based on sensory quality, intensity, and identification. As these multisensory signals converge, neuroimaging studies show compelling evidence for hedonic processing in the mid-anterior OFC. While all studies reported activation in the "common pleasure OFC area," different patterns of activation were dependent on state and type of sensory input. Sweet odors, for example, elicit a stronger reward response compared to a savory odor, illustrating an identity-specific value pattern (Howard, Gottfried, Tobler, & Kahnt, 2015). The reward of a food eaten to satiety was selectively decreased compared to foods not eaten, which correlates with subjective ratings of pleasantness (Kringelbach et al., 2003). Not only does the reward value decrease when satiation lurks, but motivation of the subject is also manipulated. During this process of eating, different groups of hedonic circuitry are recruited depending on both motivation and pleasantness, respectively (Small et al., 2001). These changes in hedonic network dynamics illustrate the many variables to be taken into consideration when conducting and interpreting research on hedonia.

The dissociation between motivational factors of pleasure (wanting) and the actual sensation of pleasure (liking) has been elucidated by neuroimaging studies mapping hedonic key areas, and by studies on neurotransmitters, accentuating the neuroanatomical parcellation and adding another important piece of the puzzle to our understanding of reward processing and the pleasure cycle. These subcomponents of pleasure processing and their respective networks are intertwined within larger, more complex brain networks, where much is still to be discovered. With current technical advances, there are high expectations for further elaboration on our understanding of pleasure in the brain and neuroscience in general. The development of whole-brain computational modeling of neuroimaging data with much better temporal resolution may help move beyond mere correlation to discover causal relationships within the pleasure systems.

The hedonic circuitry consists of a large and distributed wanting-network, and smaller clusters of hedonic hotspots, which generate liking sensations. As these networks as a whole form the hedonic circuitry and are highly interdependent, both will be described in this chapter.

4.1 Mapping the hedonic network

The neural mechanisms evaluating reward and determining which reward deserves our immediate attention have long been a topic of debate. The evolutionary contrast between the innate values of essential rewards that are needed for survival and reproduction (food, sex, shelter), and the more abstract rewards (music, art) that are only ascribed value through higher-level association, has led to the theory that essential and abstract rewards are represented in phylogenetically different brain networks (Knutson & Bossaerts, 2007). However, recent neuroimaging studies have provided a growing body of evidence against this theory, as all rewards activate a collective reward circuitry; from pleasures of odors (Howard et al., 2015), food (Kringelbach & Rolls, 2004), music (Mavridis, 2014), art (Vartanian & Skov, 2014), love (Xu et al., 2010) to sex (Georgiadis & Kringelbach, 2012).

The discovery of a pleasure center in the brain occurred in the 1950s, when James Olds and Peter Milner implanted electrodes in a rat brain and found a remarkable motivation for self-stimulation (Olds & Milner, 1954), as the rat would self-stimulate the electrode up to 2000 times per hour (Olds, 1956). This discovery led to a subsequent study, where electrodes were placed in different areas of the brain to investigate the effects on self-stimulation by neuroanatomical location of the electrode (Olds, 1958). This not only led to the finding that negative and positive behavioral reinforcement occurred dependent on location, there was also a remarkably stable quantitative difference in motivational effects dependent on electrode location. An electrode located just in front of the mammillary body resulted in self-stimulation frequency above 5000 responses per hour, while much lower rates (about 200 per hour) were found in most other parts of the limbic brain areas. This strong motivational drive to trigger stimulation based on one electrode location, compared to other locations in the limbic system, reflects the different roles the hedonic networks have when it comes to wanting pleasure.

Through studies using neuroimaging, neurostimulation, and alteration of neurotransmitters, the dynamics of hedonic networks in humans and various animal species are beginning to emerge. When dissecting the underlying processing, the reward networks include cortical neuroanatomical areas in the OFC, anterior cingulate cortices, and insula (Kringelbach et al., 2003; Kringelbach & Rolls, 2004); in addition, the pleasure networks include ventral pallidum (VP), nucleus accumbens (NAc), and amygdala (Cardinal, Parkinson, Hall, & Everitt, 2002) (see Fig. 11.3).

4.2 Key hedonic areas and the pleasure cycle

The OFC is a large and heterogeneous region, accounting for approximately 15% of the frontal lobe (Semendeferi, Damasio, Frank, & Van Hoesen, 1997), which often is subdivided into smaller regions in an equally heterogeneous manner (Carmichael, Clugnet, & Price, 1994; Haber & Knutson, 2009). However, as a general pattern, the unisensory input converges to form multisensory information in the posterior OFC with an increasing processing complexity in the more anterior parts of the OFC (Kringelbach & Rolls, 2004). The OFC is most strongly activated by olfactory and gustatory stimuli (Kringelbach & Rolls, 2004; O'Doherty, Rolls, Francis, Bowtell, & McGlone, 2001; Small et al., 1997), and involved with higher-order chemosensory processing (Gottfried & Zald, 2005; Kringelbach & Rolls, 2004), such as cognitive-affective food evaluation and appetite (Rolls et al., 1989; Seubert et al., 2015). With a high connectivity between the OFC, primary sensory cortices, and limbic structures such as the amygdala, this region is considered the key candidate for linking sensory input to form the perception of flavor, and to orchestrate reward predictions (Georgiadis & Kringelbach, 2012).

The mid-anterior OFC seems to play an essential part in reward processing, as activity related to reward value correlates negatively with satiety ratings for food being consumed, and not for other types of food. Correspondingly, both increased pleasantness ratings and increased activity in the mid-anterior OFC are found when subjects are stimulated with two tastants simultaneously compared to separate stimulations (de Araujo, Kringelbach, Rolls, & Hobden, 2003). Additionally, the lateral anterior part of the OFC is activated when retronasal olfactory stimuli were combined with gustatory stimuli, compared to little or no activation for the individual stimuli (de Araujo, Rolls et al., 2003). Besides revealing an activation pattern, where reward and variables such as satiety and multisensory synergetic enhancement are closely linked, the OFC seems to be a key region for performing predictions on future rewards. The localization of activity varies slightly between possible food sources, even if the expected reward values are rated equally (Howard et al., 2015), however much is still to be learned before the mechanism of reward predictions in the OFC is fully understood.

An important contribution to our understanding of the complex interactions of the hedonic network is the vast body of evidence showing remarkable differences in response to the two main neurotransmitters related to pleasure: dopamine (and GABA) and opioids (including endocannabinoid, orexin, and related neurotransmitters). The ability to locate and suppress or stimulate these two neurotransmitters has led to the discovery of several predominantly opioid-sensitive hedonic hotspots related to "liking" and a larger network of primarily dopamine-responding neurons related to "wanting" a reward (Berridge & Robinson, 2003; Ho & Berridge, 2013).

The small hedonic hotspots related to liking have been found in the subcortical structures, specifically in the VP, the NAc, and in the parabrachial nucleus of the pons. When stimulated with opioids, these regions are shown in rodents to amplify orofacial "liking" expressions, while dopamine had no effect on the level of pleasure (Castro & Berridge, 2014; Ho & Berridge, 2013; Peciña & Berridge, 2005). Interestingly, in a caudal region of the small opioid-sensitive part of NAc, the microinjections of opioid agonists suppress "liking" reactions and trigger "fear" reactions, acting as a hedonic cold spot (Castro & Berridge, 2014), with a gradually altering positive pleasure reaction pattern towards the rostral hotspot. The gradient of change from reward to fear is not fixed, but can be modulated by, for example, a home-like or stressful environment setting to respond predominantly with either liking or fear, respectively (Reynolds & Berridge, 2008). The hedonic hotspot of the VP has a somewhat dissimilar function, though it too has the capacity to trigger "liking," it is indispensable for preserving normal baseline levels of "liking," as excessive "disgust" will dominate in its absence (Ho & Berridge, 2013). This affective scale in the NAc and the unique structural and functional connections of the hotspots throughout the limbic system may enable differentiated hedonic sensations (Britt & McGehee, 2008), which seem central in controlling hedonic impact.

The link between dopamine and "wanting" has been illustrated with similar rodent models, where normal orofacial "liking" reactions were observed after complete destruction of dopamine systems, conversely all food-seeking behavior had been eliminated (Berridge & Robinson, 1998). Dopamine-deficient rodents become hypophagic and die of starvation by 3–4 weeks of age unless dopamine is restored medically (Hnasko, Szczypka, Alaynick, During, & Palmiter, 2004). In human dopamine-depleted brains, as seen in Parkinson disease, "liking" ratings are equally unaffected by the lack of dopamine. Conversely, the dopamine stimulation has a remarkable effect on the motivation for obtaining sucrose rewards in rodents (Peciña & Berridge, 2013).

Other brain regions can become equally important in the processing of pleasure, or lack of pleasure. In patients with chronic pain the anterior cingulate cortices are active along with the lateral OFC in response to placebo treatment, indicating the importance of these areas for pain alleviation (Kringelbach, 2005; Rolls et al., 2003).

The processes of the hedonic network, including the key aspects of "liking" and "wanting," are closely accompanied by "learning." The evaluation of reward value is intrinsically linked with prediction error, where we learn from potential differences between expected and received rewards to improve future predictions (Niv & Schoenbaum, 2008). The vast differences in hedonic input and processing call for highly differentiated learning mechanisms: a short-term memory in the OFC has been suggested (Rolls, 2005) to explain the rapid reversals in learning demonstrated here, compared to neurons in the amygdala, where a more stable plasticity encodes longer-lasting fearful or rewarding emotional associations (Namburi et al., 2015), especially the formation, consolidation, and retention of fear memories (LeDoux, 2012). However, not only structure matters in learning. Evidence suggests that dopamine and glutamate play a key role in facilitating long-term plasticity and reward learning (Kelley, 2004). As a consequence, the neural networks involved in the pleasure circuitry will learn from prior errors to enhance pleasure, avoid danger, and ensure survival.

5 Perspectives/challenges

When describing the sensory processing and topology, it is important to bear in mind that the activated areas can be limited by the neuroimaging techniques applied. Insufficient structural resolution and subsequent thresholding can result in a lack of activation of relevant areas and BOLD imaging in fMRI studies offers little appreciation of the temporal relationship between sensory stimuli (Pazart et al., 2014; Zacà, Agarwal, Gujar, Sair, & Pillai, 2014). With a lack of temporal and spatial resolution, it is possible that small differences in the processing of particular rewards have been disregarded.

The interactions between sensory perception of hedonics, quality, intensity, and identity and the subsequent influence on multisensory processing can complicate the interpretation of functional neuroimaging studies. This calls for the use of structural connectivity models in order to make a hardwired hedonic infrastructural network, which would be comparable across conditions and free from dissimilar interpersonal hedonic preferences and expectations.

A better understanding of the brain mechanisms of pleasure can not only potentially open doors to better treatment of disorders such as depression, eating disorders, addiction, gambling, and chronic pain, it can have a general applicability for the broader population. Pleasure is fundamentally linked to well-being and the good life, in numerous situations from social interactions, personal ambitions, to simpler pleasures such as eating. The profound pleasure that can emerge from eating has been an evolutionary force that ensured survival, however, food can elicit great alterations in our behavior and its effects on deeper brain functions are only beginning to be understood.

References

Arshamian, A., Iannilli, E., Gerber, J. C., Willander, J., Persson, J., Seo, H.-S., et al. (2013). The functional neuroanatomy of odor evoked autobiographical memories cued by odors and words. *Neuropsychologia, 51*, 123–131.

Bargmann, C. I. (2006). Comparative chemosensation from receptors to ecology. *Nature, 444*, 295–301.

Berridge, K. C., & Kringelbach, M. L. (2013). Neuroscience of affect: Brain mechanisms of pleasure and displeasure. *Current Opinion in Neurobiology, 23*, 1–10.

Berridge, K. C., & Kringelbach, M. L. (2015). Pleasure systems in the brain. *Neuron, 86*(3), 646–664.

Berridge, K. C., & Robinson, T. E. (1998). What is the role of dopamine in reward: Hedonic impact, reward learning, or incentive salience? *Brain Research. Brain Research Reviews, 28*, 309–369.

Berridge, K. C., & Robinson, T. E. (2003). Parsing reward. *Trends in Neurosciences, 26*, 507–513.

Blackman, J., Saliba, A., & Schmidtke, L. (2010). Sweetness acceptance of novices, experienced consumers and winemakers in Hunter Valley Semillon wines. *Food Quality and Preference, 21*, 679–683.

Bolhuis, D. P., Lakemond, C. M. M., de Wijk, R. A., Luning, P. A., & de Graaf, C. (2013). Consumption with large sip sizes increases food intake and leads to underestimation of the amount consumed. *PLoS ONE, 8* e53288.

Braude, L., & Stevenson, R. J. (2014). Watching television while eating increases energy intake. Examining the mechanisms in female participants. *Appetite, 76*, 9–16.

Breton-Provencher, V., & Saghatelyan, A. (2012). Newborn neurons in the adult olfactory bulb: Unique properties for specific odor behavior. *Behavioural Brain Research, 227*, 480–489.

Britt, J. P., & McGehee, D. S. (2008). Presynaptic opioid and nicotinic receptor modulation of dopamine overflow in the nucleus accumbens. *The Journal of Neuroscience, 28*, 1672–1681.

Buck, L., & Axel, R. (1991). A novel multigene family may encode odorant receptors: A molecular basis for odor recognition. *Cell, 65*, 175–187.

Bushdid, C., Magnasco, M. O., Vosshall, L. B., & Keller, A. (2014). Humans can discriminate more than 1 trillion olfactory stimuli. *Science, 343*, 1370–1372.

Cardinal, R. N., Parkinson, J. A., Hall, J., & Everitt, B. J. (2002). Emotion and motivation: The role of the amygdala, ventral striatum, and prefrontal cortex. *Neuroscience & Biobehavioral Reviews, 26*, 321–352.

Carmichael, S. T., Clugnet, M. C., & Price, J. L. (1994). Central olfactory connections in the macaque monkey. *The Journal of Comparative Neurology, 346*, 403–434.

Castriota-Scanderbeg, A., Hagberg, G. E., Cerasa, A., Committeri, G., Galati, G., Patria, F., et al. (2005). The appreciation of wine by sommeliers: A functional magnetic resonance study of sensory integration. *NeuroImage, 25*, 570–578.

Castro, D. C., & Berridge, K. C. (2014). Opioid hedonic hotspot in nucleus accumbens shell: Mu, delta, and kappa maps for enhancement of sweetness "liking" and "wanting". *The Journal of Neuroscience, 34*, 4239–4250.

Chaudhari, N., Yang, H., Lamp, C., Delay, E., Cartford, C., Than, T., et al. (1996). The taste of monosodium glutamate: Membrane receptors in taste buds. *Journal of Neuroscience, 16*, 3817–3826.

Chen, X., Gabitto, M., Peng, Y., Ryba, N. J. P., & Zuker, C. S. (2011). A gustotopic map of taste qualities in the mammalian brain. *Science, 333*, 1262–1266.

Cohen, L., Rothschild, G., & Mizrahi, A. (2011). Multisensory integration of natural odors and sounds in the auditory cortex. *Neuron, 72*, 357–369.

Craig, W. (1917). Appetites and aversions as constituents of instincts. *Proceedings of the National Academy of Sciences of the United States of America, 3*, 685–688.

Croy, I., Symmank, A., Schellong, J., Hummel, C., Gerber, J., Joraschky, P., et al. (2014). Olfaction as a marker for depression in humans. *Journal of Affective Disorders, 160*(C), 80–86.

de Araujo, I. E., & Simon, S. A. (2009). The gustatory cortex and multisensory integration. *International Journal of Obesity (London), 33*(Suppl. 2), S34–S43.

de Araujo, I. E. T., Kringelbach, M. L., Rolls, E. T., & Hobden, P. (2003). Representation of umami taste in the human brain. *Journal of Neurophysiology, 90*, 313–319.

de Araujo, I. E. T., Rolls, E. T., Kringelbach, M. L., McGlone, F., & Phillips, N. (2003). Taste-olfactory convergence, and the representation of the pleasantness of flavour, in the human brain. *The European Journal of Neuroscience, 18*, 2059–2068.

de Graaf, C., & Kok, F. J. (2010). Slow food, fast food and the control of food intake. *Nature Reviews Endocrinology, 6*, 290–293.

Deroy, O., Chen, Y.-C., & Spence, C. (2014). Multisensory constraints on awareness. *Philosophical Transactions of the Royal Society B: Biological Sciences, 369* 20130207.

Engen, T. (1972). The effect of expectation on judgments of odor. *Acta Psychologica, 36*, 450–458.

Erlanson-Albertsson, C. (2005). How palatable food disrupts appetite regulation. *Basic & Clinical Pharmacology & Toxicology, 97*, 61–73.

Finlayson, G., King, N., & Blundell, J. E. (2007). Is it possible to dissociate "liking" and "wanting" for foods in humans? A novel experimental procedure. *Physiology & Behavior, 90*, 36–42.

Finlayson, G., King, N., & Blundell, J. (2008). The role of implicit wanting in relation to explicit liking and wanting for food: Implications for appetite control. *Appetite, 50*, 120–127.

Frey, S., Kostopoulos, P., & Petrides, M. (2000). Orbitofrontal involvement in the processing of unpleasant auditory information. *The European Journal of Neuroscience, 12*, 3709–3712.

García-García, I., Narberhaus, A., Marqués-Iturria, I., Garolera, M., Rădoi, A., Segura, B., et al. (2013). Neural responses to visual food cues: Insights from functional magnetic resonance imaging. *European Eating Disorders Review, 21*, 89–98.

Georgiadis, J. R., & Kringelbach, M. L. (2012). The human sexual response cycle: Brain imaging evidence linking sex to other pleasures. *Progress in Neurobiology, 98*, 49–81.

Gjorgjieva, J., Sompolinsky, H., & Meister, M. (2014). Benefits of pathway splitting in sensory coding. *The Journal of Neuroscience, 34*, 12127–12144.

Gonzalez-Kristeller, D. C., do Nascimento, J. B., Galante, P. A., & Malnic, B. (2015). Identification of agonists for a group of human odorant receptors. *Frontiers in Pharmacology, 6*, 35.

Gottfried, J. A., & Zald, D. H. (2005). On the scent of human olfactory orbitofrontal cortex: Meta-analysis and comparison to non-human primates. *Brain Research Reviews, 50*, 287–304.

Grill, H. J., & Norgren, R. (1978). Chronically decerebrate rats demonstrate satiation but not bait shyness. *Science, 201*, 267–269.

Haber, S. N., & Knutson, B. (2009). The reward circuit: Linking primate anatomy and human imaging. *Neuropsychopharmacology, 35*, 4–26.

Haddad, R., Weiss, T., Khan, R., Nadler, B., Mandairon, N., Bensafi, M., et al. (2010). Global features of neural activity in the olfactory system form a parallel code that predicts olfactory behavior and perception. *Journal of Neuroscience, 30*, 9017–9026.

Hallem, E. A., & Carlson, J. R. (2006). Coding of odors by a receptor repertoire. *Cell, 125*, 143–160.

Hawkes, C. (2003). Olfaction in neurodegenerative disorder. *Movement Disorders, 18*, 364–372.

Higgs, S., & Jones, A. (2013). Prolonged chewing at lunch decreases later snack intake. *Appetite, 62*, 91–95.

Hnasko, T. S., Szczypka, M. S., Alaynick, W. A., During, M. J., & Palmiter, R. D. (2004). A role for dopamine in feeding responses produced by orexigenic agents. *Brain Research, 1023*, 309–318.

Ho, C.-Y., & Berridge, K. C. (2013). An orexin hotspot in ventral pallidum amplifies hedonic "liking" for sweetness. *Neuropsychopharmacology, 38*, 1655–1664.

Holley, A. (1999). *Éloge De L'odorant*. Paris: Odile Jacob.
Howard, J. D., Gottfried, J. A., Tobler, P. N., & Kahnt, T. (2015). Identity-specific coding of future rewards in the human orbitofrontal cortex. *Proceedings of the National Academy of Sciences of the United States of America*, *112*, 5195–5200.
Johansson, D. P., Lee, I., Risérus, U., Langton, M., & Landberg, R. (2015). Effects of unfermented and fermented whole grain rye crisp breads served as part of a standardized breakfast, on appetite and postprandial glucose and insulin responses: A randomized cross-over trial. *PLoS ONE*, *10*, e0122241.
Kayser, J., Tenke, C. E., Kroppmann, C. J., Alschuler, D. M., Ben-David, S., Fekri, S., et al. (2013). Olfaction in the psychosis prodrome: Electrophysiological and behavioral measures of odor detection. *International Journal of Psychophysiology*, *90*, 190–206.
Kelley, A. E. (2004). Ventral striatal control of appetitive motivation: Role in ingestive behavior and reward-related learning. *Neuroscience & Biobehavioral Reviews*, *27*, 765–776.
Knutson, B., & Bossaerts, P. (2007). Neural antecedents of financial decisions. *Journal of Neuroscience*, *27*, 8174–8177.
Koza, B. J., Cilmi, A., Dolese, M., & Zellner, D. A. (2005). Color enhances orthonasal olfactory intensity and reduces retronasal olfactory intensity. *Chemical Senses*, *30*, 643–649.
Kringelbach, M. L. (2004). Food for thought: Hedonic experience beyond homeostasis in the human brain. *Neuroscience*, *126*, 807–819.
Kringelbach, M. L. (2005). The human orbitofrontal cortex: Linking reward to hedonic experience. *Nature Reviews Neuroscience*, *6*, 691–702.
Kringelbach, M. L. (2015). The pleasure of food: Underlying brain mechanisms of eating and other pleasures. *Flavour*, *4*, 20.
Kringelbach, M. L., & Berridge, K. C. (2010). The functional neuroanatomy of pleasure and happiness. *Discovery Medicine*, *9*, 579–587.
Kringelbach, M. L., de Araujo, I. E. T., & Rolls, E. T. (2004). Taste-related activity in the human dorsolateral prefrontal cortex. *NeuroImage*, *21*, 781–788.
Kringelbach, M. L., O'Doherty, J., Rolls, E. T., & Andrews, C. (2003). Activation of the human orbitofrontal cortex to a liquid food stimulus is correlated with its subjective pleasantness. *Cerebral Cortex*, *13*, 1064–1071.
Kringelbach, M. L., & Rolls, E. T. (2004). The functional neuroanatomy of the human orbitofrontal cortex: Evidence from neuroimaging and neuropsychology. *Progress in Neurobiology*, *72*, 341–372.
Kringelbach, M. L., Stein, A., & Van Hartevelt, T. J. (2012). The functional human neuroanatomy of food pleasure cycles. *Physiology & Behavior*, *106*, 307–316.
Lapid, H., Shushan, S., Plotkin, A., Voet, H., Roth, Y., Hummel, T., et al. (2011). Neural activity at the human olfactory epithelium reflects olfactory perception. *Nature Neuroscience*, *14*, 1455–1461.
Lawless, H. T. (1977). The pleasantness of mixtures in taste and olfaction. *Sensory Processes*, *1*, 227–237.
LeDoux, J. (2012). Rethinking the emotional brain. *Neuron*, *73*, 653–676.
Lin, D. Y., Zhang, S.-Z., Block, E., & Katz, L. C. (2005). Encoding social signals in the mouse main olfactory bulb. *Nature*, *434*, 470–477.
Lou, H. C., Joensson, M., & Kringelbach, M. L. (2011). Yoga lessons for consciousness research: A paralimbic network balancing brain resource allocation. *Frontiers in Psychology*, *2*, 366.
Maier, J. X., Wachowiak, M., & Katz, D. B. (2012). Chemosensory convergence on primary olfactory cortex. *The Journal of Neuroscience*, *32*, 17037–17047.

Mantini, D., Hasson, U., Betti, V., Perrucci, M. G., Romani, G. L., Corbetta, M., et al. (2012). Interspecies activity correlations reveal functional correspondence between monkey and human brain areas. *Nature Methods*, *9*, 277–282.

Mattes, R. D. (2005). Fat taste and lipid metabolism in humans. *Physiology & Behavior*, *86*, 691–697.

Mavridis, I. N. (2014). Music and the nucleus accumbens. *Surgical and Radiologic Anatomy*, *37*, 121–125.

Namburi, P., Beyeler, A., Yorozu, S., Calhoon, G. G., Halbert, S. A., Wichmann, R., et al. (2015). A circuit mechanism for differentiating positive and negative associations. *Nature*, *520*, 675–678.

Nijland, R., & Burgess, J. G. (2010). Bacterial olfaction. *Biotechnology Journal*, *5*, 974–977.

Nitschke, J. B., Dixon, G. E., Sarinopoulos, I., Short, S. J., Cohen, J. D., Smith, E. E., et al. (2006). Altering expectancy dampens neural response to aversive taste in primary taste cortex. *Nature Neuroscience*, *9*, 435–442.

Niv, Y., & Schoenbaum, G. (2008). Dialogues on prediction errors. *Trends in Cognitive Sciences*, *12*, 265–272.

O'Doherty, J., Rolls, E. T., Francis, S., Bowtell, R., & McGlone, F. (2001). Representation of pleasant and aversive taste in the human brain. *Journal of Neurophysiology*, *85*, 1315–1321.

Olds, J. (1956). Pleasure centers in the brain. *Scientific American*, 105–116.

Olds, J. (1958). Satiation effects in self-stimulation of the brain. *Journal of Comparative and Physiological Psychology*, *51*, 675–678.

Olds, J., & Milner, P. (1954). Positive reinforcement produced by electrical stimulation of septal area and other regions of rat brain. *Journal of Comparative and Physiological Psychology*, *47*, 419–427.

Olender, T., Lancet, D., & Nebert, D. W. (2008). Update on the olfactory receptor (OR) gene superfamily. *Human Genomics*, *3*, 87.

Olender, T., Waszak, S. M., Viavant, M., Khen, M., Ben-Asher, E., Reyes, A., et al. (2012). Personal receptor repertoires: Olfaction as a model. *BMC Genomics*, *13*, 414.

Osterbauer, R. A., Matthews, P. M., Jenkinson, M., Beckmann, C., Hansen, P. C., & Calvert, G. A. (2005). Color of scents: Chromatic stimuli modulate odor responses in the human brain. *Journal of Neurophysiology*, *93*, 3434–3441.

Papies, E. K. (2013). Tempting food words activate eating simulations. *Frontiers in Psychology*, *4*, 838.

Pause, B. M., Raack, N., Sojka, B., Göder, R., Aldenhoff, J. B., & Ferstl, R. (2003). Convergent and divergent effects of odors and emotions in depression. *Psychophysiology*, *40*, 209–225.

Pazart, L., Comte, A., Magnin, E., Millot, J.-L., & Moulin, T. (2014). An fMRI study on the influence of Sommeliers' expertise on the integration of flavor. *Frontiers in Behavioral Neuroscience*, *8*, 358.

Peciña, S., & Berridge, K. C. (2005). Hedonic hot spot in nucleus accumbens shell: Where do mu-opioids cause increased hedonic impact of sweetness? *The Journal of Neuroscience*, *25*, 11777–11786.

Peciña, S., & Berridge, K. C. (2013). Dopamine or opioid stimulation of nucleus accumbens similarly amplify cue-triggered "wanting" for reward: Entire core and medial shell mapped as substrates for PIT enhancement. *The European Journal of Neuroscience*, *37*, 1529–1540.

Piqueras-Fiszman, B., & Spence, C. (2015). Sensory expectations based on product-extrinsic food cues: An interdisciplinary review of the empirical evidence and theoretical accounts. *Food Quality and Preference*, *40*, 165–179.

Reynolds, S. M., & Berridge, K. C. (2008). Emotional environments retune the valence of appetitive versus fearful functions in nucleus accumbens. *Nature Neuroscience*, *11*, 423–425.

Robinson, E., Aveyard, P., Daley, A., Jolly, K., Lewis, A., Lycett, D., et al. (2013). Eating attentively: A systematic review and meta-analysis of the effect of food intake memory and awareness on eating. *The American Journal of Clinical Nutrition*, *97*, 728–742.

Rolls, E. T. (2005). Taste, olfactory, and food texture processing in the brain, and the control of food intake. *Physiology & Behavior*, *85*, 45–56.

Rolls, E. T., & Baylis, L. L. (1994). Gustatory, olfactory, and visual convergence within the primate orbitofrontal cortex. *Journal of Neuroscience*, *14*, 5437–5452.

Rolls, E. T., Critchley, H. D., Browning, A. S., Hernadi, I., & Lenard, L. (1999). Responses to the sensory properties of fat of neurons in the primate orbitofrontal cortex. *Journal of Neuroscience*, *19*, 1532–1540.

Rolls, E. T., O'Doherty, J., Kringelbach, M. L., Francis, S., Bowtell, R., & McGlone, F. (2003). Representations of pleasant and painful touch in the human orbitofrontal and cingulate cortices. *Cerebral Cortex*, *13*, 308–317.

Rolls, E. T., & Rolls, J. H. (1997). Olfactory sensory-specific satiety in humans. *Physiology & Behavior*, *61*, 461–473.

Rolls, E. T., Scott, T. R., Sienkiewicz, Z. J., & Yaxley, S. (1988). The responsiveness of neurones in the frontal opercular gustatory cortex of the macaque monkey is independent of hunger. *The Journal of Physiology (Lond)*, *397*, 1–12.

Rolls, E. T., Sienkiewicz, Z. J., & Yaxley, S. (1989). Hunger modulates the responses to gustatory stimuli of single neurons in the caudolateral orbitofrontal cortex of the macaque monkey. *The European Journal of Neuroscience*, *1*, 53–60.

Roper, S. D. (2013). Taste buds as peripheral chemosensory processors. *Seminars in Cell & Developmental Biology*, *24*, 71–79.

Rowe, D. B. D., & Shepherd, G. M. (2016). The role of ortho-retronasal olfaction in mammalian cortical evolution. *The Journal of Comparative Neurology*, *524*, 471–495.

Running, C. A., Craig, B. A., & Mattes, R. D. (2015). Oleogustus: The unique taste of fat. *Chemical Senses*, *40*, 507–516.

Sáenz-Navajas, M. P., Ballester, J., Pêcher, C., Peyron, D., & Valentin, D. (2013). Sensory drivers of intrinsic quality of red wines: Effect of culture and level of expertise. *Food Research International*, *54*, 1506–1518.

Schirmer, A., & Kotz, S. A. (2006). Beyond the right hemisphere: Brain mechanisms mediating vocal emotional processing. *Trends in Cognitive Sciences*, *10*, 24–30.

Semendeferi, K., Damasio, H., Frank, R., & Van Hoesen, G. W. (1997). The evolution of the frontal lobes: A volumetric analysis based on three-dimensional reconstructions of magnetic resonance scans of human and ape brains. *Journal of Human Evolution*, *32*, 375–388.

Seo, H.-S., & Hummel, T. (2011). Auditory-olfactory integration: Congruent or pleasant sounds amplify odor pleasantness. *Chemical Senses*, *36*, 301–309.

Seo, H.-S., Song, J. T., Cheong, J. J., Lee, Y. H., Lee, Y. W., Hwang, I., et al. (2001). Jasmonic acid carboxyl methyltransferase: A key enzyme for jasmonate-regulated plant responses. *Proceedings of the National Academy of Sciences of the United States of America*, *98*, 4788–4793.

Seubert, J., Ohla, K., Yokomukai, Y., Kellermann, T., & Lundström, J. N. (2015). Superadditive opercular activation to food flavor is mediated by enhanced temporal and limbic coupling. *Human Brain Mapping*, *36*, 1662–1676.

Shakhawat, A. M., Gheidi, A., Hou, Q., Dhillon, S. K., Marrone, D. F., Harley, C. W., et al. (2014). Visualizing the engram: Learning stabilizes odor representations in the olfactory network. *Journal of Neuroscience*, *34*, 15394–15401.

Shepherd, G. M. (2004). The human sense of smell: Are we better than we think? *PLoS Biology*, *2*, E146.

Shepherd, G. M. (2005). Perception without a thalamus: How does olfaction do it? *Neuron, 46*, 166–168.

Shepherd, G. M. (2006). Smell images and the flavour system in the human brain. *Nature, 444*, 316–321.

Sherrington C. S. (1906). *The integrative action of the nervous system.* New Haven, CT: Yale University Press.

Small, D. M., Voss, J., Mak, Y. E., Simmons, K. B., Parrish, T. & Gitelman, D. (2004). Experience-dependent neural integration of taste and smell in the human brain. *Journal of Neurophysiology, 92*, 1892–1903.

Small, D. M., Gerber, J. C., Mak, Y. E., & Hummel, T. (2005). Differential neural responses evoked by orthonasal versus retronasal odorant perception in humans. *Neuron, 47*, 593–605.

Small, D. M., Jones-Gotman, M., Zatorre, R. J., Petrides, M., & Evans, A. C. (1997). Flavor processing: More than the sum of its parts. *Neuroreport, 8*, 3913–3917.

Small, D. M., Zatorre, R. J., Dagher, A., Evans, A. C., & Jones-Gotman, M. (2001). Changes in brain activity related to eating chocolate: From pleasure to aversion. *Brain, 124*, 1720–1733.

Smith, K. S., & Berridge, K. C. (2005). The ventral pallidum and hedonic reward: Neurochemical maps of sucrose "liking" and food intake. *The Journal of Neuroscience, 25*, 8637–8649.

Spence, C. (2015a). Eating with our ears: Assessing the importance of the sounds of consumption to our perception and enjoyment of multisensory flavour experiences. *Flavour, 4*, 3.

Spence, C. (2015b). On the psychological impact of food colour. *Flavour, 4*, 21.

Steiner, J. E. (1974). Discussion paper: Innate, discriminative human facial expressions to taste and smell stimulation. *Annals of the New York Academy of Sciences, 237*, 229–233.

Stevenson, R. J. (2014). Flavor binding: Its nature and cause. *Psychological Bulletin, 140*, 487–510.

Sullivan, R. M., Wilson, D. A., Ravel, N., & Mouly, A.-M. (2015). Olfactory memory networks: From emotional learning to social behaviors. *Frontiers in Behavioral Neuroscience, 9*, 36.

Suzuki, K., Simpson, K. A., Minnion, J. S., Shillito, J. C., & Bloom, S. R. (2010). The role of gut hormones and the hypothalamus in appetite regulation. *Endocrine Journal, 57*, 359–372.

Tham, W. W. P., Stevenson, R. J., & Miller, L. A. (2011). The role of the mediodorsal thalamic nucleus in human olfaction. *Neurocase, 17*, 148–159.

Tucker, R. M., Mattes, R. D., & Running, C. A. (2014). Mechanisms and effects of "fat taste" in humans. *BioFactors, 40*, 313–326.

Van Hartevelt, T. J., & Kringelbach, M. L. (2012). The olfactory system. In J. Mai & G. Paxinos (Eds.), *The human nervous system* (pp. 1219–1238) (3rd ed.). San Diego: Academic Press (Elsevier).

Vartanian, O., & Skov, M. (2014). Neural correlates of viewing paintings: Evidence from a quantitative meta-analysis of functional magnetic resonance imaging data. *Brain and Cognition, 87*, 52–56.

Velasco, C., Balboa, D., Marmolejo-Ramos, F., & Spence, C. (2014). Crossmodal effect of music and odor pleasantness on olfactory quality perception. *Frontiers in Psychology, 5*, 1352.

Vickers, Z., Peck, A., Labuza, T., & Huang, G. (2014). Impact of almond form and moisture content on texture attributes and acceptability. *Journal of Food Science, 79*, S1399–S1406.

Xu, X., Aron, A., Brown, L., Cao, G., Feng, T., & Weng, X. (2010). Reward and motivation systems: A brain mapping study of early-stage intense romantic love in Chinese participants. *Human Brain Mapping, 32*, 249–257.

Yaxley, S., Rolls, E. T., Sienkiewicz, Z. J., & Scott, T. R. (1985). Satiety does not affect gustatory activity in the nucleus of the solitary tract of the alert monkey. *Brain Research, 347*, 85–93.

Yeshurun, Y., & Sobel, N. (2010). An odor is not worth a thousand words: From multidimensional odors to unidimensional odor objects. *Annual Review of Psychology, 61*, 219–241.

Zacà, D., Agarwal, S., Gujar, S. K., Sair, H. I., & Pillai, J. J. (2014). Special considerations/technical limitations of blood-oxygen-level-dependent functional magnetic resonance imaging. *Neuroimaging Clinics of North America, 24*, 705–715.

Zheng, H., Lenard, N. R., Shin, A. C., & Berthoud, H.-R. (2009). Appetite control and energy balance regulation in the modern world: Reward-driven brain overrides repletion signals. *International Journal of Obesity and Related Metabolic Disorders, 33*, S8–S13.

The Neuroscience of Flavor

Charles Spence
Crossmodal Research Laboratory, Department of Experimental Psychology, University of Oxford, Oxford, United Kingdom

1 Introduction

It is intuitive to believe that since we experience the taste of food and drink in the mouth, our understanding of flavor perception would be advanced greatly by knowing about the different classes of receptors that can be found on the tongue. However, regardless of one's position on the question of which senses should be considered as constitutive of flavor (Spence, in press), the key point to note here is that most people now agree that flavor perception results from the multisensory integration of multiple sensory signals in the human brain (eg, Small, 2012; Spence, 2015). As if to emphasize this point, Dana Small boldly titled one of her recent review papers *"Flavor is in the brain."*[1] Hence, if you really want to know why foods and drinks taste the way that they do, then you need to understand something about the underlying neuroscience governing multisensory flavor perception, an area of research that some refer to as *"neurogastronomy"* (Shepherd, 2006, 2012; though see Spence, 2012a). While existing results are undoubtedly both fascinating and important, as will be described below, it is perhaps worth bearing in mind here that the experimental situation in which the participants find themselves in fMRI studies is just very unnatural (see Spence, submitted).

One of the striking aspects of multisensory flavor perception is the profound individual differences that exist in terms of the foods that we like/dislike: Why, for example, should it be that the foods you love I hate, and vice versa (see Prescott, 2012, and Chapter 8)? It has long been known that we live in different taste worlds (Bartoshuk, 1980; see Chapter 10). And while some of these individual differences are genetic in origin (eg, Bartoshuk, 2000; Mauer & El-Sohemy, 2012), others are based on our prior experiences with foods. Indeed, it is interesting to note how many of our food dislikes are linked to specific food textures rather than to particular tastes or flavors (see Prescott, 2012, on this point). Of course, individuals also vary in terms of how neophilic versus neophobic they are—that is, how willing they are to try new foods and flavors (eg, Henriques, King, & Meiselman, 2009; Pliner & Hobden, 1992; Veeck, 2010).

Neuroscientists have demonstrated some marked individual differences in the way in which the brains of different groups of individuals respond to one and the same

---

[1] Along similar lines, Shepherd (2012) states that: "*A common misconception is that the foods contain the flavours. Foods do contain the flavour molecules, but the flavours of those molecules are actually created by our brains*" (p. ix, emphasis in original). Shepherd continues: "*It is important to realize that flavor doesn't reside in a flavorful food any more than color resides in a colorful object*" (p. 5).

food. So, for example, Eldeghaidy et al. (2011) highlighted increased activity in the orbitofrontal cortex (OFC) amongst a group of supertasters in response to the delivery of a series of fat emulsions of increasing fat concentration as compared to a group of non-tasters exposed to exactly the same stimuli. A nascent field of neuroimaging research has also started to compare the network of brain regions that are recruited by experts and regular consumers. So, for example, Pazart, Comte, Magnin, Millot, and Moulin (2014) recently demonstrated that many of the same brain areas (eg, the insula, orbitofrontal cortex, amygdala, and frontal operculum) were activated in social wine drinkers and sommeliers when tasked with evaluating the flavor of wines that had been served to them blind. That said, reduced and more targeted sensory activity was documented in the experts (though see also Castriota-Scanderbeg et al., 2005, for a somewhat different pattern of results).

2 Flavor expectations and flavor experiences

When thinking about flavor perception, it is important to draw a distinction between flavor expectations and flavor experiences (Stevenson, 2009). Different combinations of senses are involved in the generation of flavor expectations as compared to flavor experiences: Vision and orthonasal olfaction tend to be dominant in the former case (though audition and touch can also play a role), whereas gustatory, retronasal olfactory, and oral-somatosensory/trigeminal inputs tend to dominate the latter.[2] Crucially, our expectations concerning the likely taste and flavor of those foods and beverages that we put in our mouth play an important role in determining the final experience (see Piqueras-Fiszman & Spence, 2015).

Whenever we interact with food or drink, or happen to be in a relevant consumption context, our brain interprets and integrates previously experienced (and stored) information with any newly presented cues about the food. Consequently, everything from what is known about the product prior to consumption, visual appearance cues, orthonasal olfactory cues, and, on occasion, even distal food sounds through to the context in which we happen to be eating or drinking, will set up powerful expectations in our mind about that which we are about to experience. For instance, Nitschke et al. (2006) conducted a study in which participants were sometimes informed that they would receive a very bitter tastant, while on other trials, they were told that they would taste something much less bitter. Neural changes that were observed took place at some of the earliest sites in the brain after the taste and smell signals are initially coded. In particular, activity in the middle and posterior insula was modulated by the verbal description that the participants had been given regarding the intensity of

---

[2] The *orthonasal* system is associated with the inhalation of external odors, as when we sniff. The *retronasal* system, involving the posterior nares, is associated with the detection of the olfactory stimuli emanating from the food we consume, as odors are periodically forced out of the nasal cavity when we chew or swallow food or drink. There are important differences between these two senses of smell at both the subjective/perceptual level (Bojanowski & Hummel, 2012; Diaz, 2004; Rozin, 1982), and in terms of the neural substrates involved (eg, Small, Gerber, Mak, & Hummel, 2005).

the to-be-delivered tastant (see also the follow-up work: Sarinoloulos, Dixon, Short, Davidson, & Nitschke, 2006).[3] Using fMRI, Woods et al. (2011) demonstrated that people who expected a very sweet drink, while tasting a drink that was somewhat less sweet than expected, enhanced reported sweetness and bolstered activity in the taste cortex, relative to the same drink when such an expectation was not present. Such findings provide evidence that taste expectation can modulate activity in primary taste cortex, thus implying that expectation effects do indeed impact on taste perception as opposed to merely decisional.

We not only generate expectations concerning the sensory-discriminative aspects of what a food will likely taste of, we also generate hedonic expectations that can modulate how much we end up liking the tasting experience. If our experience on actually tasting a food is not too different from the expectation (in either the sensory-discriminative or hedonic domains) then what we experience will likely end up being assimilated to what we expected (see Piqueras-Fiszman & Spence, 2015, for a review). If, however, the experience is very different from the expectation (though quite how much of a difference there needs to be is anyone's guess; see Schifferstein, 2001; Spence & Piqueras-Fiszman, 2014) then a negatively valenced disconfirmation of expectation (contrast) response is often seen instead (eg, Cardello & Sawyer, 1992; Carlsmith & Aronson, 1963; Deliza & MacFie, 1997; Zellner, Strickhouser, & Tornow, 2004).[4] Nitschke et al. (2006) also found that the OFC response changed as a function of the participants' expectations concerning the bitterness of the solutions that they were given to taste. Relevant here, separable but overlapping neural substrates have been documented as far as the anticipatory and consummatory aspects of flavor are concerned (Small, Veldhuizen, Felsted, Mak, & McGlone, 2008). In particular, the amygdala and mediodorsal thalamus exhibit preferential activation in response to food odors that predict the imminent arrival of a flavored food or drink.

3 Neural circuits underlying multisensory flavor perception

The last few years have seen a rapid growth in our understanding of the neural networks that give rise to the multisensory perception of flavor (see Shepherd, 2012; Small, 2012; Verhagen & Engelen, 2006, for reviews). The early cortical representation of visual, auditory, and somatosensory information (eg, "primary" and "secondary" areas) occurs in the so-called unisensory cortex. By contrast, the cortical representations of the chemical senses (eg, of smell and taste) are seen in the limbic and paralimbic cortex (Zatorre & Jones-Gotman, 2000). Gustatory inputs project

---

[3] However, it is not altogether clear whether this modulation was actually taking place in the primary taste cortex or in an adjacent region instead (see Grabenhorst, Rolls, & Bilderbeck, 2008, on this point).
[4] While our response to the disconfirmation of expectation is normally negatively valenced, it should be remembered that this is very much context-dependent. So, for example, if one finds oneself sitting at the table in a modernist restaurant, one may actually be positively expecting to be surprised, that is, to have one's expectations disconfirmed (see Piqueras-Fiszman & Spence, 2012; Spence & Piqueras-Fiszman, 2014).

from the tongue to the primary taste cortex (more specifically, to the anterior insula and the frontal or parietal operculum), whereas olfactory stimuli project directly to the primary olfactory (ie, piriform) cortex. From there, gustatory inputs project to the caudolateral orbitofrontal cortex (OFC), whereas olfactory inputs project to the caudomedial part of the OFC instead.

The OFC, a small (walnut-sized) part of the brain, located just behind the eyes, plays a central role in representing the pleasantness (and reward value) of food and drink (see Small, 2012, for a review). The oral-somatosensory attributes of food and drink are represented initially in the oral/primary somatosensory cortex (Wang et al., 2002), but thereafter project throughout the primary gustatory cortex (Small, 2012). Somatosensory inputs also modulate the neural activity that is seen in the OFC (eg, Cerf-Ducastel, Van de Moortele, Macleod, Le Bihan, & Faurion, 2001; Eldeghaidy et al., 2011). As Dana Small puts it: "*...while integration occurs at many, if not most, levels of the neuroaxis, the independent sensory inputs that combine to create flavour sensations are first integrated in the anterior insula. It is then proposed that these neural signatures of flavour are communicated to higher and lower-order neural relays where they are enriched and integrated with appetitive circuits to guide feeding behaviour.*" (Small, 2012, p. 540).

But what are the rules governing the integration of the various sensory signals that give rise to the multisensory perception of flavor? Certainly, the congruency between the component signals would seem to be one of the key factors here (eg, Dalton, Doolittle, Nagata, & Breslin, 2000). So, for example, the participants in one influential neuroimaging study had to rate the pleasantness and congruency of various different pairings of orthonasal olfactory and gustatory stimuli (De Araujo, Rolls, Kringelbach, McGlone, & Phillips, 2003). The olfactory stimuli in this case consisted of the aroma of strawberry and methianol, which smells like chicken broth. The tastants, delivered in solution, consisted of sucrose and monosodium glutamate (MSG). The participants received both congruent (eg, strawberry odor and sucrose) and incongruent (eg, chicken broth odor and sucrose) stimulus combinations. Intriguingly, increased OFC activity correlated with increased ratings of the pleasantness and congruency of the combined olfactory and gustatory inputs that the participants were evaluating. Thus, it would appear that the presentation of familiar, or congruent, combinations of olfactory and gustatory stimuli leads to an enhanced neural response in those parts of the brain that code for the hedonic (ie, pleasantness) and reward value of food.[5]

While De Araujo et al. (2003) presented their olfactory stimuli orthonasally, other researchers have demonstrated similar results (both behaviorally and neurally) following the presentation of retronasal olfactory stimuli (see Spence, 2012b, for a review). For example, in one such study, Dana Small and her colleagues presented familiar/unfamiliar combinations of retronasal olfactory and gustatory stimuli to their participants (Small et al., 2004). Superadditive neural interactions (see Stein, 2012) were observed in the OFC in response to familiar (or congruent, eg, a sweet taste and a vanilla aroma), but not to unfamiliar (or incongruent) combinations of stimuli

---

[5] Similar results have also been reported following the presentation of congruent combinations of visual and olfactory stimuli as well—think only of the smell of strawberries and the color red (Österbauer et al., 2005).

(such as for the combination of a salty taste with a vanilla aroma). Several other brain areas—including the dorsal insula, the frontal operculum, and the anterior cingulate cortex—also "lit up", perhaps constituting the bare bones of the brain's "flavor network" (Shepherd, 2012; Small, 2012).

In summary, congruent combinations of olfactory (both orthonasal and retronasal), gustatory, visual, and presumably also tactile/auditory stimuli give rise to increased activity in the brain's reward areas, such as the OFC. By contrast, incongruent combinations of sensory stimuli can lead to a subadditive neural response in those regions that correlate with the participant's subjective response (see also Skrandies & Reuther, 2008). Using positron emission tomography, Small, Jones-Gotman, Zatorre, Petrides, and Evans (1997) were able to demonstrate reduced blood flow in the primary gustatory and secondary gustatory and olfactory cortices (ie, OFC) when orthonasal smell and taste stimuli were delivered simultaneously rather than individually. Interestingly, incongruent combinations of smell and taste (eg, a salty taste and a strawberry aroma) have been shown to give rise to increased cerebral blood flow in the amygdala and basal forebrain, when compared to a matched or congruent condition (eg, a salty taste paired with the aroma of soy sauce).

One obvious question to ask at this point is what determines which combinations of sensory cues will be treated (by the brain) as congruent, and which incongruent. Related to the above section, this is likely determined by an individual's prior exposure, which starts while we are still in the womb (see Schaal & Durand, 2012; Schaal, Marlier, & Soussignan, 2000). In fact, by 6 months of age, infants are already sensitive to the pairing of color (of a cup) and the taste of the contents (Reardon & Bushnell, 1988). While some researchers have, on occasion, argued for the existence of innate correspondences, little convincing evidence in support of such a possibility has, as yet, been provided.

Sensory dominance: One might ask why it should be that color cues typically exert such a dramatic effect over taste/flavor perception. At one level though this is not so surprising: In particular, in terms of "cortical real estate," far more of the brain (>50%) is given over to processing what we see (Felleman & van Essen, 1991) than to processing either gustatory or olfactory information (roughly 1–2% each; see Gallace, Ngo, Sulaitis, & Spence, 2012, for a review). That said, it is worth noting how, if a food smells off, most people will not touch it, no matter how beautiful it looks. A similarly strong aversive response has also been noted when people were presented with meat that had been colored blue (see Wheatley, 1973). Results such as these can be taken to suggest that the processing of multisensory cues concerning off-flavors (ie, foods that may be potentially poisonous) may be different from the processing of the pleasant or neutral stimuli that have been presented to the participants in the majority of laboratory research (Barkow, Cosmides, & Tooby, 1992; see also Boesveldt, Frasnelli, Gordon, & Lündstrom, 2010).

The hungry brain: More generally, it is worth noting that our desire for a given food changes as a function of how hungry we are. Some years ago Dana Small and her colleagues investigated the changes in neural activity that occurred when participants were fed chocolate to satiety (Small, Zatorre, Dagher, Evans, & Jones-Gotman, 2001). As the reward value of the food decreased, activity in the insula was seen to decrease bilaterally. By contrast, posterior cingulate activity was documented in all of the conditions. When highly motivated to eat (and while still rating their preferred chocolate as very pleasant), increased activation was seen in the caudomedial OFC, the insula/operculum, the striatum, and the midbrain. While the same participants were in the satiated state, increased activation was observed in the parahippocampal gyrus, the caudolateral OFC, and prefrontal regions.

Some of the most profound increases in cerebral blood flow that have been reported in neuroimaging studies have been in hungry participants viewing images of appetizing foods (see Wang et al., 2004; and Spence, Okajima, Cheok, Petit, & Michel, in press, for a review). Van der Laan, De Ridder, Viergever, and Smeets (2011) conducted an informative meta-analysis of 17 different neuroimaging studies in which the neural activation elicited by the visual presentation of food images was assessed. While nearly 200 separate foci of activation were highlighted, a small number of key brain regions (including the bilateral posterior fusiform gyrus, the left lateral OFC, and the left middle insula) were activated in response to visual images of food (across a number of the studies). Enhanced neural activation was observed in hungry participants in the right amygdala and left lateral OFC in response to food pictures. Meanwhile, the neural response in the hypothalamus/ventral striatum tracked the expected energy content of the food.

The obese brain: Given the current obesity epidemic, there is obviously great interest in trying to understand how the brains of obese individuals, and the neural responses that appetitive food stimuli elicit in them, differ from what is seen in normal-weight individuals. Recently, Pursey et al. (2014) conducted a meta-analysis of 60 neuroimaging studies that had assessed the neural response to visual food cues as a function of the weight of their participants. Obese individuals exhibited a greater increase in neural activation in response to food as compared to nonfood images (especially for high-calorie foods), in those brain regions that are known to be associated with reward processing (eg, the insula and OFC).[6] Increased activation was also observed in those brain areas that are involved in reinforcement and adaptive learning (the amygdala, putamen, and OFC), emotional processing (the insula, amygdala, and cingulate gyrus), recollective and working memory (the amygdala, hippocampus, thalamus, posterior cingulate cortex, and caudate), executive functioning (the prefrontal cortex, caudate, and cingulate gyrus), decision making (the OFC, the prefrontal cortex, and the thalamus), visual processing (the thalamus and fusiform gyrus), and motor learning and coordination, such as hand-to-mouth movements and swallowing (the insula, putamen, thalamus, and caudate). Interestingly, obese individuals show significantly less activation of the reward-related brain areas in response to the consumption of food than do healthy-weight individuals. Such results have been taken to suggest that the latter group may anticipate more reward from the intake of food while at the same time experiencing less sensory pleasure as a result of eating (Stice, Spoor, Bohon, Veldhuizen, & Small, 2008).

---

[6]At the other end of the weight spectrum, those individuals suffering from binge-eating disorder and bulimia experience greater reward sensitivity, brain activation, and arousal, in response to viewing images of pleasant foods (Schienle, Schäfer, Hermann, & Vaitl, 2009).

4 Branding and pricing

Outside of the psychology, or food science, laboratories, we rarely taste a food or beverage without knowing something about what it is that we are eating or drinking, such as, for example, the brand name and/or perhaps the price (or approximate price range). On many occasions, of course, the food will also have some form of label or description attached. It has been known for many years that such information can change what people say about the taste, flavor, and/or aroma of a food and how much they like it, not to mention how much they are willing to pay for it (eg, Herz & von Clef, 2001; Martin, 1990; Olson & Dover, 1978). Until recently, however, it has always been unclear just how early in human information processing such effects occurred.

Over the last decade or so, a number of neuroimaging studies have demonstrated the sometimes profound changes in brain activity (both in terms of the network of brain areas that are activated and the amount of activation that is seen) that can result from the provision of such product-extrinsic information to the participant who is normally to be found lying prone in the brain scanner. What is more, the effects of such product-extrinsic cues have, on occasion, been shown to influence the neural activity at some of the earliest (ie, primary sensory) areas in the human brain. In fact, there is now good evidence to suggest that the cognitive expectations we hold regarding taste or flavor can have a profound influence on some of the earliest neural sites where olfactory and gustatory information are processed (eg, see Grabenhorst et al., 2008; McClure et al., 2004; Plassmann, O'Doherty, Shiv, & Rangel, 2008; Woods et al., 2011).

Branding: The classic experiment on the neural substrates of branding was conducted by McClure et al. (2004). The participants in this oft-cited study had their brains scanned while a cola drink (either Pepsi or Coke) was periodically squirted into their mouth while different visual information concerning the brand of cola that they were apparently tasting was projected on a screen. Under blind tasting conditions, ventromedial prefrontal cortex activity correlated with the participants' behavioral preference. Qualitatively different patterns of brain activation were observed depending on which brand the participants were told that they were tasting. In particular, on being led to believe that they were tasting Coke, activation was observed in the hippocampus, dorsolateral prefrontal cortex, and midbrain.

In a more recent follow-up, Kühn and Gallinat (2013) observed more activation in the left ventral striatum when participants believed that the cola that they were tasting was a strong rather than a weak brand (eg, Coke or Pepsi vs a national or fictitious brand). Intriguingly, this effect was stronger in those individuals who consumed cola infrequently, possibly pointing to a greater reliance on brand cues in less experienced consumers.

Pricing: A few years ago, neuroscientists in California investigated what would happen in the brain of social wine drinkers (students) when they were given different (and sometimes misleading) information about the price of red wine (Plassmann et al., 2008). A $5 bottle of wine was either correctly described or else mislabeled as a $45 wine. Another bottle of wine actually cost $90 and was either presented as a $10 or $90 wine. The third wine was correctly referred to as costing $35 a bottle. The price was displayed on a computer monitor whenever a small amount of wine was squirted into the participant's mouth. On some trials, the participants had to rate the intensity of the wine's taste on a 6-point scale, whilst on other trials they rated its pleasantness instead. Sometimes, no behavioral response was required. While flavor intensity was not influenced by the price information, participants reported liking the wine labeled as expensive more than when labeled as a cheaper wine (see also Plassmann & Weber, 2015). Crucially, analysis of the brain scans revealed increases in blood flow in the medial OFC in the trials showing higher wine prices.[7]

When the same wines were presented 8 weeks later, now without any indication as to their price (and this time away from the brain scanner), no significant differences in pleasantness were reported. As discussed earlier, the effect of the price might be stronger in normal situations, when wine can be fully enjoyed in the right context. Just imagine yourself lying flat on your back, inserted several feet down a narrow tube, with your head clamped still (in order to minimize any motion artefacts that can make it difficult to analyze the brain imaging data), and with a tube held between your teeth as wine is periodically squirted into your mouth. You have to hold the wine in your oral cavity and evaluate its taste, without swallowing, before finally having your mouth washed out with artificial saliva before the whole process starts again.

Labeling: Elsewhere, researchers have investigated whether people's response to ambiguous (ie, bivalent) odors would be influenced by the label that was provided. So, for example, verbally describing an odor as "smelly cheese" results in people rating an ambiguous aroma as more pleasant than when exactly the same odor happens to be labeled as "sweaty socks" (Herz & von Clef, 2001; see also Manescu, Frasnelli, Lepore, & Djordjevic, 2014). When De Araujo, Rolls, Velazco, Margot, and Cayeux (2005) subsequently repeated this experiment in the brain scanner, changing the label was actually shown to change the part of the OFC that was activated in their participants while sniffing this odorant (see also Djordjevic et al., 2008).

Grabenhorst et al. (2008) delivered a savory solution to participants lying in the brain scanner who were provided with one of several different labels designed to vary their expectations concerning the pleasantness, rather than the intensity, of that which they were about to taste. The descriptors included terms such as "rich and delicious taste" versus "monosodium glutamate" for an umami solution. Meanwhile, a solution to which a vegetable aroma had been added was described as "rich and delicious flavor," "boiled vegetable water," or "monosodium glutamate." Varying the verbal description modulated brain activity in the pregenual cingulate cortex, the ventral striatum, and the medial OFC.[8] On the basis of such results, Grabenhorst et al. (2008, p. 1549) went on to conclude that *"top-down language-level cognitive effects reach far down into the earliest cortical areas that represent the appetite value of taste and flavour"*. Elsewhere, Linder et al. (2010) demonstrated that labeling a food as organic

---

[7] By contrast, no change in blood flow was observed in the primary taste cortex. This null result is, however, not so surprising given that this part of the brain is more interested with deciphering the sensory-discriminative attributes of a wine's taste (eg, how sweet, sour, etc., a wine is).

[8] No such changes were observed in the insular taste cortex, presumably because people's expectations regarding the intensity of the taste were not manipulated.

led to increased activity in the ventral striatum, a part of the brain that is involved in controlling our motivation to acquire and eat food.

In fact, even reading the word salt has been shown to activate many of the same brain areas as when a salty taste is actually experienced in the mouth (see Barrós-Loscertales et al., 2012). Here, it is perhaps also worth thinking not just about flavor experiences but also flavor (mental) imagery (eg, Olivetti Belardinelli et al., 2009). Indeed, it is fascinating to note that more of our brain lights up when we merely think about (or anticipate) food than when we actually get to taste it (eg, O'Doherty, Deichmann, Critchley, & Dolan, 2002). Finally, the participants in another intriguing neuroimaging study were asked whether they would like the taste of certain unusual combinations of ingredients (Barron, Dolan, & Behrens, 2013). Do you, for example, think that you would like the taste of a raspberry and avocado smoothie, or how about a beetroot custard? Participants who performed this task exhibited increased neural activation in the medial prefrontal cortex (mPFC).

In summary, then, neuroimaging studies have enabled researchers to understand a little more about what happens in the brain when people say that a drink tastes better after having been told that it costs more (Plassmann et al., 2008; Spence, 2010). Neuroimaging research has also highlighted the way in which the provision of branding information can end up recruiting different brain networks (eg, McClure et al., 2004; see also Kühn & Gallinat 2013; and Plassmann, Ramsøy, & Milosavljevic, 2012, for a review). Finally, the way in which a food or beverage is labeled or described also has a significant impact on the way in which taste and flavor information is represented, and responded to, by the brain.

5 Conclusions

Flavor perception is one of the most multisensory of our everyday experiences involving as it does gustation (taste) and olfaction (retronasal smell) together with trigeminal inputs (see Spence, Smith, & Auvray, 2015).[9] A large body of empirical research over the last century or so has unequivocally demonstrated that the other senses, for example, oral-somatosensation, audition, and even vision can profoundly influence human flavor perception (see Spence, 2015; Stevenson, 2009, for reviews), though the jury is still out when it comes to the question of whether they are constitutive of flavor perception or merely modulate it (Spence et al., 2015). Nevertheless, regardless of where one stands on this debate, three key rules have been shown to explain the integration of signals from different sensory modalities that give rise to flavor experiences: these rules are sensory dominance, superadditivity, and subadditivity (see Spence, 2012b, 2015, for reviews).

---

[9] *"Eating is the only thing we do that involves all the senses. I don't think that we realize just how much influence the senses actually have on the way that we process information from mouth to brain."* (Heston Blumenthal, Tasting menu from 2004, The Fat Duck restaurant, Bray, UK).

As the present review has hopefully made clear, the neuroscientific study of multisensory flavor perception, and the factors that modulate it, has made great strides over the last decade or so (eg, Shepherd, 2012; Small, 2012; Verhagen & Engelen, 2006). However, that being said, it is nevertheless important not to lose sight of the fact that the situation in the brain scanner is far removed from the conditions of everyday life, and, more importantly, of everyday consumption (see Spence, 2010; Spence & Piqueras-Fiszman, 2014). One just needs to consider, for example, the participant lying passively with their head clamped still, unable to swallow until instructed to do so, as small amounts of liquid or liquified food are squirted into their mouth. Note here that the scanning environment is typically extremely noisy, such that the participant likely has to wear noise-canceling headphones, as bits of information (or visual images) are presented via mirror reflection on a computer screen, perhaps pertaining to the suggested price or brand of that which they are about to taste (Spence, submitted). Ultimately, dining is a fundamentally social activity (see Jones, 2008; Spence & Piqueras-Fiszman, 2014) and, as such, this is another aspect of our everyday experience of food and drink that it is simply very difficult to capture using the neuroimaging approach to the study of flavor perception.

References

Barkow, J. H., Cosmides, L., & Tooby, J. (Eds.), (1992). *The adapted mind: Evolutionary psychology and the generation of culture*. Oxford: Oxford University Press.

Barron, H. C., Dolan, R. J., & Behrens, T. E. J. (2013). Online evaluation of novel choices by simultaneous representation of multiple memories. *Nature Neuroscience, 16*, 1492–1498.

Barrós-Loscertales, A., González, J., Pulvermüller, F., Ventura-Campos, N., Bustamante, J. C., Costumero, V., ... Ávila, C. (2012). Reading salt activates gustatory brain regions: fMRI evidence for semantic grounding in a novel sensory modality. *Cerebral Cortex, 22*, 2554–2563.

Bartoshuk, L. (1980). Separate worlds of taste. *Psychology Today, 14*, 48–49, 51, 54–56, 63.

Bartoshuk, L. M. (2000). Comparing sensory experiences across individuals: Recent psychophysical advances illuminate genetic variation in taste perception. *Chemical Senses, 25*, 447–460.

Boesveldt, S., Frasnelli, J., Gordon, A. R., & Lündstrom, J. N. (2010). The fish is bad: Negative food odors elicit faster and more accurate reactions than other odors. *Biological Psychology, 84*, 313–317.

Bojanowski, V., & Hummel, T. (2012). Retronasal perception of odors. *Physiology & Behavior, 107*, 484–487.

Cardello, A. V., & Sawyer, F. M. (1992). Effects of disconfirmed consumer expectations on food acceptability. *Journal of Sensory Studies, 7*, 253–277.

Carlsmith, J. M., & Aronson, E. (1963). Some hedonic consequences of the confirmation and disconfirmation of expectancies. *Journal of Abnormal and Social Psychology, 66*, 151–156.

Castriota-Scanderbeg, A., Hagberg, G. E., Cerasa, A., Committeri, G., Galati, G., Patria, F., et al. (2005). The appreciation of wine by sommeliers: A functional magnetic resonance study of sensory integration. *NeuroImage, 25*, 570–578.

Cerf-Ducastel, B., Van de Moortele, P.-F., Macleod, P., Le Bihan, D., & Faurion, A. (2001). Interaction of gustatory and lingual somatosensory perceptions at the cortical level in the human: A functional magnetic resonance imaging study. *Chemical Senses, 26*, 371–383.

Dalton, P., Doolittle, N., Nagata, H., & Breslin, P. A. S. (2000). The merging of the senses: Integration of subthreshold taste and smell. *Nature Neuroscience, 3*, 431–432.

De Araujo, I. E., Rolls, E. T., Velazco, M. I., Margot, C., & Cayeux, I. (2005). Cognitive modulation of olfactory processing. *Neuron, 46*, 671–679.

De Araujo, I. E. T., Rolls, E. T., Kringelbach, M. L., McGlone, F., & Phillips, N. (2003). Taste-olfactory convergence, and the representation of the pleasantness of flavour, in the human brain. *European Journal of Neuroscience, 18*, 2059–2068.

Deliza, R., & MacFie, H. J. H. (1997). The generation of sensory expectation by external cues and its effect on sensory perception and hedonic ratings: A review. *Journal of Sensory Studies, 2*, 103–128.

Diaz, M. E. (2004). Comparison between orthonasal and retronasal flavour perception at different concentrations. *Flavour and Fragrance Journal, 19*, 499–504.

Djordjevic, J., Lundstrom, J. N., Clément, F., Boyle, J. A., Pouliot, S., & Jones-Gotman, M. (2008). A rose by any other name: Would it smell as sweet? *Journal of Neurophysiology, 99*, 386–393.

Eldeghaidy, S., Marciani, L., McGlone, F., Hollowood, T., Hort, J., Head, K., et al. (2011). The cortical response to the oral perception of fat emulsions and the effect of taster status. *Journal of Neurophysiology, 105*, 2572–2581.

Felleman, D. J., & Van Essen, D. C. (1991). Distributed hierarchical processing in primate cerebral cortex. *Cerebral Cortex, 1*, 1–47.

Gallace, A., Ngo, M. K., Sulaitis, J., & Spence, C. (2012). Multisensory presence in virtual reality: Possibilities & limitations. In G. Ghinea, F. Andres, & S. Gulliver (Eds.), *Multiple sensorial media advances and applications: New developments in MulSeMedia* (pp. 1–40). Hershey, PA: IGI Global.

Grabenhorst, F., Rolls, E. T., & Bilderbeck, A. (2008). How cognition modulates affective responses to taste and flavor: Top-down influences on the orbitofrontal and pregenul cortices. *Cerebral Cortex, 18*, 1549–1559.

Henriques, A. S., King, S. C., & Meiselman, H. L. (2009). Consumer segmentation based on food neophobia and its application to product development. *Food Quality and Preference, 20*, 83–91.

Herz, R. S., & von Clef, J. (2001). The influence of verbal labelling on the perception of odors: Evidence for olfactory illusions? *Perception, 30*, 381–391.

Jones, M. (2008). *Feast: Why humans share food*. Oxford: Oxford University Press.

Kühn, S., & Gallinat, J. (2013). Does taste matter? How anticipation of cola brands influences gustatory processing in the brain. *PLoS ONE, 8*(4), e61569.

Linder, N. S., Uhl, G., Fliessbach, K., Trautner, P., Elger, C. E., et al. (2010). Organic labeling influences food valuation and choice. *NeuroImage, 53*, 215–220.

Manescu, S., Frasnelli, J., Lepore, F., & Djordjevic, J. (2014). Now you like me, now you don't: Impact of labels on odor perception. *Chemical Senses, 39*, 167–175.

Martin, D. (1990). The impact of branding and marketing on perception of sensory qualities. *Food Science & Technology Today: Proceedings, 4*(1), 44–49.

Mauer, L., & El-Sohemy, A. (2012). Prevalence of cilantro (*Coriandrum sativum*) disliking among different ethnocultural groups. *Flavour, 1*, 8.

McClure, S. M., Li, J., Tomlin, D., Cypert, K. S., Montague, L. M., & Montague, P. R. (2004). Neural correlates of behavioral preference for culturally familiar drinks. *Neuron, 44*, 379–387.

Nitschke, J. B., Dixon, G. E., Sarinopoulos, I., Short, S. J., Cohen, J. D., Smith, E. E., et al. (2006). Altering expectancy dampens neural response to aversive taste in primary taste cortex. *Nature Neuroscience, 9*, 435–442.

Olivetti Belardinelli, M., Palmiero, M., Sestieri, C., Nardo, D., Di Matteo, R., Londei, A., et al. (2009). An fMRI investigation on image generation in different sensory modalities: The influence of vividness. *Acta Psychologica, 132*, 190–200.

Olson, J. C., & Dover, P. A. (1978). Cognitive effects of deceptive advertising. *Journal of Marketing Research, 15*, 29–38.

Österbauer, R. A., Matthews, P. M., Jenkinson, M., Beckmann, C. F., Hansen, P. C., & Calvert, G. A. (2005). Color of scents: Chromatic stimuli modulate odor responses in the human brain. *Journal of Neurophysiology, 93*, 3434–3441.

O'Doherty, J., Deichmann, R., Critchley, H. D., & Dolan, R. J. (2002). Neural responses during anticipation of a primary taste reward. *Neuron, 33*, 815–826.

Pazart, L., Comte, A., Magnin, E., Millot, J.-L., & Moulin, T. (2014). An fMRI study on the influence of sommeliers' expertise on the integration of flavor. *Frontiers in Behavioral Neuroscience, 8*, 358.

Piqueras-Fiszman, B., & Spence, C. (2012). Sensory incongruity in the food and beverage sector: Art, science, and commercialization. *Petits Propos Culinaires, 95*, 74–118.

Piqueras-Fiszman, B., & Spence, C. (2015). Sensory expectations based on product-extrinsic food cues: An interdisciplinary review of the empirical evidence and theoretical accounts. *Food Quality & Preference, 40*, 165–179.

Plassmann, H., O'Doherty, J., Shiv, B., & Rangel, A. (2008). Marketing actions can modulate neural representations of experienced pleasantness. *Proceedings of the National Academy of Sciences of the USA, 105*, 1050–1054.

Plassmann, H., Ramsøy, T. Z., & Milosavljevic, M. (2012). Branding the brain: A critical review and outlook. *Journal of Consumer Psychology, 22*, 18–36.

Plassmann, H., & Weber, B. (2015). Individual differences in marketing placebo effects: Evidence from brain imaging and behavioural experiments. *Journal of Marketing Research, LII*, 493–510.

Pliner, P., & Hobden, K. (1992). Development of a scale to measure the trait of food neophobia in humans. *Appetite, 19*, 105–120.

Prescott, J. (2012). *Taste matters: Why we like the foods we do*. London: Reaktion Books.

Pursey, K. M., Stanwell, P., Callister, R. J., Brain, K., Collins, C. E., & Burrows, T. L. (2014). Neural responses to visual food cues according to weight status: A systematic review of functional magnetic resonance imaging studies. *Frontiers in Nutrition, 1*, 7.

Reardon, P., & Bushnell, E. W. (1988). Infants' sensitivity to arbitrary pairings of color and taste. *Infant Behavior and Development, 11*, 245–250.

Rozin, P. (1982). "Taste-smell confusions" and the duality of the olfactory sense. *Perception & Psychophysics, 31*, 397–401.

Sarinoloulos, I., Dixon, G. E., Short, S. J., Davidson, R. J., & Nitschke, J. B. (2006). Brain mechanisms of expectation associated with insula and amygdala response to aversive taste: Implications for placebo. *Brain, Behavior, and Immunity, 20*, 120–132.

Schaal, B., & Durand, K. (2012). The role of olfaction in human multisensory development. In A. J. Bremner, D. Lewkowicz, & C. Spence (Eds.), *Multisensory development* (pp. 29–62). Oxford: Oxford University Press.

Schaal, B., Marlier, L., & Soussignan, R. (2000). Human foetuses learn odours from their pregnant mother's diet. *Chemical Senses, 25*, 729–737.

Schienle, A., Schäfer, A., Hermann, A., & Vaitl, D. (2009). Binge-eating disorder: Reward sensitivity and brain activation to images of food. *Biological Psychiatry, 65*, 654–661.

Schifferstein, H. N. J. (2001). Effects of product beliefs on product perception and liking. In L. Frewer, E. Risvik, & H. Schifferstein (Eds.), *Food, people and society: A European perspective of consumers' food choices* (pp. 73–96). Berlin: Springer Verlag.

Shepherd, G. M. (2006). Smell images and the flavour system in the human brain. *Nature, 444,* 316–321.
Shepherd, G. M. (2012). *Neurogastronomy: How the brain creates flavor and why it matters.* New York, NY: Columbia University Press.
Skrandies, W., & Reuther, N. (2008). Match and mismatch of taste, odor, and color is reflected by electrical activity in the human brain. *Journal of Psychophysiology, 22,* 175–184.
Small, D. M. (2012). Flavor is in the brain. *Physiology & Behavior, 107,* 540–552.
Small, D. M., Gerber, J. C., Mak, Y. E., & Hummel, T. (2005). Differential neural responses evoked by orthonasal versus retronasal odorant perception in humans. *Neuron, 47,* 593–605.
Small, D. M., Jones-Gotman, M., Zatorre, R. J., Petrides, M., & Evans, A. C. (1997). Flavor processing: More than the sum of its parts. *Neuroreport, 8,* 3913–3917.
Small, D. M., Veldhuizen, M. G., Felsted, J., Mak, Y. E., & McGlone, F. (2008). Separable substrates for anticipatory and consummatory food chemosensation. *Neuron, 57,* 786–797.
Small, D. M., Voss, J., Mak, Y. E., Simmons, K. B., Parrish, T., & Gitelman, D. (2004). Experience-dependent neural integration of taste and smell in the human brain. *Journal of Neurophysiology, 92,* 1892–1903.
Small, D. M., Zatorre, R. J., Dagher, A., Evans, A. C., & Jones-Gotman, M. (2001). Changes in brain activity related to eating chocolate: From pleasure to aversion. *Brain, 124,* 1720–1733.
Spence, C. (submitted). The neuroscience of behaviour. *Organizational Research Methods.*
Spence, C. (in press). Multisensory flavour perception. In C. Korsmeyer (Ed.), *The taste culture reader* (2nd Ed.). Oxford: Bloomsbury.
Spence, C. (2010). The price of everything—The value of nothing? *The World of Fine Wine, 30,* 114–120.
Spence, C. (2012a). Book review: "Neurogastronomy: How the brain creates flavor and why it matters" by Gordon M. Shepherd. *Flavour, 1,* 21.
Spence, C. (2012b). Multi-sensory integration & the psychophysics of flavour perception. In J. Chen & L. Engelen (Eds.), *Food oral processing—Fundamentals of eating and sensory perception* (pp. 203–219). Oxford: Blackwell.
Spence, C. (2015). Multisensory flavour perception. *Cell, 161,* 24–35.
Spence, C., Okajima, K. Cheok, A. D., Petit, O., & Michel, C. (in press). Eating with our eyes: From visual hunger to digital satiation. *Brain & Cognition.*
Spence, C., & Piqueras-Fiszman, B. (2014). *The perfect meal: The multisensory science of food and dining.* Oxford: Willey-Blackwell.
Spence, C., Smith, B., & Auvray, M. (2015). Confusing tastes and flavours. In D. Stokes, M. Matthen, & S. Biggs (Eds.), *Perception and its modalities* (pp. 247–274). Oxford: Oxford University Press.
Stein, B. E. (Ed.). (2012). *The new handbook of multisensory processing.* Cambridge, MA: MIT Press.
Stevenson, R. J. (2009). *The psychology of flavour.* Oxford: Oxford University Press.
Stice, E., Spoor, S., Bohon, C., Veldhuizen, M. G., & Small, D. M. (2008). Relation of reward from food intake and anticipated food intake to obesity: A functional magnetic resonance imaging study. *Journal of Abnormal Psychology, 117,* 924–935.
Van der Laan, L. N., De Ridder, D. T. D., Viergever, M. A., & Smeets, P. A. (2011). The first taste is always with the eyes: A meta-analysis on the neural correlates of processing visual food cues. *NeuroImage, 55,* 296–303.
Veeck, A. (2010). Encounters with extreme foods: Neophilic/neophobic tendencies and novel foods. *Journal of Food Products Marketing, 16,* 246–260.

Verhagen, J. V., & Engelen, L. (2006). The neurocognitive bases of human multimodal food perception: Sensory integration. *Neuroscience and Biobehavioral Reviews, 30*, 613–650.

Wang, G.-J., Volkow, N. D., Felder, C., Fowler, J. S., Levy, A. V., Pappas, N. R., et al. (2002). Enhanced resting state activity of the oral somatosensory cortex in obese subjects. *Neuroreport, 13*, 1151–1155.

Wang, G.-J., Volkow, N. D., Telang, F., Jayne, M., Ma, J., Rao, M., et al. (2004). Exposure to appetitive food stimuli markedly activates the human brain. *NeuroImage, 212*, 1790–1797.

Wheatley, J. (1973). Putting colour into marketing. *Marketing* October, 24–29, 67.

Woods, A. T., Lloyd, D. M., Kuenzel, J., Poliakoff, E., Dijksterhuis, G. B., & Thomas, A. (2011). Expected taste intensity affects response to sweet drinks in primary taste cortex. *Neuroreport, 22*, 365–369.

Zatorre, R. J., & Jones-Gotman, M. (2000). Functional imaging of the chemical senses. In A. W. Toga & J. C. Mazziota (Eds.), *Brain mapping: The systems* (pp. 403–424). San Diego, CA: Academic Press.

Zellner, D., Strickhouser, D., & Tornow, C. (2004). Disconfirmed hedonic expectations produce perceptual contrast, not assimilation. *American Journal of Psychology, 117*, 363–387.

Responses of the Autonomic Nervous System to Flavors

13

René A. de Wijk[1] and Sanne Boesveldt[2]
[1]Consumer Science & Health, WUR, Wageningen, The Netherlands, [2]Division of Human Nutrition, Wageningen University, Wageningen, The Netherlands

1 The problem

Up to 80% of all new food products fail in the marketplace, despite the fact that they are typically subjected to a large number of sensory and consumer tests before their market introduction (Crawford, 1977). This suggests that the "standard" sensory and consumer tests, which typically include sensory analytical profiling and liking tests, have a low predictive validity with respect to general product performance. Possibly, consumer food choice outside the laboratory may be less based on cognitive information processing and rational reasoning, and more on unarticulated/unconscious motives and associations (Wansink, 2004). Reasons for likes or dislikes of specific foods are typically difficult to articulate but may determine much of our food choice. Unarticulated/unconscious motives and associations are not very well captured by traditional tests based on conscious cognitive processes, and may be better captured by physiological and behavioral measures (eg, facial expressions) of the autonomic nervous system (ANS) which do not require conscious processes (Greenwald, 2009).

2 The central and autonomic nervous systems

Subconscious processing can be monitored at the central nervous system (CNS), the brain, or in the ANS. The brain is the place where all consumers' planning takes place and actions are initiated. These actions are then executed by the ANS resulting in the appropriate behaviors that can be roughly categorized into fight or flight reactions: The flight system is mainly active in those situations involving threat, with the resulting behaviors aimed at withdrawal, attack, and escape. In contrast, the fight system is mainly active in situations "that promote survival including sustenance, procreation, and nurturance, with a basic behavioural repertoire of ingestion, copulation, and caregiving" (Bradley, Codispoti, Cuthbert, & Lang, 2001, p. 276; Bradley, Codispoti, Sabatinelli, & Lang, 2001). Other terms for these dual responses are defensive/appetitive and approach/avoidance. Emotions are an essential link between specific situations and the appropriate behavior. A situation where one is confronted with a hostile person can evoke scared emotions resulting in avoidance behavior. Similarly, a situation where one is confronted with spoiled food can evoke disgust emotions, resulting in the same avoidance behavior. In contrast, an encounter with a loved one, or with a delicious food, can evoke happiness emotions resulting in approach behavior (see Piqueras-Fiszman, Kraus, & Spence, 2014).

ANS activity is not only determined by appetitive/defensive behaviors that are of interest for this chapter but also by all other non-CNS processes that are required to function properly, including motor activity, attention, homeostasis, and digestion (Berntson & Cacioppo, 2000). These processes can be divided into those processes that are related to activation, determined by sympathetic activation, and relaxation, determined by parasympathetic activation. Given the broad range of processes affecting ANS activity, of which only a limited number may be relevant to consumer decision-making, and the fact that ANS activity reflects the execution part of consumer behavior rather than the planning and preparations that takes place in the brain, ANS activity may seem an odd target for consumer tests.

1. ANS processes do not "merely" follow and execute the results of CNS processing, but also provide extensive feedback to the CNS where it becomes an important contributor for CNS decision-making. Or, as Critchley and Nagai (2012, p. 163) put it: "bodily states shape mental content: if we are dehydrated we experience thirst, if nutritionally deprived, hunger. These basic motivational changes influence how we interact mentally and physically with our environment. Motivational changes are elicited by other external threats to the physiological integrity and by external cues that may ensure the body's integrity. Changes in the physiological state of the body accompany the processing of such aversive and appetitive cues to facilitate and anticipate appropriate behaviors." Emotions play a central role as mediators between bodily changes and appropriate actions. One could say that a stimulus that does not result in a bodily change cannot be viewed as emotive (Critchley & Nagai, 2012). Following this line of reasoning, James (1994) suggested that bodily responses precede emotions and that emotions are in fact "feelings" of those responses. Damasio, Tranel, and Damasio (1991) continued along this line with the "somatic marker hypothesis" that assumes that arousal responses reflect learning and unconsciously facilitate decision-making, resulting in appropriate behavior. Others have pointed out that differentiation between bodily responses is too crude to account for the broad range of emotional experiences, and that different bodily responses can trigger the same emotion (Cannon, 1927).
2. CNS consumer processes may not always be localized, but may be organized in networks of activation that spread across hemispheres. Consequently, tests that focus on localized brain activity, such as functional magnetic resonance imaging (fMRI) may be of limited use. A more general measure of brain activity such as electroencephalography (EEG) may be more appropriate. For example, emotions that affect consumer decisions as to whether to approach or avoid a food product seem to be associated with asymmetrical activation of the two hemispheres, whereby more activity in the left hemisphere is associated with "approach" emotions, such as happy, surprised, and angry, and more activity in the right hemisphere is associated with "avoidance" emotions, such as scared, sad, and disgusted.
3. Consumer processes change rapidly over time. Current theories of product emotions, a powerful driver of consumer decisions, assume that emotional responses are not static but dynamic reflecting sequential judgements—or appraisals—of the products of aspects such as its pleasantness and familiarity (eg, Scherer, 2001). These appraisals may take place within seconds after the product is first encountered whereby an appraisal may take as little as a few tenths of a second. Any successful measuring tool for consumer processes should therefore have a temporal resolution that is good enough to monitor the results of this dynamic appraisal. Established CNS tests such as fMRI have extremely poor temporal

resolution, often exceeding 10 s, and are therefore not well suited. The exception, once again, is EEG, which offers a sufficient degree of temporal resolution.
4. Consumer behavior is often habitual and automatic, and often irrational. Most of us consumers have a pretty good idea that certain foods such as vegetables are good for us, while others, such as hamburgers, are not. Still, the first McDonalds restaurant serving only vegetable dishes has yet to open. Despite their rational thoughts, many consumers find the temptations fuelled by the aromas of the fast food restaurant difficult to resist, which is a good example of appetitive ANS reactions literally guiding the consumers towards the desired hamburger.

3 Emotions and specificity of responses

Most current theories of emotions assume that emotional response starts with appraisal of the personal significance of an event (eg, Lazarus, 1991; Mauss & Robinson, 2009; Scherer, 1984), which then initiates an emotional response based on subjective experience, physiology, and behavior (eg, Frijda, 1999). A crucial and much debated question is whether emotional responses, as reflected for example in the ANS system, are discrete or dimensional. The discrete emotions perspective assumes that each emotion corresponds to a specific "profile" or pattern in experience, physiology, and behavior (eg, Ekman, 1992). With respect to ANS activity, this would mean that anger would be associated with a different ANS pattern than, for example, happiness. Ekman's results (1983), based on actors and scientists who were instructed to show specific emotional expressions, suggest at least some specificity of ANS patterns. For example, anger was associated with increased heart rate and increased skin temperature, fear with increased heart rate and decreased temperature, and disgust with reduced heart rate and reduced temperature (see Fig. 13.1).

In contrast, the dimension perspective of emotions assumes the existence of a limited number of dimensions that organize all emotional responses. The most commonly assumed dimensions are valence and arousal, as well as approach-avoidance (eg, Davidson, 1999; Lang, Bradley, & Cuthbert, 1997; Russell & Barrett, 1999). Russell's circumplex model of affect, shown in Fig. 13.2, organizes emotions along two dimensions related to valence and activation or arousal. Examples of negatively valenced emotions with low and high arousal are fatigued and tense, respectively, whereas examples of positive emotions with low and high arousal are calm and alert, respectively (Russell & Barrett, 1999).

Even within the dimensionality community, there continue to be debates about the specific dimensions. For example, Russell's view that negative and positive emotions are inversely related, that is, are part of the same dimension, is opposed by others who assume that negative and positive emotions are, in fact, relatively independent of each other (eg, Lewis, Critchley, Rotshtein, & Dolan, 2007; Tellegen, Watson, & Clark, 1999).

Others have speculated that ANS patterns lack any specificity or dimensionality (eg, Feldman-Barrett & Wager, 2006) or that specificity is only found for specific emotions such as fear and anger (eg, Stemmler, 2004). In between these two extremes,

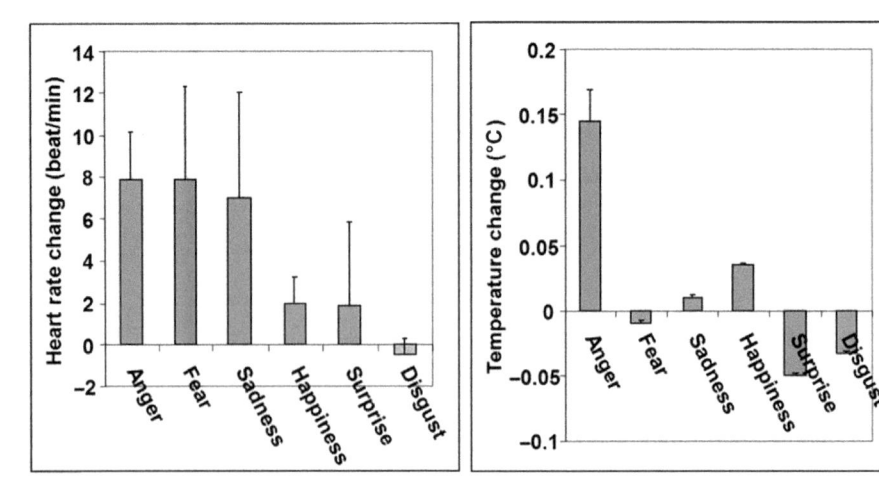

Figure 13.1 Changes in heart rate (left) and skin temperature (right) during directed facial tasks. Error bars are standard errors.
Source: Adapted from Ekman et al. (1983).

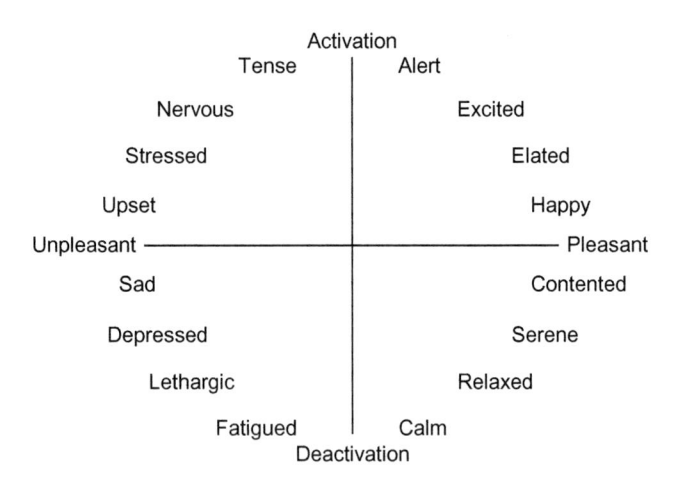

```
                        Activation
                 Tense  |  Alert
              Nervous   |   Excited
             Stressed   |    Elated
             Upset      |    Happy
  Unpleasant ───────────┼─────────── Pleasant
              Sad       |   Contented
            Depressed   |   Serene
            Lethargic   |   Relaxed
                Fatigued|  Calm
                       Deactivation
```

Figure 13.2 Circumplex modal of affect organizing emotions along two independent valence and activation dimensions.
Source: From Russell and Barrett (1999).

Cacioppo, Berntson, Klein, and Poehlmann (1997) and Cacioppo, Berntson, Larsen, Poehlmann, and Ito (2000) claim that there is at least some degree of emotional-specificity to ANS patterns. They also note that valence-specific patterns (reflecting combinations of positive vs negative emotions) tend to be more consistent than emotion-specific patterns (De Wijk, Kooijman, Verhoeven, Holthuysen, & de Graaf, 2012).

4 Measures of autonomic nervous system activity

Measures of ANS activity can be separated into cardiovascular measures, respiratory measures, and electrodermal measures. An extensive review of 134 ANS studies by Kreibig (2010) revealed that heart rate measurement is the most popular cardiovascular measure, followed by systolic and diastolic blood pressure, heart rate variability, and finger temperature. Most popular measures with regard to respiration are respiratory rate, followed by respiratory depth and period, tidal volume and respiratory variability, and with regard to electrodermal measures, skin conductance level, followed by skin conductance response rate, and skin conductance amplitude. Overall, heart rate is the most commonly used index for ANS activity. Different measures reflect different types of activity. For example, skin conductance reflects primarily sympathetic (action-related) activity, whereas heart rate and blood pressure primarily reflect a combination of sympathetic and parasympathetic activity. In contrast, heart rate variability reflects primarily parasympathetic (relaxation-related) activity (Cacioppo et al., 2000; see Mauss and Robinson, 2009, for a review). A more complete set of 33 cardiovascular measures, 20 respiratory measures, and 6 electrodermal measures has been provided by Kreibig (2010) (see Table 13.1).

5 Patterns of ANS activity

In an effort to link ANS patterns to specific emotions, Kreibig (2010) provided an extensive review of the literature, where responses of the ANS were linked to various types of emotional stimuli, such as film clips and personalized recall of specific situations. The emotions that were investigated most often were fear, sadness, anger, disgust, and happiness (see Kreibig, 2010). For each type of ANS response measure, consensus of the relevant studies was represented as either an increase, a decrease, or no change compared to baseline of that measure in relation to a specific emotion. The results shown in Table 13.1 suggest a considerable degree of ANS specificity between positive and negative emotions; anger was associated with increased heart rate and reduced finger temperature (and increased skin conductance), whereas happiness was associated with increased heart rate and increased finger temperature. More impressively, ANS specificity is also observed between negative (and between positive) emotions: increased heart rate, reduced finger temperature, and increased skin conductance were observed for the negative anger emotions, whereas reduced heart rate, reduced finger temperature, and reduced skin conductance were observed for acute sadness.

Without going into the specifics of the overview, a number of things stand out: (1) the majority of measures show a response associated with a specific emotion and (2) the direction of the response associated with a specific emotion is not unequivocal but may vary between studies. For example, the heart rate associated with pride or amusement may increase as well as decrease. Similarly, the finger temperature associated with disgust related to mutilation or contamination may increase or decrease

Table 13.1 Overview of ANS responses found for reviewed emotions by Kreibig (2010)

	Anger	Anxiety	Disgust contamination	Disgust mutilation	Embarrassment	Fear	Fear imminent threat	Sadness crying	Sadness noncrying	Sadness anticipatory	Sadness acute	Affection	Amusement	Contentment	Happiness	Joy	Antic pleasure visual	Antic pleasure imagery	Pride	Relief	Surprise	Suspense
Cardiovascular																						
Heart rate	↑	↑	↓-	↓	↑	↑	↓	↑	↓	↑	↓	↓	↑↓	↓	↑	↑	↓	↑	↑↓	↓-	↑	↓
Heart rate variability (HRV)	↓	↑	↑	-	↓	↓	↑	-	↑	↑	↑		↑	↑↓	↓	↑	↑		↑			
Low frequency spectral HRV		↑		↑		↑									-							
Low frequency/High frequency ratio		↑		↑		↓							↑									
P-wave amplitude	↓					↑																
T-wave amplitude	↓			↑	↑	↓				↑↓					↑	↑↓						
Left ventricular ejection time	↑					↓				↓	↓				↑	↓						
H	↓					↑				↓	↓											
Preejection period	↓	↓	↓	↑	↑	↑	↑			↓	↓		↑	↑	↑				↑			
Stroke volume	↑↓		↑	↑	↑	↑	↑			↓	↓		↑	↑	↑	↑						
Cardiac output	↑↓	↑	↑	↑	↑	↑	↑			↓	↓		↓	↑	↑	-						
Systolic blood pressure	↑	↑	↑	↑		↑				↑	↑		↓	↑	↑	↑			↑			
Diastolic blood pressure	↑		↑	↑		↑					↓		↓	↑	↑	-						
Mean arterial pressure	↑		↑	↑	↑	↑				↑	↓		↑	↑	↑	↑						
Total peripheral resistance	↑		↑	↓		↑				↑			↑	↑	↑							
Finger pulse amplitude	↑	↓	↓	↓	↓	↓	↓	↓		↑	↑		↑	↑	↑↓	↑						
Finger pulse transit time	↓	↓	↑↓	↑↓		↓	↑			↑	↓		↑↓	↑	↑↓				↓			
Ear pulse transit time		↓	↑↓	↑↓		↓	↑			↑	↑		↑↓	↑	↓				↓			
Finger temperature	↓	↑	↑	↑		↓		↓	↓	↑	↓		↑		↓							
Forehead temperature	↑↓					↓	↑	↓	↓		↓				↓		↑				↑↓	

Electrodermal

Variable																
Skin conductance response (amplitude, evoked)	↑	↑	↑	↑	↑	↑	↑	(−)	↑	↑	→	↑			(↑)	
Nonspecific skin conductance response rate	↑	↑	↑	↑	↑	→	↑	→	↑	↑	(↓)	↑	(↑)		↓	(↑)
Skin conductance level	↑	→	↑	(↑)	↑	↑	↑	↑	−	↑	→	↑	(↑)			(↑)

Respiratory

Variable																
Respiration rate	↑	↑	↑	↑	↑	↑	↑	↑	↑	(↑)	↑	(↑)	↑−	(↑)	(↑)	
Inspiratory time	(↓)	→	→	−	↓−	−	(↓)	(↓)	↑↓	(↓)	(↑)	(↑)	→	(↑)	↓	(↓)
Expiratory time	(↓)	→	↓	−	→→	−	−	→	→→	(↓)	(↓)	(↓)	−	(↓)	(↓)	(↓)
Post-inspiratory pause time	(↑)		(↑)					(↓)								
Post-expiratory pause time					(↑)		→		(↓)	↑↓						
Inspiratory duty cycle		(↓)		−	↑↓			↑↓	↑↓	↑↓	(↑)	↑		(↑)	↑−	
Tidal volume	↑↓	→	→	(↓)	(↑)	↑	↑	(↓)	↑	(↑)	↑	(↓)	(↑)			(↓)
Inspiratory flow rate or inspiratory drive			↓	−	→		↓							(↑) ↑↓		
V (rhyth)	(↑)	(↑)	(↑)	−	(↑)	−		(↑)	(↑)	(−)	(↓)	(↑)			(↑)	
V (vol)	(↑)	↑↓	(↑)	−	(↑)	←		(↑)	(↑)	(↓)						
Sighing	(↑)	(↑)	(↑)	−	↑			→								
Oscillatory resistance	(↑)	→	(↑)		→				(?)	(−)						
End-tidal carbon dioxide partial pressure					↓			↓	↑	↑	(↓)					

ANS activation components

Variable																
α-adrenergic	↑	(↑)	(↑)	(−)	↑	↑	↑	(↑)	−	(↑)	↑−	(↑)	−	→	→	(↑)
β-adrenergic	↑	(↑)	(↑)	(↑)	↑	↑	↑	↑↓	↑↓	(↓)	(↓)	→	(↑)	↓	→	(↑)
Cholinergic	↑	→	↑	(↑)	↑	→	↑	↑	↑	→	↓−	↑−	(↑)		↑	↓exp
Vagal	→	→	↓	(↓)	→	→	↓	(↓)	(↓)	(↓)	→	→	(↓)			
Respiratory	↑	↑	↑	−	↑	↑	↑	(↓) depth	(↓) depth	(↓) depth	→	→	↑ depth	(↓)	↑ pause	
	↑pause	↓depth	↑exp		↑insp	↓depth	↑depth	↓depth	↓var	↑var			↓depth			

Note: *Modal responses were defined as the response direction reported by the majority of studies (unweighted), with at least three studies indicating the same response direction. Arrows indicate increased (↑), decreased (↓), or no change in activation from baseline (−), or both increases and decreases between studies (↓↑). Arrows in parentheses indicate tentative response direction, based on fewer than three studies. Abbreviations: pause—respiratory pause time; depth—respiratory depth; exp—respiratory expiration time; insp—respiratory inspiration time; var—respiratory variability.

depending on the specific study. These different and apparently contradictory patterns may reflect differences between studies with respect to organ activity, motor activity, digestion, and/or metabolism that may be unrelated to the experimental questions but that are known to affect ANS activity, as discussed earlier. Fig. 13.3 shows examples of this phenomenon from our laboratory. Visual inspection of five well-liked commercial breakfast drinks resulted in moderate decreases or increases in heart rate (see Fig. 13.3A), whereas tasting of the same drinks increased heart rate by as much as 10 beats/min (see Fig. 13.3B). These apparently contradictory results are probably caused by the increased motor activity, and concomitant increase in heart rate, required for the oral processing of the breakfast drinks.

5.1 Specificity of ANS activity in food, drinks, and aromas

Given the inconsistent association between ANS measures and emotions, probably as a result of the wide range of stimuli, conditions, and tasks used in the various studies, we will limit our discussion now to ANS studies that use stimuli that are especially relevant to foods and drinks, namely odors and tastants, either presented individually or as part of a food or drink.

In a previous study, Alaoui-Ismaïli, Robin, Rada, Dittmar, and Vernet-Maury (1997) and Alaoui-Ismaïli, Vernet-Maury, Dittmar, Delhomme, and Chanel (1997) related various autonomic parameters to the pleasantness of five odorants. They found that unpleasant odors were associated with increased heart rate and longer skin conductance responses as compared to pleasant odors. Bensafi et al. (2002) related ANS measures to rated pleasantness, arousal, intensity, and familiarity for a set of six odorants and found that their results could be explained by two main factors: pleasantness, inversely related to heart rate (similar to Alaoui-Ismaïli, Robin et al., 1997; Alaoui-Ismaïli, Vernet-Maury et al., 1997) and arousal, positively related to skin conductance and rated intensity. Delplanque et al. (2009), Glass, Lingg, and Heuberger (2014), and He, Boesveldt, de Graaf, and de Wijk (2014) found stronger skin conductance responses and a higher heart rate for unpleasant as compared to pleasant odors (see also van den Bosch, van Delft, de Wijk, de Graaf, & Boesveldt, 2015, who found changed skin conductance after pairing initially neutral odors with unpleasant tastes or pictures). Similarly, Rousmans, Robin, Dittmar, and Vernet-Maury (2000) found

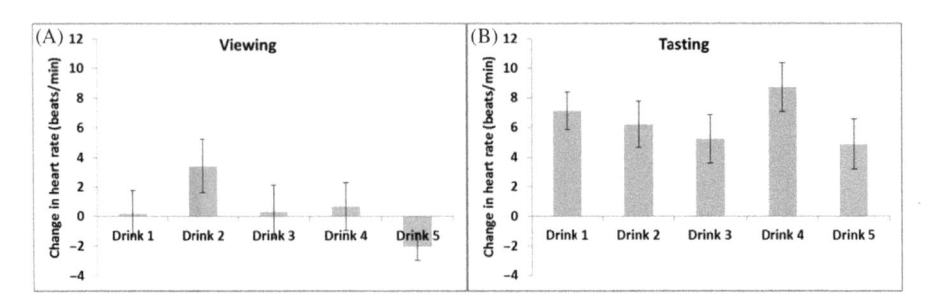

Figures 13.3 Changes in heart rate as a result of viewing (A) or tasting (B) five breakfast drinks (de Wijk, unpublished results).

stronger ANS responses for unpleasant as compared to pleasant primary tastes, but Leterme, Brun, Dittmar, and Robin (2008) suggested that the weak ANS responses for pleasant (sweet) tastes reflected habituation rather than pleasure. Delplanque et al. also established that heart rate differences between pleasant and unpleasant odors occurred relatively late in the deceleration phase, approximately 5–8 s after odor presentation (Delplanque et al., 2009, but see also Møller & Dijksterhuis, 2003). Meanwhile, He et al. (2014) observed that heart rate increased with odor intensity. No intensity effects were found for skin conductance and skin temperature. Taken together, the results show consistencies as well as inconsistencies and relations between specific ANS measures and factors such as pleasantness/valence and arousal are not unambiguous. Other factors, such as state of hunger/satiation, may also be relevant, as suggested by Delplanque et al. (2009), but have been virtually unexplored.

Various authors have established decision trees to relate specific ANS patterns to specific emotions. Examples of decision trees were developed by Collet, Vernet-Maury, Delhomme, and Dittmar (1997) for pictures, by Alaoui-Ismaïli, Robin et al. (1997), Alaoui-Ismaïli, Vernet-Maury et al. (1997), and Vernet-Maury, Alaoui-Ismaïli, Dittmar, Delhomme, and Chanel (1999) for odors, and by Robin, Rousmans, Dittmar, and Vernet-Maury (2003) for basic tastes.

De Wijk et al. (2012) found that Robin's decision tree for tastants (see Fig. 13.4) also seems to work for those ANS patterns that are associated with the sight of foods.

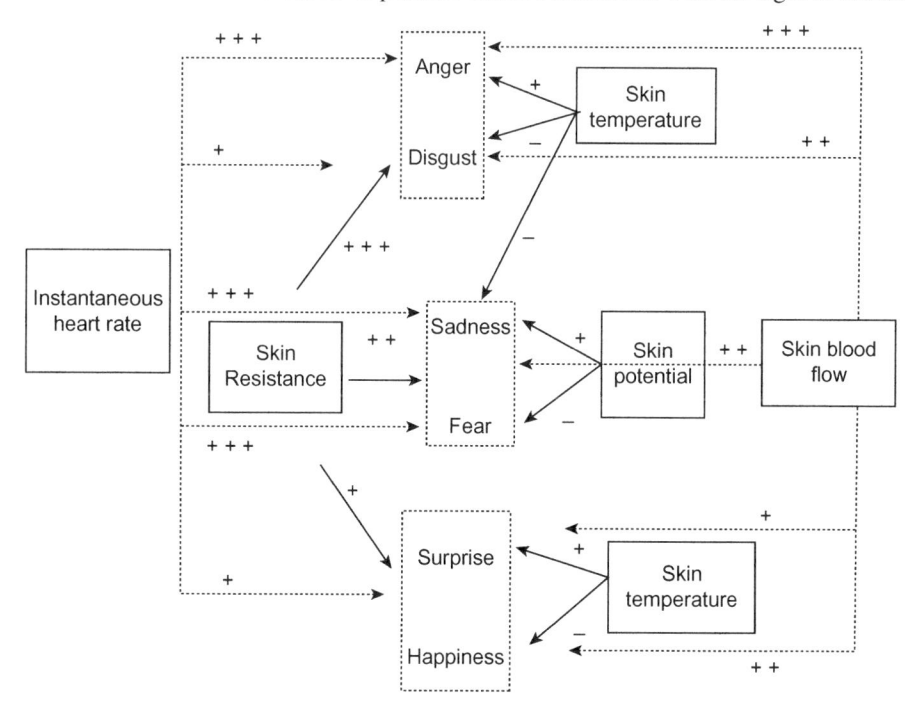

Figure 13.4 Example of decision tree that couples effects of basic tastes on selected ANS measures (skin resistance, skin temperature and heart rate) to specific emotions.
Source: From Robin et al. (2003).

The sight of unpleasant foods resulted in small heart rate changes, large skin conductance changes, and reduced skin temperature. The sight of pleasant foods also resulted in small heart rate changes, but smaller skin conductance changes and increased skin temperature.

According to Robin's decision tree, the observed patterns for unpleasant foods corresponded well with the disgust, and the observed patterns for pleasant foods corresponded well to a positive surprise emotion (De Wijk et al., 2012). Results of studies from the same laboratory (De Wijk, He, Mensink, Verhoeven, & de Graaf, 2014) and from other laboratories (Danner, Haindl, Joechl, & Duerrschmid, 2014), where foods were actually consumed, have been more difficult to associate with specific emotions. The muscle activity during consumption results in large increases in, for example, heart rate. Food specific effects are relatively subtle, especially when foods are very similar, and are superimposed on these large increases, which makes it difficult to determine whether food-specific heart rate effects on heart rate are positive, negative, or, in fact, nonexistent relative to baseline. In summary, the results of studies with consumed foods demonstrate a high degree of sensitivity of ANS patterns to subtle differences between foods that are not picked up by other measures such as preferences. However, interpretation of these differences has been difficult so far.

5.2 *Temporal dynamics of food-related ANS activity*

As was discussed above ANS activity not only reflects emotions but also many other factors, such as the specific experimental task, the type of stimulus, and the physiological state of the consumer. For example, heart rate during consumption of semisolid *foods* typically increases by as many as 10–12 beats/min (see Fig. 13.3B). Heart rate in response to *odors* presented independent of respiration shows either no effect or a moderate slowly evolving increase in heart rate, depending on the odor's valence (see Fig. 13.5A). In contrast, heart rate in response to combinations of *odors* and liquid *tastants* shows a marked increase prior to and immediately following stimulation, followed by a relatively large decrease (see Fig. 13.6). Similar variations, albeit not as extreme, were observed in other ANS measures. The differences in heart rate patterns probably reflect differences in the types of stimulus and stimulation. Foods are introduced into the mouth and require muscle activity during oral processing, which results in an increased heart rate. Systematic differences between foods are relatively small and are superimposed on these gross changes in HR. Odors presented with an injection olfactometer require no active inhalation—and no additional muscle activity—in order to be perceived. Consequently, changes in heart rate are small and probably primarily attributable to the odor's valence. The pattern found for odor/liquid combinations is more difficult to explain because the odors were again presented with an injection olfactometer, and the liquid still evoked some muscle activity that should have increased rather than decreased heart rate. The pattern is, however, also observed in other studies whereby the initial increase in heart rate is typically linked to orienting/novelty responses and the decreasing part to the valence of the stimulus. Indeed, the decrease in heart rate for our odor/liquid combinations varied with the valence of the stimulus combinations.

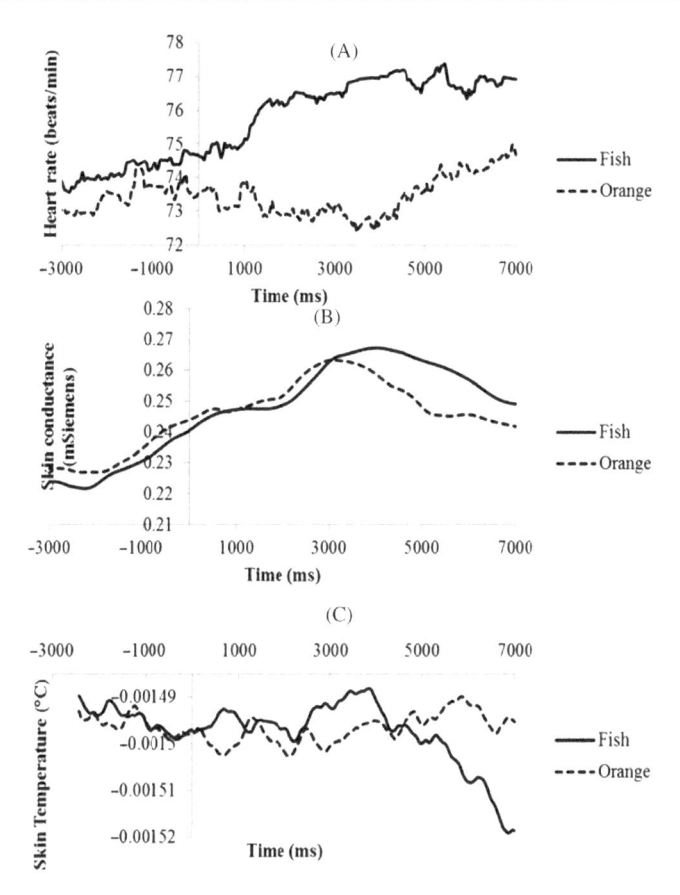

Figures 13.5 Temporal responses of (A) heart rate, (B) skin conductance, and (C) skin temperature for a pleasant (orange) and unpleasant (fish) odor.
Source: From He et al. (2014).

The dynamic features over time of physiological responses and facial expressions have typically fallen outside of the scope of most studies, even though they play a key role in several modern theories on emotion, the so-called componential appraisal models (see Ellsworth & Scherer, 2003, for an overview). The appraisal models of emotions assume that the elicitation and the differentiation of emotions are determined by appraisals, the continuous, recursive evaluations of events. Delplanque et al. (2009) investigated the appraisal of odor novelty and pleasantness and consequent emotional responses by measuring facial muscle activity and heart rate. They demonstrated that odors were detected as novel or familiar before being evaluated as either pleasant or unpleasant (Distel et al., 1999; Royet et al., 1999). In addition, their results also argued in favor of a dynamic construction of facial expressions providing support for sequential appraisal theories (Ellsworth & Scherer, 2003). For example, early reactions, such as raising the eyebrows and opening the eyes, were related to the detection of a novel or unexpected stimulus, which is associated with increased

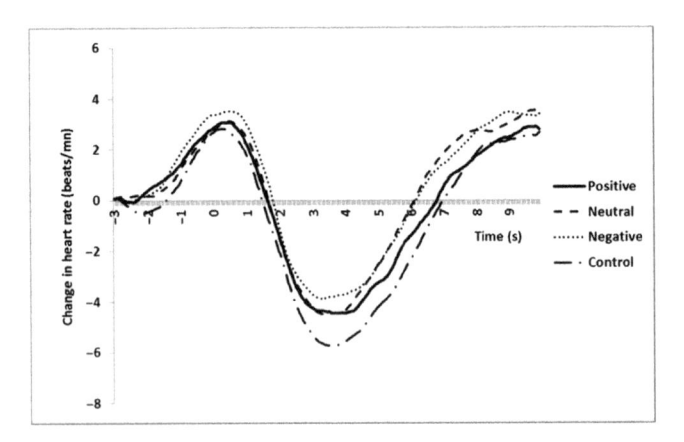

Figure 13.6 Change in heart rate to neutral odors and pleasant (positive) and unpleasant (negative) liquids.
Source: Unpublished data from Boesveldt et al. (2015).

alertness and attention. After this novelty detection, the assessment of pleasantness may lead to avoidance when the stimulus is aversive or threatening, or approach when a pleasant response is activated. Similarly, De Wijk et al. (2014), studying ANS response patterns in response to commercial breakfast drinks, related fast heart rate and skin temperature patterns (after 1 s) to the food's pleasantness and somewhat slower patterns (after 2 s) to the intensity of the food's flavor (De Wijk et al., 2014).

He et al. (2014) measured the temporal characteristics of ANS responses to odors and found that a very fast temporal differentiation between pleasant and unpleasant odors in heart rate (within 400 ms after the start of odor presentation) and skin temperature (within 500–600 ms) and a somewhat slower differentiation in skin conductance (within 4 s; see also Fig. 13.5). The results also confirm previous studies demonstrating longer-lasting heart rate responses for unpleasant tastants (eg, Horio, 2000) and odors (Delplanque et al., 2008) and showed that similar effects are also found for skin conductance and skin temperature.

5.3 Other potential applications of ANS measures in consumer research

ANS measures in consumer research do not have to be limited to the interaction between the consumer and the taste, smell, or texture of the product, but can, in principle, be applied to all aspects of consumer interactions. As was shown earlier, ANS patterns differentiate not only between the actual tastes of foods, but also between the sight of foods (see Fig. 13.3A). Similarly, ANS activity can be used on packaging and the effect of packaging and labeling on consumer product experiences. Unpublished data demonstrate that branding, the combination of package plus brand name, affects ANS activity during the visual inspection phase as well as during the tasting phase. In this study, participants were either exposed to a glass with a breakfast drink

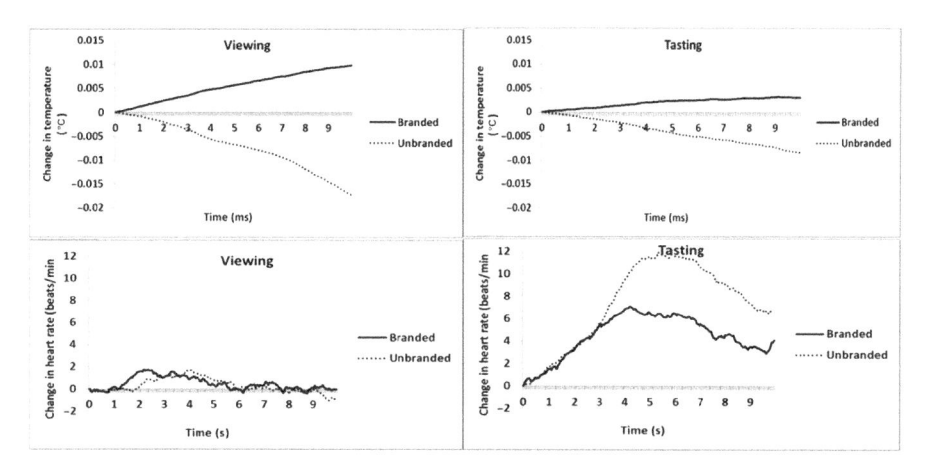

Figure 13.7 Effect of visual inspection (left figures) and tasting (right figures) of branded and unbranded breakfast drinks on skin temperature (top figures) and heart rate (bottom figures). Unpublished resulted by de Wijk et al. (2015).

(unbranded condition) or to a glass with the breakfast drink plus the package with the brand name (branded condition) of the drink. Following the 10-s visual exposure, the breakfast drinks were tasted by the participants. ANS activity (heart rate, skin conductance, and skin temperature) was recorded both during visual inspection and during tasting. The results showed clear effects of task (visual inspection vs tasting) and of branding on ANS patterns. The results are shown in Fig. 13.7.

The effects of visual inspection were in general stronger than the effects of tasting, except for heart rate, which is probably again related to the muscle activity during oral processing, and the effects were product-specific. Obviously, visual inspection of the food provides important anticipatory information about the upcoming taste, which may explain the relatively strong ANS effects of visual inspection. Branding provides additional important anticipatory information. Not only did the participant know that he/she was about to taste a breakfast drink, but also which specific breakfast drink. This additional anticipatory information has relatively large effects on skin temperature and skin conductance during visual inspection and on all three ANS measures during tasting. Our results, that also included facial expressions, suggested that—at least in our case—branding intensified ANS responses (as well as facial expressions) and that ANS activity during visual inspection was associated with ANS activity for the same breakfast drink during tasting. The latter result suggests that anticipation during visual inspection may actually involve a perceptual re-enactment of the upcoming taste, triggering similar ANS patterns. When expectations are not met, for example, when a different breakfast drink is tasted than the one that was visually inspected, the ANS activity, especially skin conductance, becomes markedly different (see Fig. 13.8), as would be expected.

So far, ANS results have been limited to single sniffs or odors and single "bites" of foods but ANS patterns may also contribute to insights in processes during repeated

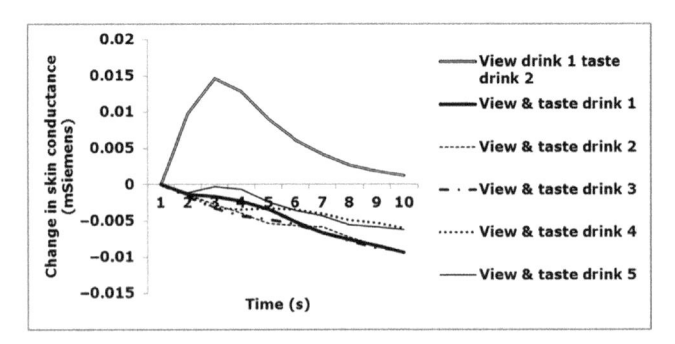

Figure 13.8 Skin conductance responses during tasting of breakfast drinks after either the corresponding or a different package and brand name was visually inspected. Unpublished results from de Wijk et al. (2015).

exposures to the same stimulus, a process that is commonly known as "eating." Eating behavior, and especially deviations from "normal" eating behavior, leading to over- or under-consumption, has clear health implications, and critical insights in the mechanism underlying meal termination are still missing. What determines when a consumer stops eating depends on a myriad of factors, some related to the consumption environment (eg, the food is gone, no more time, the social company) and some related to the food perception (the food is no longer as well-liked, a phenomenon known as "sensory specific satiety" (Rolls, Rolls, Rowe, & Sweeney, 1981)). Other factors are related to the physiological state (eg, hormonal and mechanical signals from the periphery to the brain of the consumer signaling that the point of satiation has been reached) and psychological state (cognitions such as "in view of my new year weight goals I should REALLY stop eating now") of the consumer. Some of these factors, such as the cognitions, are really easy to capture with questionnaires, whereas other factors play a more unconscious role and are probably captured with other techniques, such as ANS. Preliminary results from our laboratory indicate that sensory-specific satiety, the decrease in sensory liking of a consumed food but not of other foods, not only shows up in preference scores but also in ANS activity, that is, ANS patterns after consumption differentiate between the consumed sweet or savory foods, a nonconsumed food that also tastes sweet or savory, and a nonconsumed food with a different taste (eg, savory after sweet).

ANS measurements may also provide insights in consumer segmentation. As marketers and consumer scientists have known for years, the consumer population is not homogeneous and products that are wanted by certain consumers are not wanted by others. Hence, consumer segmentation into segments that share similar perceptions and attitudes towards consumer products is essential to tailor consumer products to the consumer that actually needs and wants them. Examples of segments that are easy to identify include males versus females, children versus middle aged versus elderly, free living versus institutionalized elderly, and consumers from different cultures. The relatively few ANS consumer studies aimed at comparing different consumer segments demonstrate differences, for example, between males and females with regard to their heart rate and skin conductance responses to emotional images (Bradley, Codispoti, Cuthbert et al., 2001; Bradley, Codispoti, Sabatinelli et al., 2001). Others, however, have failed to find systematic gender differences, for example, in response to tastants (Robin et al., 2003). Effects of aging are similarly inconsistent. De Wijk et al. (2012) failed to find systematic differences between ANS patterns of children and young adults in response to foods. Others, using wider ranges of older participants, found diminished physiological responses in the elderly. Elderly participants who discussed emotional conflicts (Kunzmann, Kupperbusch, & Levenson, 2005; Levenson, Carstensen, & Gottman, 1994); who relived emotional experiences (Labouvie-Vief, DeVoe, & Bulka, 1989); or who watched emotional films (Kunzmann et al., 2005; Tsai, Levenson, & Carstensen, 2000) demonstrated reduced heart rate responses (see also Smith, Hillman, & Duley, 2005). In addition, a diminished skin conductance and respiratory response was reported when elderly participants watched disgusting films (Kunzmann et al., 2005). However, several other studies report no age difference in terms of the skin conductance response (Kunzmann & Grühn, 2005; Neiss, Leigland, Carlson, & Janowsky, 2009; Tsai et al., 2000), startle-blink responses (Smith et al., 2005), and cardiovascular and respiratory responses (Kunzmann & Grühn, 2005). Finally, food neophobics (that is, people who are reluctant to try novel foods) showed stronger ANS responses to food stimuli than neophilics, whereas no differences were found for nonfood stimuli (Raudenbush & Capiola, 2012). Similarly, Nederkoorn, Smulders, Havermans, and Jansen (2004) observed that finger pulse amplitude related positively to the urge to binge eat.

These ANS consumer studies use relatively crude segments of consumers, and future studies may focus on more specific segments, for example, based on household status, income, food phobia, or specific segments of elderly with a comparable physiological and psychological condition.

5.4 ANS measurements in the real consumer world

So far, ANS studies have primarily been conducted in laboratories, whereas actual consumer behavior takes place in real-life supermarkets and other consumer settings. The main reason for this discrepancy is that ANS responses are technically (still) challenging to measure and are prone to artefacts and all types of interferences of factors that are not of relevance for the study. These artefacts and interferences are more likely to occur in real-life consumer situations which are almost by definition noisy and not well-controlled. As a result, most researchers prefer to conduct their consumer ANS studies in the well-controlled laboratory environment. Unfortunately, the so-called "noise" in the real-life consumer environment may well be a major determining factor in consumer behavior, and by eliminating this factor from laboratory studies, the outcome of these studies may lose their relevance for consumer science. The solution may be to incorporate ANS measurements in real-life consumer situations and real consumers, despite all the potential problems discussed above, or to bring the "noisy" real-life consumer situation to the laboratory, for example, with virtual reality.

With regard to the former solution, considerable advances have been made in developing ANS wireless sensors that allow on- or offline recordings of ANS responses in free-moving consumers. These sensors are typically noninvasive and interfere only minimally with "normal" consumer behavior. A potential problem is to relate these ANS recordings to meaningful events during the trip of the free-moving consumer. For example, how to link a specific response, say a skin conductance response signaling a sudden increase in arousal, to a specific event, for example, a sign that a particular item is for sale? This problem can partly be eliminated by combining ANS recordings with eye-tracking recordings. Modern head- or even glass-mounted eye-trackers can be used to record the gaze behavior of consumers during their trip. This can subsequently be linked to ANS recordings, thereby linking attention to ANS responses. With the solving of the technical difficulties, the potential problems shift towards data analysis.

Analysis of ANS data under laboratory conditions is already far from routine and involves labor-intensive cleaning of the raw data from artefacts, linking them to timestamps that indicate specific events, as well as the analyses of the resulting data. Analyses of real-world ANS data will be even more labor-intensive, and interpretation of the results will be even more challenging. A possible alternative is to bring the real-life consumer situation to the laboratory. This can be done by actually recreating (part of) the environment in the laboratory, such as a supermarket cabinet with shelves and products. Recently, more high-tech approaches have been introduced in which the real-life consumer situation is virtually re-created with virtual-reality techniques. This so-called "immersive" technology allows the "consumer" to virtually explore, for example, a supermarket, while sitting behind a computer monitor. The consumer is more or less static, whereas the situation is dynamic as a result of the consumer's virtual movement pattern. The fact that the consumer is more or less static allows the use of the well-established ANS laboratory techniques in combination with recordings of virtual movement patterns. Obviously, the success of these advanced laboratory techniques depends on the degree to which the experimental results resemble those obtained in real-life situations. The definitive answer to this question is not yet known, but the results so far look promising.

Despite the fact that ANS recordings are cumbersome and analysis and interpretation are not straightforward, the consumer scientist and marketer have really limited alternatives because most traditional tests, based on questionnaires and conscious processing, do not provide sufficient insight into mechanisms underlying consumers' reactions to products, as demonstrated by the high failure rates of new market introductions.

6 Concluding remarks

Autonomic nervous system responses offer insights into consumers' responses that are probably highly relevant for behavior but that remain largely unexplored with standard tests due to their highly dynamic nature. Responses within 500 ms may go consciously unnoticed, but may play an important role in filtering out most of the information to which the consumers are exposed. A typical supermarket environment offers consumers information in the form of 30,000 items, displays, and PA systems.

In order to be able to finish shopping within a reasonable time, consumers need an effective filtering mechanism. Part of the filtering may come from the items that are selected habitually every time that the consumer enters the supermarket. However, most of the product selection is nonhabitual and is based on in-store decisions (Inman, Winer, & Ferraro, 2009). Especially these type of decisions need an effective filtering mechanism that minimizes the—conscious—burden on the consumer, and still provides him/her with the wanted items. This filtering mechanism probably reflects many factors related to the consumer's personal shopping history, experiences, and preferences as well as factors related to the shopping environment, including packaging, brand names, price, and so on. Obviously, the specific factors as well as their interactions will be extremely complex, and the resulting shopping behavior will therefore be difficult to predict by any test. The fact that interpretation of ANS results is often not straightforward may well reflect the complexity of the underlying process rather than uncertainties about the ANS measures themselves.

References

Alaoui-Ismaïli, O., Robin, O., Rada, H., Dittmar, A., & Vernet-Maury, E. (1997). Basic emotions evoked by odorants: Comparison between autonomic responses and self-evaluation. *Physiology & Behavior, 62*(4), 713–720.

Alaoui-Ismaïli, O., Vernet-Maury, E., Dittmar, A., Delhomme, G., & Chanel, J. (1997). Odor hedonics: Connection with emotional response estimated by autonomic parameters. *Chemical Senses, 22*, 237–248.

Bensafi, M., Rouby, C., Farget, V., Bertrand, B., Vigouroux, M., & Holley, A. (2002). Autonomic nervous system responses to odours: The role of pleasantness and arousal. *Chemical Senses, 27*, 703–709.

Berntson, G. G., & Cacioppo, J. T. (2000). From homeostasis to allodynamic regulation. In J. T. Cacioppo, L. G. Tassinary, & G. G. Berntson (Eds.), *Handbook of psychophysiology* (pp. 459–481) (2nd Ed.). New York, NY: Cambridge University Press.

Bradley, R. M., Codispoti, M., Cuthbert, B. N., & Lang, P. J. (2001). Emotion and motivation I: Defensive and appetitive reactions in picture processing. *Emotion, 1*(3), 276–298.

Bradley, M. M., Codispoti, M., Sabatinelli, D., & Lang, P. J. (2001). Emotion and motivation II: Sex differences in picture processing. *Emotion, 1*(3), 300–319.

Cacioppo, J., Berntson, G. G., Klein, D. J., & Poehlmann, K. M. (1997). The psychophysiology of emotion across the lifespan. *Annual Review of Gerontology and Geriatrics, 17*, 27–74.

Cacioppo, J. T., Berntson, G. G., Larsen, J. T., Poehlmann, K. M., & Ito, T. A. (2000). The psychophysiology of emotion. In M. Lewis & J. M. Haviland-Jones (Eds.), *The handbook of emotion* (pp. 173–191). New York, NY: Guildford Press.

Cannon, W. B. (1927). The James-Lange theory of emotions: A critical examination and an alternative theory. *The American Journal of Psychology, 39*(1/4), 106–124.

Collet, C., Vernet-Maury, E., Delhomme, G., & Dittmar, A. (1997). Autonomic nervous system response patterns specificity to basic emotions. *Journal of the Autonomic Nervous System, 62*, 45–57.

Crawford, C. (1977). Marketing research and the new product failure rate. *Journal of Marketing, 41*, 55–61.

Critchley, H. D., & Nagai, Y. (2012). How emotions are shaped by bodily states. *Emotion Review*, *4*(2), 163–168.

Damasio, A. R., Tranel, D., & Damasio, H. C. (1991). Ch. 11: Somatic markers and the guidance of behaviour: Theory and preliminary testing. In H. S. Levin, H. M. Eisenberg, & A. L. Benton (Eds.), *Frontal lobe function and dysfunction* (pp. 217–229). New York, NY: Oxford University Press.

Danner, L., Haindl, S., Joechl, M., & Duerrschmid, K. (2014). Facial expressions and autonomous nervous system responses elicited by tasting different juices. *Food Research and International*, *64*, 81–90.

Davidson, R. J. (1999). Neuropsychological perspectives on affective styles and their cognitive consequences. In T. Dalgleish & M. J. Power (Eds.), *Handbook of cognition and emotion* (pp. 103–123). New York, NY: Wiley.

Delplanque, S., Grandjean, D., Chrea, C., Aymard, L., Cayeux, I., LeCalvé, B., et al. (2008). Emotional processing of odors: Evidence for a nonlinear relation between pleasantness and familiarity evaluations. *Chemical Senses*, *33*(5), 469–479.

Delplanque, S., Grandjean, D., Chrea, C., Coppin, G., Aymard, L., Cayeux, I., et al. (2009). Sequential unfolding of novelty and pleasantness appraisals of odors: Evidence from facial electromyography and autonomic reactions. *Emotion*, *9*, 316–328.

De Wijk, R. A., He, W., Mensink, M. M., Verhoeven, R., & de Graaf, C. (2014). ANS responses and facial expressions differentiate between repeated exposures of commercial breakfast drinks. *PLoS One*, *9*(4). http://dx.doi.org/10.1371/journal.pone.0093823.

De Wijk, R. A., Kooijman, V., Verhoeven, R. H. G., Holthuysen, N. T. E., & de Graaf, C. (2012). Autonomic nervous system responses on and facial expressions to the sight, smell, and taste of liked and disliked foods. *Food Quality and Preference*, *26*, 196–203.

Distel, H., Ayabe-Kanamura, S., Martínez-Gómez, M., Schicker, I., Kobayakawa, T., Saito, S., et al. (1999). Perception of everyday odors—Correlation between intensity, familiarity and strength of hedonic judgement. *Chemical Senses*, *24*(2), 191–199.

Ekman, P. (1992). Facial expressions of emotion: An old controversy and new findings. *Philosophical Transactions B of the Royal Society, London*, *335*, 63–69.

Ekman, P., Levenson, R. W., & Friesen, W. V. (1983). Autonomic nervous system activity distinguishes among emotions. *Science*, *221*(4616), 1208–1210.

Ellsworth, P. C., & Scherer, K. R. (2003). Appraisal processes in emotion. In R. J. Davidson, K. R. Scherer, & H. H. Goldsmith (Eds.), *Handbook of affective sciences* (pp. 572–595). New York, NY: Oxford University Press.

Feldman-Barrett, L., & Wager, T. D. (2006). The structure of emotion evidence from neuroimaging studies. *Current Directions in Psychological Science*, *5*(2), 79–83.

Frijda, N. H. (1999). Emotions and hedonic experience. In D. Kahneman, E. Diener, & N. Schwarz (Eds.), *Well-being: The foundations of hedonic psychology* (pp. 190–210). New York, NY: Russell sage foundation.

Glass, S. T., Lingg, E., & Heuberger, E. (2014). Do ambient urban odors evoke basic emotions. *Frontiers in Psychology*, *5*, 340. <http://dx.doi.org/10.3389/fpsyg.2014.00340>.

Greenwald, A. (2009). Supplemental material for understanding and using the implicit association test: III. Meta-analysis of predictive validity. *Journal of Personality and Social Psychology*, *97*, 17–41.

He, W., Boesveldt, S., de Graaf, C., & de Wijk, R. A. (2014). Dynamics of autonomic nervous system responses and facial expressions to odors. *Frontiers in Psychology*, *5*, 110.

Horio, T. (2000). Effects of various taste stimuli on heart rate in humans. *Chemical Senses*, *25*(2), 149–153.

Inman, J. J., Winer, R. S., & Ferraro, R. (2009). The interplay among category characteristics, customer characteristics, and customer activities on in-store decision making. *Journal of Marketing, 73*(5), 19–29.

James, W. (1994). Physical basis of emotion. *Psychological Review, 101*, 205–210.

Kreibig, S. (2010). Autonomic nervous system activity in emotion: A review. *Biological Psychology, 84*, 394–421.

Kunzmann, U., & Grühn, D. (2005). Age differences in emotional reactivity: The sample case of sadness. *Psychology and Aging, 20*, 47–59. http://dx.doi.org/10.1037/0882-7974.20.1.47.

Kunzmann, U., Kupperbusch, C. S., & Levenson, R. W. (2005). Behavioral inhibition and amplification during emotional arousal: A comparison of two age groups. *Psychology and Aging, 20*, 144–158. http://dx.doi.org/10.1037/0882-7974.20.1.144.

Labouvie-Vief, G., DeVoe, M., & Bulka, D. (1989). Speaking about feelings: Conceptions of emotion across the life span. *Psychology and Aging, 4*, 425–437. http://dx.doi.org/10.1037//0882-7974.4.4.425.

Lang, P. J., Bradley, M. M., & Cuthbert, B. N. (1997). Motivated attention: Affect, activation, and action. In P. J. Lang, R. F. Simons, & M. T. Balaban (Eds.), *Attention and orienting: Sensory and motivational processes* (pp. 97–135). Mahwah, NJ: Lawrence Erlbaum Associates.

Lazarus, R. S. (1991). *Emotion and adaptation*. Oxford: Oxford University Press.

Leterme, A., Brun, L., Dittmar, A., & Robin, O. (2008). Autonomic nervous system responses to sweet taste: Evidence for habituation rather than pleasure. *Physiology & Behavior, 93*, 994–999.

Levenson, R. W., Carstensen, L. L., & Gottman, J. M. (1994). Influence of age and gender on affect, physiology, and their interrelations: A study of long-term marriages. *Journal of Personality and Social Psychology, 67*, 56.

Lewis, P. A., Critchley, H. D., Rotshtein, P., & Dolan, R. J. (2007). Neural correlates of processing valence and arousal in affective words. *Cerebral Cortex, 17*(3), 742–748.

Mauss, I. B., & Robinson, M. D. (2009). Measures of emotion: A review. *Cognition & Emotion, 23*(2), 209–237.

Møller, P., & Dijksterhuis, G. (2003). Differential human electrodermal responses to odours. *Neuroscience Letters, 346*, 129–132.

Nederkoorn, C., Smulders, F., Havermans, R., & Jansen, A. (2004). Exposure to binge food in bulimia nervosa: Finger pulse amplitude as a potential measure of urge to eat and predictor of food intake. *Appetite, 42*, 125–130.

Neiss, M. B., Leigland, L. A., Carlson, N. E., & Janowsky, J. S. (2009). Age differences in perception and awareness of emotion. *Neurobiology of Aging, 30*, 1305–1313. http://dx.doi.org/10.1016/j.neurobiolaging.2007.11.007.

Piqueras-Fiszman, B., Kraus, A., & Spence, C. (2014). "Yummy" versus "yucky"! Explicit and implicit approach—Avoidance motivations toward appealing and disgusting foods in normal eaters. *Appetite, 78*, 193–202.

Raudenbush, B., & Capiola, A. (2012). Physiological responses of food neophobics and food neophilics to food and non-food stimuli. *Appetite, 58*, 1106–1108.

Robin, O., Rousmans, S., Dittmar, A., & Vernet-Maury, E. (2003). Gender influence on emotional responses to primary tastes. *Physiology & Behavior, 78*, 385–393.

Rolls, B. J., Rolls, E. T., Rowe, E. A., & Sweeney, K. (1981). Sensory specific satiety in man. *Physiology & Behavior, 27*(1), 137–142.

Rousmans, S., Robin, O., Dittmar, A., & Vernet-Maury, E. (2000). Autonomic nervous system responses associated with primary tastes. *Chemical Senses, 25*(6), 709–718.

Royet, J. P., Koenig, O., Gregoire, M. C., Cinotti, L., Lavenne, F., Le Bars, D., et al. (1999). Functional anatomy of perceptual and semantic processing for odors. *Journal of Cognitive Neuroscience*, *11*(1), 94–109.

Russell, J. A., & Barrett, L. F. (1999). Core affect, prototypical emotional episodes, and other things called emotion: Dissecting the elephant. *Journal of Personality and Social Psychology*, *76*(5), 805–819.

Scherer, K. R., Scherer, K. R., Schorr, A., & Johnstone, T. (2001). Appraisal considered as a process of multilevel sequential checking. In Scherer, A. Schorr, & T. Johnstone (Eds.), *Appraisal processes in emotion: Theory, methods, research*. New York, NY: Oxford University Press, p. 92–120.

Scherer, K. R. (1984). Emotion as a multicomponent process: A model and some cross-cultural data. *Review of Personality and Social Psychology*, *5*, 37–63.

Smith, D. P., Hillman, C. H., & Duley, A. R. (2005). Influences of age on emotional reactivity during picture processing. *Journal of Gerontology-Series B Psychological Sciences and Social Sciences*, *60*, P49–P56.

Stemmler, G. (2004). Physiological processes during emotion. In P. Philippot & R. S. Feldman (Eds.), *The regulation of emotion* (pp. 33–70). Mahwah, NJ: Lawrence Erlbaum Associates, Inc.

Tellegen, A., Watson, D., & Clark, L. A. (1999). On the dimensional and hierarchical structure of affect. *Psychological Science*, *10*(4), 297–303.

Tsai, J. L., Levenson, R. W., & Carstensen, L. L. (2000). Autonomic, subjective, and expressive responses to emotional films in older and younger Chinese Americans and European Americans. *Psychology and Aging*, *15*, 684–693. http://dx.doi.org/10.1037//0882-7974.15.4.684.

van den Bosch, I., van Delft, J. M., de Wijk, R. A., de Graaf, C., & Boesveldt, S. (2015). Learning to (dis)like: The effect of evaluative conditioning with tastes and faces on odor valence assessed by implicit and explicit measurements. *Physiology & Behavior*, *151*, 478–484.

Vernet-Maury, E., Alaoui-Ismaïli, O., Dittmar, A., Delhomme, G., & Chanel, J. (1999). Basic emotions induced by odorants: A new approach based on autonomic pattern results. *Journal of the Autonomic Nervous System*, *75*, 176–183.

Wansink, B. (2004). Environmental factors that increase the food intake and consumption volume of unknowing consumers. *Annual Review of Nutrition*, *24*, 455–479.

Assessing the Influence of the Drinking Receptacle on the Perception of the Contents

Charles Spence[1] and Xiaoang Wan[2]
[1]Crossmodal Research Laboratory, Department of Experimental Psychology, University of Oxford, Oxford, United Kingdom, [2]Department of Psychology, School of Social Sciences, Tsinghua University, Beijing, P.R. China

Drinking nearly always involves exposure to, and physical contact with, the receptacle in which those liquids are contained.[1] One might ask, therefore, just how much of an influence over our perception of a beverage the glasses, cups, mugs, cans, and bottles from which we drink actually have. Furthermore, and separate from the perceptual question, one can also ask whether people's consumption behaviors are affected by the receptacle in which drinks are served as well. The emerging body of scientific evidence reviewed in this chapter supports the view that the vessels we drink from exert a far greater influence over our perception and behavior than most people realize. In this chapter, we look first at the question of whether the shape of the drinking receptacle influences people's consumption behavior. We then go on to review the literature regarding how the color, weight, and texture of the receptacle influence people's perception of a drink. "Sensation transference" is introduced as one possible mechanism to explain certain of these results. Third, we review the emerging evidence showing that the containers we drink from affects everything from how much we like a drink through to how refreshing we find it. Finally, we highlight some of the opportunities for marketers in terms of the introduction of signature branded glassware, as well as signposting what we see as some of the most exciting future directions for research and development in this area. Ultimately, we argue that drinking vessels should be designed to convey some functional and/or perceptual benefit when it comes to the consumer's experience of the contents (eg, Stead, Angus, Macdonald, & Bauld, 2014).

---

[1] For some, the receptacles themselves have become a vital part of human drinking behavior. As a case in point, Li Bai, arguably China's greatest poet, the so-called Immortal Poet (see https://en.wikipedia.org/wiki/Li_Bai), wrote many poems about wine while using delicate language to describe the containers, closely linking them to Chinese culture, such as, for example, the Lanling wine that was served in a jade cup, and the clear liquor that was supposed to be served in a golden goblet.

1 The shape of the receptacle

1.1 Expectations regarding type/flavor of drink

On occasion, the shape of the container influences the flavor that people associate with a drink of a particular color. For example, in a recent series of cross-cultural studies (Wan et al., 2014; Wan, Woods, Seoul, Butcher, & Spence, 2015), the same set of seven colored drinks were presented in three different types of glass to participants from mainland China, the United States, the United Kingdom, India, and South Korea (see Fig. 14.1A). The participants were instructed to indicate the flavor that first came to mind on seeing each of the colored drinks. The results revealed that a given beverage color could elicit either similar or different flavor expectations depending on both the type of receptacle and the cultural background of the participant. For example, all five groups of participants expected the green liquid to taste of kiwi when it was presented in the wine or cocktail glass. Furthermore, all except the Indians also associated the green color with lime flavor (see Fig. 14.1B). However, the different cultural groups had very different flavor expectations when the green drink was presented in the water glass instead: in particular, the Chinese expected the drink to taste of apple, those from the United States and Korea expected it to taste of melon, the Indians expected mint/pineapple, and the British expected a pear flavor. When the green drink was presented in the cocktail glass, the Koreans expected that the drink would taste of melon, whereas the Indians expected it to taste of mint or grape instead.

The latest research suggests that different groups of consumers may associate the same glass with different types of drinks and/or different drinking occasions (see Wan et al., 2014; Wan, Woods, et al., 2015). In fact, one important question to investigate in

Figure 14.1 (A) Pictures of drinks in seven different colors presented in three types of glasses in Wan et al.'s (2014) study to participants from three different countries. (B) An illustration of the three most popular flavors that the participants from mainland China, the United States, the United Kingdom, India, and South Korea associated with the green drinks presented in the wine and cocktail glasses.

Figure 14.1 (Continued)

this area concerns the consequences of serving a drink from a container that is deemed "appropriate" versus inappropriate by the consumer/participant for a given drink (eg, Raudenbush, Meyer, Eppich, Corley, & Petterson, 2002). Note that what counts as "appropriate" will normally be determined by an individual's previous experience of the consumption of particular beverages from specific receptacles. Presumably, serving a drink from the appropriate vessel may, over time, come to enhance the drinker's experience of the contents by a process of associative learning. Furthermore, according to the literature on the "mere exposure effect" (eg, Gallace & Spence, 2014; Harrison, 1977; Jakesch & Carbon, 2012; Pliner, 1982; Suzuki & Gyoba, 2008; Zajonc, 1968, 1980, 2001), those combinations of drinks and associated drinking vessels that we are exposed to on a regular basis will end up becoming more familiar to us, and hence, over time, more liked.

Research on the free sorting of wine glasses has assessed the influence of the shape of the glass. In particular, Faye, Courcoux, Giboreau, and Qannari (2013) conducted a study in which more than 200 participants were classified into one of three groups based on their wine knowledge. The groups so formed included connoisseurs, intermediates, and nonconnoisseurs. The participants viewed photographs of 30 different wine glasses of varying size, volume, and overall design. The participants had to freely sort these glasses into mutually exclusive categories and then use whatever language they wanted in order to describe the categories so formed. The results revealed that the connoisseurs tended to sort the glasses on the basis of their usage and technical characteristics (eg, angular or wide based), whereas the nonconnoisseurs sorted the glasses based primarily on more basic properties of the designs and glass shapes. Such results, then, already highlight the influence of prior knowledge and experience on people's perception of wine glasses.

1.2 What is the "best" shape for a wine glass?

Over the years, more research has been directed at studying the impact of the glass in which wine is served than any other drink (see Spence, 2011, for a review). It has been suggested that the perceived intensity of a wine's odor increases as the ratio between the opening of the glass and its largest diameter increases (Fischer & Loewe-Stanienda, 1999). On the one hand, a larger surface area may well promote the oxidation of the wine. On the other, if a glass has an opening that is too small, a wine taster may simply find it impossible to sniff the odor while simultaneously tasting the wine. This might well be expected to have an adverse effect on their overall perception of a wine's aroma, since one and the same odor can appear very different as a function of whether it happens to be experienced orthonasally (ie, when sniffing) or retronasally (eg, when the odor reaches the posterior nares after swallowing; Rozin, 1982; though see also Diaz, 2004; Pierce & Halpern, 1996). Furthermore, it should also be remembered here that a large part of what people call taste is derived not from what is going on on their tongue but actually from olfactory stimulation.[2]

---

[2] Some have been tempted to argue that as much as 80–95% of what we call taste is actually derived from olfactory inputs (eg, see Martin, 2004; though see Spence, 2015b, for a critical review).

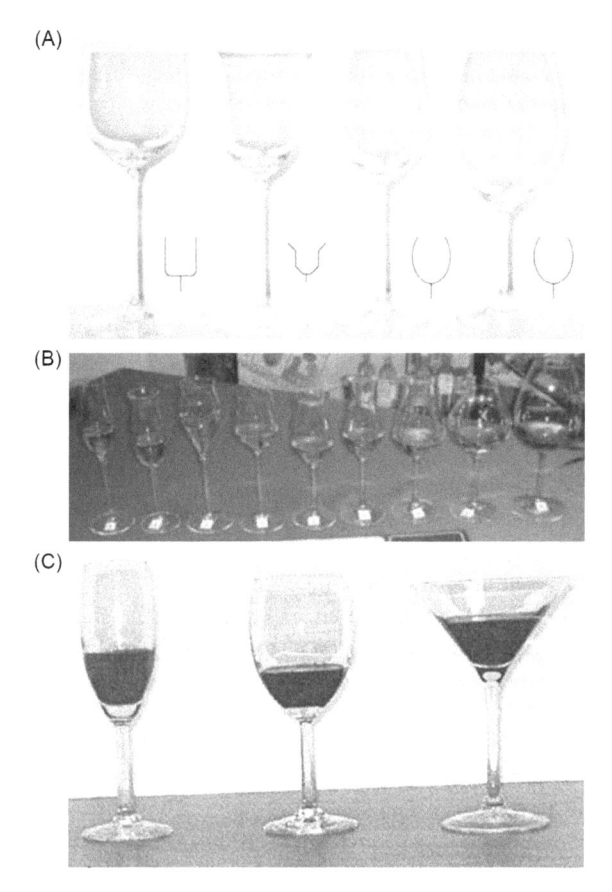

Figure 14.2 (A) Four wine glasses (straight-sided, v-shaped, and bulbous, small and large) used in Hummel et al. (2003). (B) The nine glasses used in Vilanova et al.'s (2008) study of the effect of glass shape on sweet wine. (C) The three glasses (Flute, Bordeaux, or Martini) in which the wine was kept prior to testing in Russell et al.'s (2005) study.

In a study conducted by Hummel, Delwiche, Schmidt, and Hüttenbrink (2003), 200 untrained participants had a red wine (a Cabernet Sauvignon) and a white wine (a Chardonnay) to evaluate from one of four different glasses (see Fig. 14.2A). The glasses were all matched in terms of their opening diameter and height. The participants first had to rate the intensity of the odor by sniffing the wine in the glass and then indicate how much they liked it. Next, they tasted the wine, and rated it once again. The results revealed that glasses with a bulbous shape generally gave rise to the highest ratings of odor intensity, both before and after tasting. There was also a trend in the data toward wines being preferred when they were tasted in bulbous glasses. Intriguingly, those wines that were presented in straight-sided glasses were rated as having a sourer taste than those evaluated in any of the other glasses. However, some of the strongest support for the claim that the shape of the wine glass affects the taster's perception of wine aroma came from the fact that the participants generated significantly more verbal descriptors when evaluating the wine served in the bulbous glasses. This suggests that the wine's aroma was perceived as somewhat more complex. Intriguingly, when the participants were asked how many different wines they had tasted, most said two or three. That is, they were often fooled into believing that they had tasted a different wine when exactly the same wine had been served in a different glass.

Cliff (2001) tested 18 students on an oenology and viticulture degree course using three glasses: a tulip-shaped Chardonnay glass, a bowl-shaped Burgundy glass, and a standard egg-shaped tasting glass (ISO 3591) that had been approved by the International Standards Organization). The students worked in pairs, one swirling the glass under the nose of their blindfolded accomplice. Their task was to rate the fruitiness (for white) or vegetativeness (for red), volatility, and overall intensity of one red (Cabernet Franc) and one white (Riesling) wine without being able to either see or touch the glasses.[3] The shape of the glass had no effect on the students' perception of the fruitiness (white), vegetativeness (red), or volatile acidity of the wines. However, ratings of the total aroma intensity for the wines were as much as 10% higher in the red Burgundy glass, and lowest in the Chardonnay glass.

Research conducted by Vilanova, Vidal, and Cortes (2008) shows that more experienced wine tasters are also affected by the shape of the glass. In one study, for example, six sweet wines were evaluated in nine different glasses (see Fig. 14.2B). A group of professional wine tasters had to rate the intensity and quality of the aroma and taste of each glass (of wine) on an 11-point scale. Even though the tasters were aware that they would be evaluating exactly the same wine in each session, their aroma ratings still varied by as much as 1.8 points on a 10-point rating scale as a function of the glass that the wine was served in. Once again, bulbous glasses gave rise to the highest wine ratings. Interestingly, the glasses exerted more of an influence on the wine tasters' aroma judgments than on their taste judgments. Thus, it would appear that social drinkers', oenology students', and professional wine tasters' ratings of a wine's aroma, flavor, and, to a lesser extent, its taste can all be affected by the dimensions of the glass in which the wine happens to be served.

1.3 Mixed evidence regarding the influence of the wine glass

While a number of studies have demonstrated that the same wine will be rated very differently when served in different wine glasses (eg, Anon., 1973; Fischer, 1996; see also Anon., 2011c; Cloake, 2012; Gawel, 2010; McCarthy, 2006), others have reported that people are unable to detect any perceptible difference between samples of the same wine when evaluated in different wine glasses (Postgate, 1951). Typically, under those strictly controlled laboratory conditions in which a participant's

---

[3] Note that 200 parts per million of ethyl acetate was added to some of the wines in order to synthetically create a defect known as volatile acidity. The wines were poured into the glasses an hour before their evaluation.

awareness of the glass in which a wine is served, or has been allowed to breath, is removed, then no detectable difference in people's ratings of the sensory properties of the wine is reported (see Delwiche & Pelchat, 2001, 2002; Russell, Zivanovic, Morris, Penfield, & Weiss, 2005).

So, for instance, Russell et al. (2005) conducted a study examining the effect of the glass on the evolution of wine, but in this case the results failed to show any effect of glass shape on the perception of wine when all of the known psychological cues had been removed. Twelve participants had to try to pick the odd one out from three beakers of wine in a triangle test. The wine, a Merlot, was kept in one of three differently shaped glasses (see Fig. 14.2C) for 0, 15, or 30 minutes. In order to prevent any bias that might have been induced by seeing/handling the glasses themselves, the tasters evaluated 10-mL wine samples taken from the various glasses but served to them in identical 50-mL breakers. The tasters swirled the wines in their mouths for at least 15 seconds before evaluating them, and expectorating. While high-performance liquid chromatography did indeed reveal measurable differences between the wines that had been kept in the different glasses, the participants themselves were unable to pick the odd wine out. This was true even if one sample was from a newly opened bottle, whereas the other two samples were from wines that had been stored in the glass for 30 minutes. That said, the much smaller sample size of participants in this study as compared to certain of the other studies discussed earlier, might perhaps raise concerns that the latter study may have simply been underpowered in terms of detecting a significant effect.

Elsewhere, Delwiche and Pelchat (2002) utilized a wine-tasting procedure that enabled them to focus solely on their participants' experience of the contents of the wine glass. They had blindfolded, naïve, nonexpert wine-drinkers sniff a red and a white wine. The wines were presented in four different types of glass in order to evaluate the wine's aroma, while the distance between them and the wine was kept constant with each participant's head being positioned in a headrest (see Fig. 14.3). The results revealed that the type of glass had no influence on most aspects of the perceived wine aroma, though the Bordeaux glass did reduce the perceived intensity of the wine's aroma somewhat.

How should the seeming contradiction between research demonstrating a significant effect of glass shape/size on wine perception on the one hand, and those studies showing that people cannot tell any difference be resolved? The solution here requires us to take a closer look at the methods used in the various studies that have been summarized thus far. It turns out that the literature can be more or less neatly summarized by saying that people find it much more difficult to distinguish between wines served/stored in different glasses if they are not allowed to see/touch them (Cliff, 2001; Delwiche & Pelchat, 2002; Russell et al., 2005). By contrast, in pretty much every study in which the participants have been allowed to hold the wine glasses, much larger (and typically significant) effects of the glass on the sensory properties of the wine have normally been observed (eg, Fischer & Loewe-Stanienda, 1999; Vilanova et al., 2008).

Importantly, whenever a person can see and/or physically interact with the wine glass (which is the normal situation when drinking) when rating the sensory

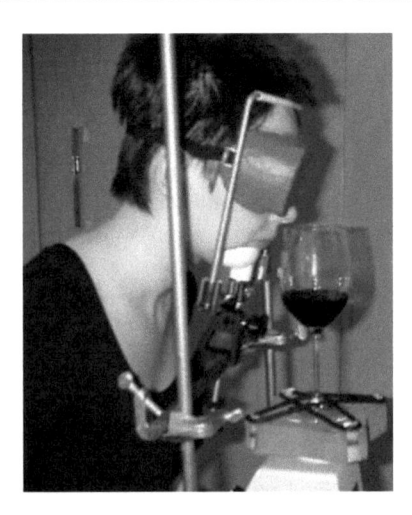

Figure 14.3 The scientific approach to the study of the impact of the wine glass on aroma perception. The participants in a study by Delwiche and Pelchat (2002) were not allowed to see or feel the glass that they were sniffing from. Note how the wine glass is mounted on a mechanical swirling device to enhance the release of volatile odor molecules. Consequently, any effects of glass shape on aroma perception could be unequivocally attributed to the physical (rather than the psychological) effect of the glassware.

properties of the wine contained within, then the shape and/or size of the glass impacts their ratings of the contents (Cliff, 2001; Hummel et al., 2003; Russell et al., 2005; Vilanova et al., 2008).

1.4 The physical versus psychological influence of the shape of the wine glass

One simple question here concerns whether any change in the size/shape of the glass influences the release, and/or the retention, of volatiles at the liquid's surface? The answer is undoubtedly, yes. For example, Arakawa et al. (2015) recently developed a sniffer-camera to visualize the concentration distribution of ethanol vapor that was emitted from a wine served in a wine glass, a cocktail glass, and a water glass. The ethanol vapor acquired a ring shape around the rim of the wine glass when a wine was served at 13°C, while no such result was obtained for the other two glasses. Here, however, it is important to note that there is a separate question as to whether the drinker's nose or palate is also sensitive enough to be able to pick up on such differences. Of course, one might also wonder whether there are any differences between the trained and naive palate. Intriguingly, however, the single biggest factor determining just how much of an influence the glass has turns out to be the drinker's knowledge concerning the glass that they happen to be evaluating the drink from.

Contrast this with previous suggestions that differing glass shapes may result in a differing spread of the wine over the surface of the taster's tongue (Peynaud, 1987).

Instrumental measurements have demonstrated differences in the volatiles in the headspace of a glass, in the case of both still (eg, Hirson, Heymann, & Ebeler, 2012) and sparkling wine (Liger-Belair, Bourget, Pron, Polidori, & Cilindre, 2012). These differences, will, of course, be influenced by the amount of time that a given drink happens to have been sitting in a particular glass (Hirson et al., 2012). Once again, though, the question of whether such differences are perceptible to the consumer remains open. Relevant here, Russell et al. (2005) revealed an influence of the exposure to air (0, 15, or 30 minutes) on the concentration of gallic acid, even though, as we saw earlier, there was no effect of glass shape on their participants' perception of the wine.

Therefore, we would argue that the various results that have been reported in this section are most consistent with the view that the influence of the glass on the perception of the wine has more of a psychological, rather than a chemical or physical, origin. It may be reasonable to speculate that the influence of the glass on the subjective rating of the wine should be conceptualized as a conscious multisensory process involving the direct (visual and/or haptic) perception of the glass as well as any awareness that the participant may have about the meaning and associations of the glass itself (based on prior experience and/or influenced by marketing communications). And given that the consumer will nearly always be aware of the particular glass from which he or she happens to be drinking, spending on the "right" glassware is probably justified. This is especially so when it is realized that the differences in the rating of the aroma of the very same wine in excess of 150% have, on occasion, been reported when a wine is assessed in different types of wine glass (Fischer & Loewe-Stanienda, 1999; see also Vilanova et al., 2008).[4]

In summary, the studies that have been reviewed in this section highlight that the shape of a wine glass can influence the drinker's perception of the taste and flavor of wine. To date, the available research suggests that the shape and dimensions of the glass primarily affect the overall intensity of the aroma, rather than necessarily enhancing or masking any specific aroma notes. Across a number of studies, the highest aroma intensities have typically been documented where the wines concerned have been served in glasses where the maximum diameter/opening diameter ratio is highest. Put simply, a large diameter encourages volatilization of the odor compounds, while a narrow opening helps to retain them in the glass (Fischer & Loewe-Stanienda, 1999).

2 Sensation transference

The results of numerous studies now demonstrate that the consumer's perception of, and feelings about, the sensory properties of the receptacle in which a drink is presented (eg, its color, texture, weight, etc.) influence a consumer's perception of the

---

[4] It could, though, be argued that educating the drinker that much of the effect of different glassware on their perception and enjoyment of wine may be squarely in their head (ie, psychological, rather than physiochemical, in origin) would also be a worthwhile endeavor.

contents by means of what is often referred to as "sensation transference" (Cheskin, 1957; Spence & Piqueras-Fiszman, 2012) or "affective ventriloquism" (when referring specifically to the transfer of hedonic properties; Spence & Gallace, 2011). In the following section, we review the evidence concerning the impact of varying different aspects of the drinking vessel on people's perception of the contents. Note that these studies appear to be driven by a much more basic response to the sensory attributes of the container or drinking vessel.

2.1 On the color of the containers

2.1.1 Colored glasses

Changing the color of the glass in which a drink is served has been shown to influence the consumer's perception of the drink's thirst-quenching properties. In one representative study, Guéguen (2003) reported that people rated a soda drink as tasting significantly more thirst-quenching when served from a "cold-colored" blue glass (nearly 47.5% of the participants chose this as the most thirst-quenching) than when the same drink was served from a green, yellow, or red glass instead. Only 15% of the participants rated the drink from the "warm-colored" red glass to be more thirst-quenching. Researchers have demonstrated that the color of the wine glass exerts an influence over people's perception of the contents. For instance, Ross, Bohlscheid, and Weller (2008) found that a trained panel preferred two red wines (a Syrah and a Pinot Noir) when they were tasted from a blue wine glass than from a clear wine glass (under normal white-light illumination). Even trained assessors have been reported to change their sensory evaluations of ports and Bordeaux red wines when served in red rather than clear glass (Williams, Langron, & Noble, 1984; Williams, Langron, Timberlake, & Bakker, 1984). Given such results, one might, of course, be right to worry about the influence of the black tasting glasses that are used in so many professional tastings (eg, see Harrar, Smith, Deroy, & Spence, 2013, for one such example).

2.1.2 Colored cups and mugs

Changing the color of the cups or mug in which hot drinks are often served also influences people's perception of hot beverages. For instance, Piqueras-Fiszman and Spence (2012a) noted that people's perception of the flavor of a hot chocolate drink changed as a function of the color of the plastic cup (red, orange, white, and dark cream) in which it was served (see Fig. 14.4). Interestingly, the beverage was liked more when served in the red cup than when served in the white cup, whereas the orange cup (with a white interior) enhanced the ratings of chocolate flavor when compared to either the red or white cup. Finally, the dark-cream cup subtly enhanced the sweetness and chocolate aroma when compared to the red cup.

It is important here to be clear about exactly which aspects of the drinker's response to the drinking vessel is being transferred to their ratings of the contents (or to the overall drinking experience). It would seem unlikely that the participants in the hot chocolate study simply liked the red color more than any of the other three cup colors (see Palmer & Schloss, 2010, for evidence concerning people's color

Figure 14.4 Four differently colored vending cups used in Piqueras-Fiszman and Spence's (2012b) study. Cups from the left to the right were white, cream, orange, and red. A drink of hot chocolate was rated as tasting significantly more chocolatey when sampled from the orange cup than from either the red or white cups.

preferences, which generally tend toward the cooler colors). Alternatively, however, it could be that red is the "warmest" color (see Ho, Van Doorn, Kawabe, Watanabe, & Spence, 2014).

Elsewhere, Van Doorn, Wuillemin, and Spence (2014) have demonstrated that people's perception of a café latte drink is affected by the visual appearance of the mug in which the drink is served. In this case, people rated the coffee's flavor as approximately 25% more intense when the beverage was served from a white porcelain mug than when exactly the same drink was served from a clear glass mug (or from an opaque blue mug) instead. It is at present unclear whether such results are best conceptualized in terms of yet another example of sensation transference. The alternative here is that they may reflect a more perceptual color contrast effect between the color of the coffee as it is perceived by the drinker against the background of the cup. Given that changing the color of a beverage can change its perceived sweetness, and that color contrast effects can also affect the perceived color of a food or beverage, it will undoubtedly require careful future research in order to figure out the underlying cause(s) of this effect (eg, see Bruno, Martani, Corsini, & Oleari, 2013; Spence, 2015a).

Cup color can also influence the perceived warmth of coffee (Guéguen & Jacob, 2012; see also Favre & November, 1979; Ho et al., 2014). More than 100 people tasted coffee from four different-colored cups (ie, blue, green, yellow, and red) in Guéguen and Jacob's study. The participants indicated which coffee was the warmest (in terms of its temperature). 38.3% of them reported that the coffee served from the red cup was the warmest, followed by yellow (28.3%), green (20.0%) and, finally, blue (13.3%). In the latest research from Alberto Gallace and his colleagues over in Milan, cup color has also been shown to exert a significant influence over the perception of carbonation in a drink (Risso, Maggiono, Olivero, & Gallace, 2015). The participants in the latter study had to rate natural, slightly carbonated, and carbonated water samples that were served in red, blue, or white cups. When the participants actually tasted the water samples, they rated the same water presented in the red or blue cups as more carbonated than those in the white cup. By contrast, when they just looked at the water samples without actually tasting them, they expected the samples presented in the red or white cups to be more carbonated than those seen in the blue cups.

2.1.3 Colored cans

A figure that one often sees quoted is that something like 30% of all the food products we eat or drink are consumed direct from the packaging. If anything, one could imagine that this figure is likely to be higher when it comes specifically to beverages. As such, one might want to know just how much of an influence the color of the can or bottle in which the vast majority of our beverages are sold, and hence presumably also consumed, has on perception. There is certainly an older literature on this topic hinting that the color of the packaging may influence people's perception of the taste/flavor of the contents (eg, Cheskin, 1957; Cheskin & Ward, 1948). It was Louis Cheskin who first reported that the consumers in one of his studies rated 7-Up as tasting more lemony/limey when they were given a can that happened to be yellower than normal to drink from (see also Becker, van Rompay, Schifferstein, & Galetzka, 2011; Esterl, 2011; Hine, 1995).

More recently, we conducted a study in which several hundred people were invited to drink a bottle of beer and to fill out a questionnaire about their experience (see Barnett & Spence, submitted for publication). One group of participants sampled the beer from an unlabeled brown beer bottle. Another group sampled the same beer from a brown bottle with a brown label. A third group sampled the beer from a bottle with a label that contained both yellow and green elements (designed to bring out the citrus notes in the beer; see Fig. 14.5). The results of this study, which was conducted at an international science festival in Scotland revealed that those who sampled the beer from the green bottle rated the citrusy notes as significantly more intense than those in either of the other two conditions (see Table 14.1). The increase in citrus ratings as a function of the color of the label was reported to be in the order of 10%.

2.2 On the texture of the drinking vessel

Thus far, we have focused our review on those studies that have varied the visual appearance of the drinking vessel. That being said, a recent search has also revealed the impact of the feel of the container on people's perception. Of course, in many cases, the consumer may well drink a beverage direct from the bottle or can. Under such conditions, any branding effects associated with the packaging may also be expected to play a substantial role on the perception of the contents (eg, see Gates, Copeland, Stevenson, & Dillon, 2007).

Krishna and Morrin (2008) conducted an intriguing between-participants study in which they investigated the impact of the feel of the container (a plastic cup) on people's perception of water mixed with Sprite. Their study was conducted in a university cafeteria. A total of 180 people evaluated the drink after having tasted it through a straw. Half of them touched the flimsy cup in which the drink was contained with their hand before evaluating the drink, whereas the others did not. Those participants who exhibited less of a need for touch were affected in their evaluation of the drink by the feel (ie, the firmness vs flimsiness) of the cup, whereas those participants who scored lower in terms of their need for touch were not. The participants rated the drink as being lower in quality when they felt its flimsiness. These results clearly suggest

Figure 14.5 Two of the labels used in Barnett and Spence's (submitted for publication) recent study of the effect of the label on the taste of the beer: (A) Genius Loki, (B) Liquid Hop Chemistry.

that changes in the haptic qualities of the receptacle in which a drink is served might have different effects on different people depending on their general liking for haptic input (see also Becker et al., 2011).

Consistent with these findings, a recent study by Tu, Yang, and Ma (2015, Experiment 1) demonstrated that participants rated the iciness and coldness of a cup of Chinese tea as higher when presented in a glass container than in a plastic or paper

Table 14.1 **Mean scores for taste, quality, citrus flavor, purchase intention, and price in Barnett and Spence's (submitted for publication) study of labeling effects on beer**

Bottle/label	Taste (1–9)	Quality* (1–9)	Citrus flavor* (1–9)	Purchase intent (1–9)	Price (£)
LHC	6.9	7.2	6.4^A	6.4^A	3.3
Genius Loki	6.5	6.9	5.7^{AB}	5.9^{AB}	3.4
Blind	6.5	6.8	5.9^B	5.8^B	3.4

Attributes marked with an * revealed a significant difference between bottle labels at the $P<0.05$ level. Different letters within columns indicate significant differences between conditions. Higher values indicate better (for taste), higher (for quality), more intense (for citrus flavor), and more likely (for purchase intent).

container of approximately the same size. In order to rule out any influence of differences in the weight of the container (ie, these glass, plastic, and paper containers had different weights), Tu et al. also conducted a second experiment in which they added a hidden glass base to the plastic container so that it had the same weight as the glass container, and replicated their results.

Elsewhere, Schifferstein (2009) conducted a study in which he presented participants with several cups made of various different materials and finishes, including glass, ceramic, opaque plastic, translucent plastic, and melamine. The cups were either empty or filled with different drinks (hot tea or cold soda). The participants rated their drinking experience on the basis of a number of scales (such as good–bad, beautiful–ugly, etc.). The participants' experience of drinking from the various cups was significantly affected by the cup. So, for example, the participants reported that they enjoyed drinking the soda from the plastic cup more than from the ceramic cup. Those drinks consumed from the pinkish cups were rated as tasting significantly sweeter than when exactly the same drink was evaluated from a transparent cup instead. Such results suggest that the participants may unconsciously have been transferring their experience of (or expectations related to) the pinkness of the cups (or rather, their intuitions about pink foodstuffs being sweet) to their judgments of the drinks themselves (see Spence et al., 2015).

Given that a number of companies have recently started to give their packaging a distinctive feel (see the Heineken tactile can; Anon., 2011a; Murray, 2011), it is interesting to consider what effect changing the surface texture of a drinking receptacle might have on the consumer's perception of the contents (see Spence & Piqueras-Fiszman, 2012, for a review). While we are not aware of anyone having addressed this question directly in the case of the perception of drinks, related research can be taken to suggest that such textural sensation transfer effects can be observed (see Piqueras-Fiszman & Spence, 2012c).[5] One glassware manufacturer has recently started to

---

[5] One might also expect the thermal properties of the drinking vessel to influence people's perception. Intriguingly, Williams and Bargh (2008) have shown that the warmth of a briefly held cup of hot, versus iced, coffee affected people's judgments of a target person without their being aware of this influence. Should these intriguing results be replicated then the implications for glassware could be significant.

manufacture glasses with a textured rim or lip (see Fig. 14.6A for an example). Innovative cocktail makers have also started to play with the outer texture of their drinking vessels. Tony Conigiaro, for instance, has wrapped cord around the stems of certain of his cocktail glasses in order to give them a distinctive feel (see http://www.69colebrookerow.com/#about; see Fig. 14.6B). Elsewhere, at the House of Wolf restaurant in London's Islington (which closed in 2014), one of the cocktails came in a glass that had been covered with bubble wrap.

Figure 14.6 (A) An illustration of glasses with textured rim or lip (manufactured by Roberts Wineware Inc.). (B) An example of the cocktail maker Tony Conigliaro from 69 Colebrooke Row, adding texture to the stem of one of his glasses.

2.3 On the weight of the drinking vessels

To date, very little research has been published on the question of the impact that the weight of the drinking vessel has on the drinker's experience. The available evidence from a growing number of studies suggests the existence of a significant sensation transference effect resulting from the weight of the receptacle (see Gatti, Spence, & Bordegoni, 2014; Piqueras-Fiszman, Harrar, Roura, & Spence, 2011; Piqueras-Fiszman & Spence, 2012b; Spence & Piqueras-Fiszman, 2011, for related examples; cf. Lin, 2013). However, one relevant question here concerns whether it is weight per se that matters, or rather it is "heavier than expected" that is crucial to eliciting these weight-related transference effects. In a recent conference presentation, Maggioni, Risso, Olivero, and Gallace (2014) reported a study where the participants had to rate the pleasantness, freshness, lightness, and level of carbonation of still and sparkling water samples on a series of visual analogue scales. The water samples were served from plastic cups that were light (2 g), medium, (11 g), or heavy (30 g). In this case, participants rated the water as more carbonated when served from the heaviest cup, but as most pleasant when served from the lightest cup. Meanwhile, Kampfer, Leischnig, Ivens, and Spence (submitted for publication) conducted a study in which people evaluated the taste of a can of soft drink that they were given to open. The results demonstrated that the increased weight of the packaging (287 g with half having a 60-g weight added to the bottom of their can) had a significant positive effect on perceived flavor intensity ($M = 4.64$, vs 3.90, on a 7-point Likert scale) which, in turn, influenced taste evaluations.

In summary, the studies reviewed in this section highlight how sensation transference can be used to account for the observed effects; be it the induction of a "cold" feeling from the color of the cup, the low quality from its material properties, or the feeling of "lightness" from the weight of the cup, or packaging for those drinks consumed direct from the packaging (see Anon., 2011b; Poulter, 2010). Very often, then, the attribute (be it sensory, emotional, or evaluative) that people associate with the drinking vessel appears to be transferred to their rating of the drink, at least under laboratory conditions (see Spence & Piqueras-Fiszman, 2014, for a review). Nevertheless, it should be noted that, at present, sensation transference is more of a descriptive rather than an explanatory concept. In the future, it would certainly be helpful to get a more mechanistic understanding of the conditions under which it occurs and how the effect occurs neurally.

3 Drinking vessels and consumption behavior

It is interesting to consider whether the physical properties of the drinking vessels from which drinks are consumed may also affect people's consumption behavior. Suggestive evidence in this regard comes from the work of Wansink and Van Ittersum (2003, 2005). These researchers demonstrated how the shape of a glass plays a role in how much people pour and presumably also how much they drink. In one study, people were found to drink as much as 88% more when served in a short, wide glass than when the same drink was served in a tall, narrow glass instead (Wansink & Van Ittersum, 2003). This difference is all the more surprising given that the two glasses in their study had the same volume. Wansink and Van Ittersum (2005) found that even experienced bartenders would pour more than 25% extra alcohol into tumblers than highball glasses when measuring out a shot of spirits (see also Gill & Donaghy, 2004; White, Kraus, McCracken & Swartzwelder, 2003).

The size of the receptacle that a person drinks from can also influence sip volume (Lawless, Bender, Oman, & Pelletier, 2003). In particular, elongated receptacles are perceived to contain more of a drink than is actually the case (Yang & Raghubir, 2005), suggesting that people use the height of the container in order to estimate the volume of the drinks. That said, in another study, elongated receptacles were found to result in people underestimating the volume of liquid that they had consumed. They ended up drinking more (Raghubir & Krishna, 1999). Taken together, then, these results suggest that everyone, expert and novice alike, falls prey to the well-known vertical–horizontal illusion (Avery & Day, 1969).

Recently, Attwood, Scott-Samuel, Stothart, and Munafò (2012) conducted an intriguing between-participants laboratory-based study in the United Kingdom in which social drinkers were shown to consume a drink of lager 60% more slowly from a filled 12 fl oz straight-sided glass than from a flute glass of the same volume that had curved sides. What is more, the participants underestimated the true half-way point of the curved glass to a greater extent than when drinking from the straight glass.[6]

Intriguing research has revealed that the shape of the glass influences people's subjective ratings of, and willingness-to-pay for, alcoholic drinks. People like alcoholic drinks more when they consider the glassware and drink to be "congruent" (Raudenbush et al., 2002). For instance, the 61 participants in one study rated hot chocolate, beer, and orange juice served in containers (a cup, bottle, or glass) that were deemed either congruent or incongruent with the participants' everyday experience of drinking these beverages (Raudenbush et al., 2002). The hot chocolate drink was rated as tasting significantly more pleasant when served in a ceramic cup (ie, in an appropriate receptacle) than when served in either the glass or the bottle (inappropriate). While the change was not large (c. 4% change), it is worth remembering that the participants in this study were informed what they would be drinking at the start of each trial. It would seem likely, then, that the detrimental consequences of serving a hot beverage in a product-incongruent container might be larger if the consumer was to be surprised to find that the contents did not match up to the expectations set by the container (see Piqueras-Fiszman & Spence, 2015, for a review).

In another recent study conducted by Wan, Zhou, Woods, and Spence (2015), participants from China and the United States were instructed to rate their liking, familiarity, and congruency (between the drink and the glassware), as well as how much they would be willing to pay for red wine, white wine, beer, whiskey, and Chinese baijiu drinks presented in different glassware (see Fig. 14.7A). Intriguingly,

---

[6] Interestingly, no such effect was observed in those participants who consumed a carbonated soft drink (7-Up) from the same pair of glasses instead, thus suggesting a rather more complex interaction between the glassware and the contents.

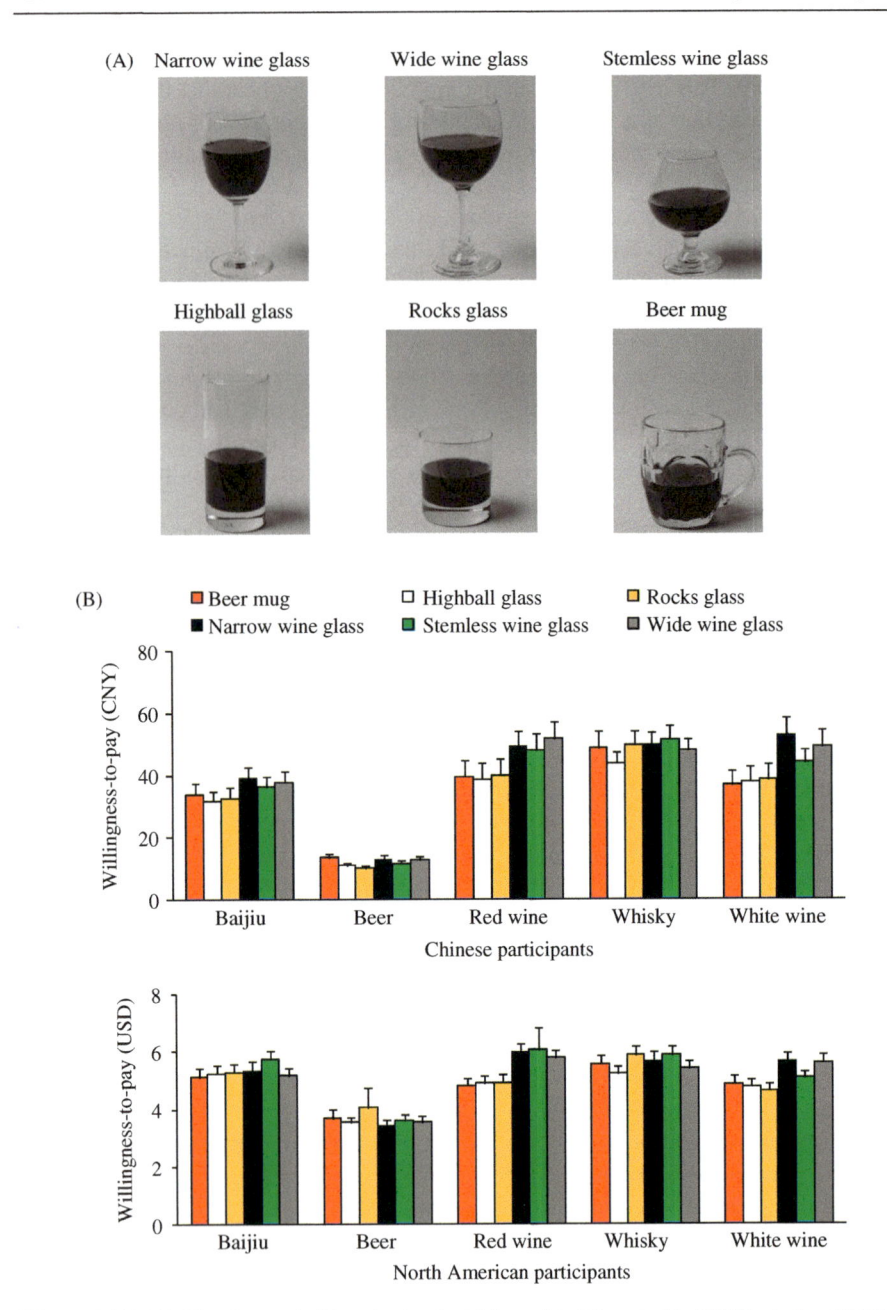

Figure 14.7 (A) Six glasses (with red wine inside) used in Wan et al.'s (2015) study. (B) The price that the Chinese and North American participants reported that they would be willing to pay for each type of alcoholic drink served in each glass (in CNY and USD, respectively). Error bars show the standard errors of the means.

the type of drink and the shape/type of glassware influenced participants' subjective ratings of and willingness-to-pay for these drinks. In particular, the participants from both countries liked the red and white wine more, and were willing to pay significantly more for if they thought that the glassware was congruent with the drinks (see Fig. 14.7B).

Finally, here, it is worth noting that one might also expect the weight of the glassware/cup to influence expected satiety following the consumption of the contents (cf. Piqueras-Fiszman & Spence, 2012b).

4 Implications for marketing practice

Advertising and marketing play a not inconsiderable role in the psychological influence of containers on consumers. For example, in the case of wine, Cliff (2001, p. 40) suggests that: *In fact, the advertising/marketing of wine glasses has been so effective that most wine connoisseurs have a distinct bias towards the appropriateness of a particular wine glass.* Indeed, a highly profitable industry has developed around the matching of specific glass shapes to particular types of wine or even grape varieties (eg, Cloake, 2012; Stead et al., 2014). Even beer glasses have started to appear in a greater variety of forms than ever before (Peyton, 2015). Recently, carefully designed drinking vessels (with strategic curves) made of high-end glass for beers have started to appear (Moynihan, 2014). Elsewhere, in order to influence people's multisensory processing of the flavor, Diageo's Johnnie Walker even went so far as to develop glassware to deliver music to the drinkers via vibrations directed to their inner ears when their teeth happened to be touching the glass (Baker, 2015).

There is also growing interest amongst the marketing community in the opportunities associated with branded glassware and signature packaging for their drinks brands. Such moves make sense given evidence that: *According to research by Interbrew, 50 per cent of consumers prefer beer to be served in a branded glass and 37 per cent are prepared to pay more for it,* (McFarland, 2002). Indeed, as noted by Attwood et al. (2012), the last few years have seen an increase in the range and availability of branded drinking glasses. It is interesting to note how many of these glasses include shape as a distinguishing feature: think only of chalice glasses, curved beer flutes, tulip glasses, tankards, and novel curved beer glasses, now used by a number of alcohol brands (Bob the Brit, 2010; Reis, 2012). In the United Kingdom, we have also seen the return of the dimpled pint glass (Barford & Rohrer, 2014). In contrast to this "male" glassware, part of the push toward novel glass designs in recent years has been toward the "girl glass" (eg, see Black, 2010; Hook, 2009). For some female drinkers, it has been reported that they may prefer a glass design that does not result in their exposing their neck by having to tilt their head back too far.

Branded glassware has, on occasion, been shown to lead to a dramatic increase in sales (Foottit, 2010). So, for example, Carlsberg UK apparently found that sales of their beer went up by 14% (and, more importantly, profit margins by 37%) after having introduced their own branded glass. The marketing opportunities associated with the introduction of heavier glassware have not gone unnoticed. Think here only of the distinctively heavy Hoegaarden beer glass.[7] On the other hand, the copa de balon, the large-bowled, long stemmed glass in which the Spanish Latin-American market like to serve/drink their gin and tonic also feels very heavy in the hand once filled with ice. Once again, it may be that the overall weight of the glass plus contents in hand, especially when held via the long narrow stem of the glass that can help to elevate a drinker's perception of the contents in this particular case (see also Chabanis, 2014).

5 Future research directions and conclusions

We believe that there are a number of interesting questions still to be addressed in this area. The first one would be to examine the cross-cultural similarities and differences in the impact of containers on people's perception of the contents. The findings of such research could well have direct implications for international marketers interested in breaking into new markets, for example, the red wine market of mainland China (Culliney, 2011). For example, if one takes a drink such as gin and tonic, then the appropriate glassware in the United Kingdom, not to mention in many other markets, is the highball or collins glass. In Spain and Latin America, by contrast, the balloon copa glass is considered to be most appropriate instead (eg, Anon., 2013). Hence people in different parts of the world may well have very different ideas about what the most "appropriate" glassware for this drink is. Hence, it is quite possible that consumers from different cultures and/or age groups may also hold very different views about the appropriateness of a given container for a particular drink (see also Barford & Rohrer, 2014; cf. Zhou, Wan, Mu, Du, & Spence, 2015, for a similar study of plateware).

Second, if one contrasts the numerous research studies that have been published on the influence of the wine glass, as well as a much smaller selection of studies of hot beverages such as coffee and hot chocolate, more intensive research is undoubtedly called for to investigate the influence of the container on the perception of other popular beverages around the world such as tea (Varma, 2013). Given tea's global popularity, it is surprising that more research has not been conducted on the optimum receptacle for drinking this ubiquitous beverage. Tea drinking is not only a daily habit for many people around the world, but also associated with a certain lifestyle or occasion in countries such as China, the United Kingdom, and Japan. Interestingly, as shown in Fig. 14.8, the containers used to serve the same type of tea (eg, green tea) might be dramatically different in different countries (Wan, Yu, Zhao & Spence, in preparation). It is, therefore, going to be very interesting to examine the cross-cultural similarities and differences in the impact of containers on the perception of tea.

---

[7] Another example of the heavy beer glass might be that the dimpled pint glass that has recently started to regain its former popularity in British pubs, after having vanished from use in many pubs about a decade ago (Barford & Rohrer, 2014).

| British tea cup | Chinese tea cup | Japanese tea cup |
| (with matching plate) | (with matching plate and cover) | |

Figure 14.8 Photos of the same green tea beverage served in the British tea cup (with matching plate), Chinese tea cup (with matching plate and cover), and Japanese tea cup, were shown to the participants in Wan et al.'s (in preparation) study.

In conclusion, the results of the various studies that have been reviewed in this chapter clearly demonstrate the profound impact that the drinking receptacle has on people's perception of beverages, from wine through to coffee, and from hot chocolate through to soda and beer. The research outlined here supports a number of conclusions: (1) the drinking vessel impacts both the sensory-discriminative and hedonic response of consumers (or participants) to the contents; and (2) the dimensions of the drinking vessel influence everything from the amount that people pour through to how much they consume. It is also worth bearing in mind that the glassware not only impacts a person's impression of whatever happens to be in the glass, but it can potentially also affect their whole meal experience should they, for example, happen to be drinking wine with dinner (Billing, Öström, & Lagerbielke, 2008).

While the majority of the research published to date has tended to focus on the physical properties of the drink itself, and/or on labeling, branding, packaging, and/or pricing (eg, Gates et al., 2007; Spence & Piqueras-Fiszman, 2012), the research that has been summarized in this chapter also argues that sensory scientists and companies alike should be thinking much more carefully about how to optimize the sensory and conceptual properties of the drinking vessel (be it the glass, can, bottle, mug, or cup) on the experience of the consumer, since it is such an intrinsic part of the overall multisensory drinking experience. Some marketers have already started to realize the marketing opportunities associated with the introduction of innovative branded glassware (Stead et al., 2014). Certainly, there is growing interest online from those who are curious to know just how much of an influence the glassware really makes to the consumer's experience of the contents (D'Costa, 2011; Reis, 2012). The era of the neuroscience-inspired drinking vessel is surely only just around the corner.[8]

---

[8] A final area of important research concerns the impact of the size and sensory properties of the straw (eg, Lawless et al., 2003; Lin, Lo, & Liao, 2013), although this is a topic that we will have to leave for another day.

References

Anon, (1973). The ideal wine glass. *Search: Science Technology and Society*, 4(1/2), 4.
Anon. (2011a). *The Dutch touch*. Downloaded from <http://175proof.com/triedandtested/the-dutch-touch/> Accessed 10.05.14.
Anon. (2011b). Arla foods target super lightweight polybottles. Downloaded from <http://wwwpackagingwastecompliancecouk/latest-news/article/29/> Accessed 01.07.11.
Anon. (2011c). Glass act. *The Cocktail Lovers*, July 16. Downloaded from <http://www.thecocktaillovers.com/2011/07/glass-act/> Accessed 11.05.14.
Anon. (2013). Gin tonic served Spanish style. Downloaded from <http://eatsdrinksandsleeps.com/2013/05/06/gin-tonic-served-in-a-spanish-style-copa/> Accessed 11.05.14.
Arakawa, T., Ititani, K., Wang, X., Kajiro, T., Toma, K., Yano, K., et al. (2015). A sniffer-camera for imaging of ethanol vaporization from wine: The effect of wine glass shape. *The Analyst*, 140, 2881–2886.
Attwood, A. S., Scott-Samuel, N. E., Stothart, G., & Munafò, M. R. (2012). Glass shape influences consumption rate for alcoholic beverages. *PLoS ONE*, 7(8), e43007.
Avery, G. C., & Day, R. H. (1969). Basis of the horizontal-vertical illusion. *Journal of Experimental Psychology*, 81, 376–380.
Baker, N. (2015). Johnnie Walker reveals futuristic glass. Downloaded from <http://www.thedrinksbusiness.com/2015/04/johnnie-walker-reveals-futuristic-glass/> Accessed 07.07.15.
Barford, V., & Rohrer, F. (2014). The return of the dimpled pint glass. *BBC News Magazine*, April 30. Downloaded from <http://www.bbc.co.uk/news/magazine-27188915/> Accessed 15.05.14.
Barnett, A., & Spence, C. When changing the label (of a bottled beer) modifies the taste (submitted for publication).
Becker, L., van Rompay, T. J. L., Schifferstein, H. N. J., & Galetzka, M. (2011). Tough package, strong taste: The influence of packaging design on taste impressions and product evaluations. *Food Quality and Preference*, 22, 17–23.
Billing, M., Öström, Å., & Lagerbielke, E. (2008). The importance of wine glasses for enhancing the meal experience from the perspectives of craft, design, and science. *Journal of Foodservice*, 19, 69–73.
Black, R. (2010). Woodpecker has a glass for the girls. *Publican's Morning Advertiser*, September 26. Downloaded from <http://www.morningadvertiser.co.uk/content/view/print/444423/> Accessed 16.05.14.
Bob the Brit (2010). *Beer glass styles*. Downloaded from <http://thebrewclub.com/2010/05/28/beer-glass-styles/> Accessed 11.05.14.
Bruno, N., Martani, M., Corsini, C., & Oleari, C. (2013). The effect of the color red on consuming food does not depend on achromatic (Michelson) contrast and extends to rubbing cream on the skin. *Appetite*, 71, 307–313.
Chabanis, J. (2014). Downloaded from <http://cargocollective.com/jocelynchabanis/Pastis-51-Piscine/> Accessed 11.05.14.
Cheskin, L. (1957). *How to predict what people will buy*. New York, NY: Liveright.
Cheskin, L., & Ward, L. B. (1948). Indirect approach to market reactions. *Harvard Business Review*, 26, 572–580.
Cliff, M. A. (2001). Influence of wine glass shape on perceived aroma and colour intensity in wines. *Journal of Wine Research*, 12, 39–46.
Cloake, F. (2012). Can the shape of your glass enhance the taste of the wine? Do you need to change your glass depending on what you're drinking? *The Guardian Online*, April 25.

Downloaded from <http://www.theguardian.com/lifeandstyle/2012/apr/25/shape-of-glass-enhance-wine/> Accessed 11.05.14.

Culliney, K. (2011). *Red wine for 'joy and strength' in China: Analyst.* Downloaded from <http://www.beveragedaily.com/Markets/Red-wine-for-joy-and-strength-in-China-Analyst/?c=NBddb94xF1vUW8AloPKEXQ%3D%3D&utm_source=newsletter_daily&utm_medium=email&utm_campaign=Newsletter%2BDaily/> Accessed 07.07.15.

D'Costa, K. (2011). Does your beer glass matter? *Scientific American.* Downloaded from <http://blogs.scientificamerican.com/anthropology-in-practice/2011/08/22/does-your-beer-glass-matter/> Accessed 11.05.14.

Delwiche, J. F., & Pelchat, M. L. (2001). Influence of glass shape on wine aroma. *Chemical Senses, 26,* 804–805.

Delwiche, J. F., & Pelchat, M. L. (2002). Influence of glass shape on wine aroma. *Journal of Sensory Studies, 17,* 19–28.

Diaz, M. E. (2004). Comparison between orthonasal and retronasal flavour perception at different concentrations. *Flavour and Fragrance Journal, 19,* 499–504.

Esterl, M. (2011). A frosty reception for Coca-Cola's white Christmas cans. *The Wall Street Journal,* December 1. Downloaded from <http://online.wsj.com/article/SB10001424052970204012004577070521211375302.html/> Accessed 28.10.12.

Favre, J. P., & November, A. (1979). *Colour and communication.* Zurich: ABC-Verlag.

Faye, P., Courcoux, P., Giboreau, A., & Qannari, E. M. (2013). Assessing and taking into account the subjects' experience and knowledge in consumer studies. Application to the free sorting of wine glasses. *Food Quality and Preference, 28,* 317–327.

Fischer, U. (1996). Weinglasser—Aesthetik oder sesorische Eignung? [Wine glass—aesthetic and sensory suitability?]. *Deutsche-Weinbau, 22,* 22–26.

Fischer, U., & Loewe-Stanienda, B. (1999). Impact of wine glasses for sensory evaluation. *International Journal of Vine and Wine Sciences, Wine Tasting, Special Edition, 33*(Suppl. 1), 71–80.

Foottit, L. (2010). Carlsberg launches glassware promotion. *Publican's Morning Advertiser,* February 26. Downloaded from <http://www.morningadvertiser.co.uk/content/view/print/440048/> Accessed 14.05.14.

Gallace, A., & Spence, C. (2014). *In touch with the future: The sense of touch from cognitive neuroscience to virtual reality.* Oxford, UK: Oxford University Press.

Gates, P. W., Copeland, J., Stevenson, R. J., & Dillon, P. (2007). The influence of product packaging on young people's palatability ratings for RTDs and other alcoholic beverages. *Alcohol and Alcoholism, 42,* 138–142.

Gatti, E., Spence, C., & Bordegoni, M. (2014). Investigating the influence of colour, weight, & fragrance intensity on the perception of liquid bath soap. *Food Quality and Preference, 31,* 56–64.

Gawel, R. (2010). *Does the type of wine glass effect the "taste" of wine?* Downloaded from <http://www.aromadictionary.com/articles/wineglass_article.html/> Accessed 29.06.15.

Gill, J. S., & Donaghy, M. (2004). Variation in the alcohol content of a 'drink' of wine and spirit poured by a sample of the Scottish population. *Health Education Research, 19,* 485–491.

Guéguen, N. (2003). The effect of glass colour on the evaluation of a beverage's thirst-quenching quality. *Current Psychology Letters Brain Behaviour and Cognition, 11*(2), 1–6.

Guéguen, N., & Jacob, C. (2012). Coffee cup color and evaluation of a beverage's "warmth quality. *Color Research and Application, 39,* 79–81.

Harrar, V., Smith, B., Deroy, O., & Spence, C. (2013). Grape expectations: How the proportion of white grape in Champagne affects the ratings of experts and social drinkers in a blind tasting. *Flavour, 2,* 25.

Harrison, A. A. (1977). Mere exposure. In L. Berkowitz (Ed.), *Advances in experimental social psychology* (Vol. 10, pp. 39–83). New York, NY: Academic.

Hine, T. (1995). *The total package: The secret history and hidden meanings of boxes, bottles, cans, and other persuasive containers*. New York, NY: Little Brown.

Hirson, G. D., Heymann, H., & Ebeler, S. E. (2012). Equilibration time and glass shape effects on chemical and sensory properties of wine. *American Journal of Enology and Viticulture, 63*, 515–521.

Ho, H.-N., Van Doorn, G. H., Kawabe, T., Watanabe, J., & Spence, C. (2014). Colour-temperature correspondences: When reactions to thermal stimuli are influenced by colour. *PLoS ONE, 9*, e91854.

Hook, S. (2009). Girl glasses to be designed. *Publican's Morning Advertiser*, October 8. Downloaded from <http://www.morningadvertiser.co.uk/content/view/print/437544/> Accessed 16.05.14.

Hummel, T., Delwiche, J. F., Schmidt, C., & Hüttenbrink, K.-B. (2003). Effects of the form of glasses on the perception of wine flavors: A study in untrained subjects. *Appetite, 41*, 197–202.

Jakesch, M., & Carbon, C. C. (2012). The mere exposure effect in the domain of haptics. *PLoS ONE, 7*, e31215.

Kampfer, K., Leischnig, A., Ivens, B. S., & Spence, C. Touch-taste-transference: Assessing the effect of the weight of product packaging on flavor perception and taste evaluation. *PLoS ONE* (submitted for publication).

Krishna, A., & Morrin, M. (2008). Does touch affect taste? The perceptual transfer of product container haptic cues. *Journal of Consumer Research, 34*, 807–818.

Lawless, H. T., Bender, S., Oman, C., & Pelletier, C. (2003). Gender, age, vessel size, cup vs. straw sipping, and sequence effects on sip volume. *Dysphagia, 18*, 196–202.

Liger-Belair, G., Bourget, M., Pron, H., Polidori, G., & Cilindre, C. (2012). Monitoring gaseous CO_2 and ethanol above champagne glasses: Flute versus coupe, and the role of temperature. *PLoS ONE, 7*(2), e30628.

Lin, H.-M. (2013). Does container weight influence judgments of volume? *International Journal of Research in Marketing, 30*, 308–309.

Lin, H.-M., Lo, H.-Y., & Liao, Y.-S. (2013). More than just a utensil: The influence of drinking straw size on perceived consumption. *Marketing Letters, 24*, 381–386.

Maggioni, E., Risso, P., Olivero, N., & Gallace, A. (2014). The effect of the weight of the container on people's perception of mineral water. In *Poster presented at the 15th annual meeting of the international multisensory research forum, Amsterdam, June 11–14*.

Martin, G. N. (2004). A neuroanatomy of flavour. *Petits Propos Culinaires, 76*, 58–82.

McCarthy, E. (2006). *The incredible importance of the wine glass*. Downloaded from <http://www.winereviewonline.com/mccarthy_on_glasses.cfm/> Accessed 09.05.14.

McFarland, B. (2002). A glass of their own. *Publican's Morning Advertiser*. April 9. Downloaded from <http://www.morningadvertiser.co.uk/content/view/print/582085/> Accessed 03.07.14.

Moynihan, T. (2014). *A high-end beer glass that makes your stouts taste better*. Downloaded from <http://www.wired.com/2014/04/spiegelau-stout-glass/?mbid=social_twitter/> Accessed 07.07.15.

Murray, F. (2011). New unified global identity for Heineken. *TheDrinksReport.com*. January 1. Downloaded from <http://www.thedrinksreport.com/news/2011/13543-new-unified-global-identity-for-heineken.html/> Accessed 03.07.14.

Palmer, S. E., & Schloss, K. B. (2010). An ecological valence theory of human color preference. *Proceedings of the National Academy of Sciences of the USA, 107*, 8877–8882.

Peynaud, E. (1987). *The taste of wine: The art and science of wine appreciation* (M. Schuster, Trans.). London: Macdonald & Co.
Peyton, J. (2015). *Drinker's guide to choosing the right glassware.* Unpublished poster.
Pierce, J., & Halpern, B. P. (1996). Orthonasal and retronasal odorant identification based upon vapor phase input from common substances. *Chemical Senses, 21*, 529–543.
Piqueras-Fiszman, B., Harrar, V., Roura, E., & Spence, C. (2011). Does the weight of the dish influence our perception of food? *Food Quality and Preference, 22*, 753–756.
Piqueras-Fiszman, B., & Spence, C. (2012a). Does the color of the cup influence the consumer's perception of a hot beverage? *Journal of Sensory Studies, 27*, 324–331.
Piqueras-Fiszman, B., & Spence, C. (2012b). The weight of the container influences expected satiety, perceived density, and subsequent expected fullness. *Appetite, 58*, 559–562.
Piqueras-Fizman, B., & Spence, C. (2012c). The influence of the feel of product packaging on the perception of the oral-somatosensory texture of food. *Food Quality and Preference, 26*, 67–73.
Piqueras-Fizman, B., & Spence, C. (2015). Sensory expectations based on product-extrinsic food cues: An interdisciplinary review of the empirical evidence and theoretical accounts. *Food Quality and Preference, 40*, 165–179.
Pliner, P. (1982). The effects of mere exposure on liking for edible substances. *Appetite, 3*, 283–290.
Postgate, R. (1951). *The plain man's guide to wine.* Berkeley, CA: University of California Press.
Poulter, S. (2010). Raise a glass of wine to the Dragon-slayer. *Daily Mail, June 14*, 13.
Raghubir, P., & Krishna, A. (1999). Vital dimensions in volume perception: Can the eye fool the stomach? *Journal of Marketing Research, 36*, 313–326.
Raudenbush, B., Meyer, B., Eppich, W., Corley, N., & Petterson, S. (2002). Ratings of pleasantness and intensity for beverages served in containers congruent and incongruent with expectancy. *Perceptual and Motor Skills, 94*, 671–674.
Reis, M. (2012). *Beer glassware: Does it really matter?* Downloaded from <http://drinks.seriouseats.com/2012/06/beer-glasses-best-glass-for-craft-beer-taste-test.html/> Accessed 11.05.14.
Risso, P., Maggiono, E., Olivero, N., & Gallace, A. (2015). The association between the colour of a container and the liquid inside: An experimental study on consumers' perception, expectations and choices regarding mineral water. *Food Quality and Preference, 44*, 17–25.
Ross, C. F., Bohlscheid, J., & Weller, K. (2008). Influence of visual masking technique on the assessment of 2 red wines by trained and consumer assessors. *Journal of Food Science, 73*, S279–S285.
Rozin, P. (1982). "Taste-smell confusions" and the duality of the olfactory sense. *Perception and Psychophysics, 31*, 397–401.
Russell, K., Zivanovic, S., Morris, W. C., Penfield, M., & Weiss, J. (2005). The effect of glass shape on the concentration of polyphenolic compounds and perception of Merlot wine. *Journal of Food Quality, 28*, 377–385.
Schifferstein, H. N. J. (2009). The drinking experience: Cup or content? *Food Quality and Preference, 20*, 268–276.
Spence, C. (2011). Crystal clear or gobbletigook? *The World of Fine Wine, 33*, 96–101.
Spence, C. (2015a). Eating with our eyes: On the colour of flavour. In A. Elliott & M. Fairchild (Eds.), *The handbook of color psychology.* Cambridge, UK: Cambridge University Press.
Spence, C. (2015b). Just how much of what we taste derives from the sense of smell? *Flavour, 4*, 30.

Spence, C., & Gallace, A. (2011). Multisensory design: Reaching out to touch the consumer. *Psychology and Marketing*, *28*, 267–308.

Spence, C., & Piqueras-Fiszman, B. (2011). Multisensory design: Weight and multisensory product perception. In G. Hollington (Ed.), *Proceedings of RightWeight2* (pp. 8–18). London, UK: Materials KTN.

Spence, C., & Piqueras-Fiszman, B. (2012). The multisensory packaging of beverages. In M. G. Kontominas (Ed.), *Food packaging: Procedures, management and trends* (pp. 187–233). Hauppauge NY: Nova Publishers.

Spence, C., & Piqueras-Fiszman, B. (2014). *The perfect meal: The multisensory science of food and dining*. Oxford, UK: Wiley-Blackwell.

Spence, C., Wan, X., Woods, A., Velasco, C., Deng, J., Youssef, J., et al. (2015). On tasty colours and colourful tastes? Assessing, explaining, and utilizing crossmodal correspondences between colours and basic tastes. *Flavour*, *4*, 23.

Stead, M., Angus, K., Macdonald, L., & Bauld, L. (2014). Looking into the glass: Glassware as an alcohol marketing tool, and the implications for policy. *Alcohol and Alcoholism*, *49*, 317–320.

Suzuki, M., & Gyoba, J. (2008). Visual and tactile cross-modal mere exposure effects. *Cognition and Emotion*, *22*, 147–154.

Tu, Y., Yang, Z., & Ma, C. (2015). Touching tastes: The haptic perception transfer of liquid food packaging materials. *Food Quality and Preference*, *39*, 124–130.

Van Doorn, G., Wuillemin, D., & Spence, C. (2014). Does the colour of the mug influence the taste of the coffee? *Flavour*, *3*, 10.

Varma, S. (2013). Tea innovation around the world. *New Food*, *16*(5), 20–22.

Vilanova, M., Vidal, P., & Cortés, S. (2008). Effect of the glass shape on flavor perception of "toasted wine" from Ribeiro (NW Spain). *Journal of Sensory Studies*, *23*, 114–124.

Wan, X., Velasco, C., Michel, C., Mu, B., Woods, A. T., & Spence, C. (2014). Does the shape of the glass influence the crossmodal association between colour and flavour? A cross-cultural comparison. *Flavour*, *3*, 3.

Wan, X., Woods, A. T., Seoul, K.-H., Butcher, N., & Spence, C. (2015). When the shape of the glass influences the flavour associated with a coloured beverage: Evidence from consumers in three countries. *Food Quality and Preference*, *39*, 109–116.

Wan, X., Yu, T., Zhao, H., & Spence, C. *The influence of the receptacle on the consumer's perception of British, Chinese, and Japanese green tea* (in preparation).

Wan, X., Zhou, X., Woods, A. T., & Spence, C. (2015). Influence of the glassware on the perception of alcoholic drinks. *Food Quality and Preference*, *44*, 101–110.

Wansink, B., & van Ittersum, K. (2003). Bottoms up! The influence of elongation on pouring and consumption. *Journal of Consumer Research*, *30*, 455–463.

Wansink, B., & van Ittersum, K. (2005). Shape of glass and amount of alcohol poured: Comparative study of the effect of practice and concentration. *British Medical Journal*, *331*, 1512–1514.

White, A. M., Kraus, C. L., McCracken, L. A., & Swartzwelder, H. S. (2003). Do college students drink more than they think? Use of a free-pour paradigm to determine how college students define standard drinks. *Alcohol Clinical and Experimental Research*, *27*, 1750–1756.

Williams, A. A., Langron, S. P., & Noble, A. C. (1984). Influence of appearance on the assessment of aroma in Bordeaux wines by trained assessors. *Journal of the Institute of Brewing*, *90*, 250–253.

Williams, A. A., Langron, S. P., Timberlake, C. F., & Bakker, J. (1984). Effect of colour on the assessment of ports. *International Journal of Food Science and Technology*, *19*, 659–671.

Williams, L. E., & Bargh, J. A. (2008). Experiencing physical warmth promotes interpersonal warmth. *Science, 322*, 606–607.

Yang, S., & Raghubir, P. (2005). Can bottles speak volumes? The effect of package shape on how much to buy. *Journal of Retailing, 81*, 269–281.

Zajonc, R. B. (1968). Attitudinal effects of mere exposure. *Journal of Personality and Social Psychology, 9*, 1–28.

Zajonc, R. B. (1980). Feeling and thinking: Preferences need no inferences. *American Psychologist, 35*, 151–175.

Zajonc, R. B. (2001). Mere exposure: A gateway to the subliminal. *Current Directions in Psychological Science, 10*, 224–228.

Zhou, X., Wan, X., Mu, B., Du, D., & Spence, C. (2015). Examining colour-receptacle-flavour interactions for Asian noodles. *Food Quality and Preference, 41*, 141–150.

The Roles of the Senses in Different Stages of Consumers' Interactions With Food Products

15

Hendrik N.J. Schifferstein
Department of Industrial Design, Delft University of Technology, Delft, The Netherlands

1 Introduction

Although many studies on the sensory properties of foods tend to focus on what people perceive when they insert a food product into their mouth, in everyday life consumers who want to consume a food product typically go through a number of events, or stages, before they actually do so. For instance, if I want to eat chocolate now, I can go to my neighborhood store, select a chocolate bar from the shelf, buy it and take it home, and then unwrap it before I can eventually take a bite. After eating the chocolate, I may wrap the remaining chocolate and store it, or if I finished it I will dispose of the packaging materials. During the course of these events, I will perceive different types of information by means of my senses, which may generate expectations for the sensations that I am about to come across, and that affect how I will evaluate the chocolate.

Food products are unique in that sensory experiences during our interaction with them can involve all of the senses: vision, touch, audition, smell, and taste (Schifferstein, 2006). The perception of sensory information is the starting point for how the product is experienced: whether it is pleasing or not, the cognitive associations and meanings it evokes, the actions it triggers, and the emotional responses that it may elicit (Brakus, Schmitt, & Zarantonello, 2009; Hekkert & Schifferstein, 2008; Vyas & van der Veer, 2006).

The example described earlier shows that experiences are not static during user–product interactions, but change over time; different types of experiences are evoked consecutively. The role of the senses in these different stages of our interaction with a food product may vary, and in each stage different meanings and cognitive associations may be evoked, various action tendencies may be activated, and several emotions may be elicited. An experience does not only result from the interaction, but also accompanies and guides the interaction, and thus affects the qualities of the interaction (Desmet & Hekkert, 2007). In addition, user–product interactions in daily life always take place within a physical, social, cultural, political, and economical context that affects the experience (Law, Roto, Hassenzahl, Vermeeren, & Kort, 2009).

This chapter focuses on the different roles that the senses play in the various stages of user–product interactions with food products. In this discussion, I will focus primarily on the sensory aspects of the product experience, although, in some cases, links with aesthetic, semantic, and emotional aspects are described as well.

2 Research approaches to the study of dynamic experiences

In the area of food perception, research on the dynamics of sensory perception has mainly been focused on variation during food consumption. For instance, temporary dynamics in a single bite or sip have been assessed using the time–intensity methodology. This involves recording the evolution of the intensity of a given sensory attribute over time during the tasting of a single product (eg, Larson-Powers & Pangborn, 1978). Alternatively, in the temporal dominance of sensations method (Pineau et al., 2009), a set of attributes is presented on a computer screen while a product is tasted. The panelist indicates what the dominant sensory perception is by selecting the corresponding attribute and by rating its intensity during oral processing. However, in order to assess consumer perception and acceptance at the key moments of a packaged food experience (not only during eating, but also including buying, storage, and food preparation) other methods are required.

Typically, the food industry has studied food experiences by conducting a number of separate, consecutive tests for the different key moments. For instance, consumer buying behavior has been studied by testing different packaging designs in a virtual display or on the shelf (eg, Burke, Harlam, Kahn, & Lodish, 1992; Garber, Hyatt, & Boya, 2008). A broad range of sensory profiling techniques have been developed to describe the sensory characteristics of foods during eating in a laboratory setting (eg, Meilgaard, Civille, & Carr, 1991). Furthermore, home use tests have been conducted to evaluate product storage, preparation, consumption, and disposal (eg, Lawless & Heymann, 1998). However, it may be difficult to relate the outcomes of all these very different types of tests to each other and, therefore, I recommend using a more unified approach, in which the different stages of the food experience can be linked together directly.

In the area of industrial design, the dynamics of consumer experience has been investigated by asking participants to describe their experience during different stages of the user–product interaction using a single method. This approach focuses on entire usage stages (eg, eating the product) instead of singular events (eg, taking a bite) and requires participants to quantify the contributions of all sensory modalities simultaneously. Fenko, Schifferstein, and Hekkert (2010) instructed 243 participants to evaluate the importance of the sensory modalities in determining their experiences with durable consumer products while buying a product, after the first week, the first month, and the first year of usage. Although subjective estimates of perceived importance may deviate from estimates determined in controlled experiments (Fenko, Schifferstein, Huang, & Hekkert, 2009; van Ittersum, Pennings, Wansink, & van Trijp, 2007; Verlegh, Schifferstein, & Wittink, 2002), they do provide a first approximation of how people perceive the roles of the senses in their subjective experiences.

Fenko et al. (2010) showed that the dominant sensory modality depends on the period of product usage. At the moment of purchase, when people often have to compare multiple slightly different products to make an optimal decision, vision is the most important modality. However, during usage the other senses gain importance, depending on their roles in user–product interactions. When users become acquainted with the product, motor skills are required to operate or use the product, which typically

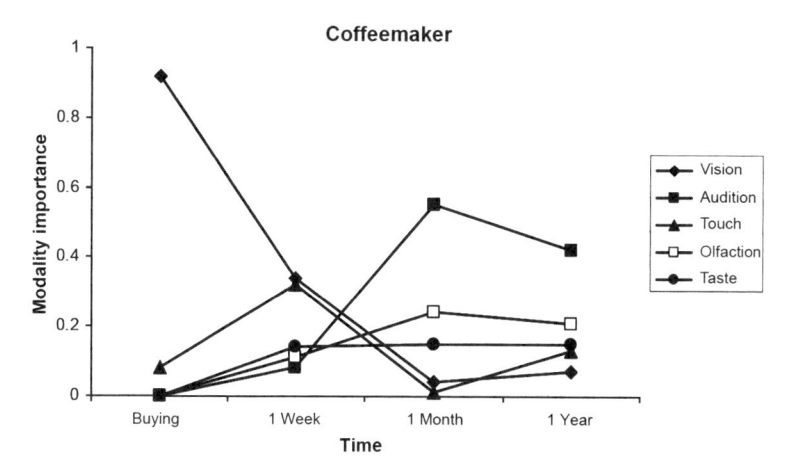

Figure 15.1 Changes in modality importance over time for a coffeemaker (Fenko et al., 2010)

increases the importance of touch. After users become familiar with their appliances, they do not need to pay close attention to visual and tactile attributes any longer, but start to notice the sounds the product makes, which they often describe as irritating and annoying (Özcan & van Egmond, 2012). For an electric coffeemaker ($N = 24$), Fenko et al. (2010) showed that vision and touch were judged to be about equally important after one week of usage. After 1 month, the sounds of the coffeemaker became most important for users, followed by the smell (see Fig. 15.1). Please note that in the case of the coffee machine, it is unlikely that users will describe the loud noise of the motor and the water pump as unpleasant, because in their memory this sound is associated with the pleasant smell and taste of freshly brewed coffee.

Schifferstein, Fenko, Desmet, Labbe, and Martin (2013) used a similar approach to investigate how important the sensory modalities are during the experience of a dehydrated food product at four different stages of product usage: choosing a product on a supermarket shelf, opening a package, cooking, and eating the food. The data were analyzed both qualitatively and quantitatively in order to determine the changes in product experience at different stages of product usage and the influence of package design on the overall product experience. These researchers found that vision was the most important sense at the buying stage. Smell was dominant at the cooking stage, and taste was the most important sensation while eating the food (see Fig. 15.2). The comments mostly reflected responses to sensory qualities (good or bad appearance, taste, smell, texture), usability aspects (easy or difficult to open package or to prepare the product), and the nature of the product itself. At the purchase stage, preexisting attitudes and stereotypes toward the product group seemed to play a major role in affective reactions, while in the other stages when multiple modalities were actively involved, participants' emotional judgments mainly reflected their direct sensory experiences.

In a similar vein, Labbe, Ferrage, Rytz, Pace, and Martin (2015) studied the impact of the motivation to drink coffee (enjoyment or being stimulated) on consumer responses for the importance of the five senses and for the satisfaction with each of

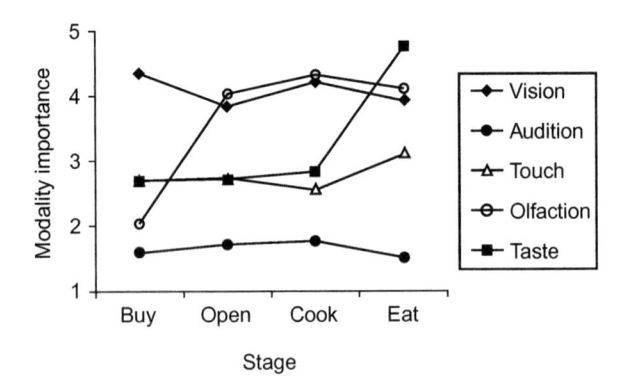

Figure 15.2 Perceived importance of the senses at four stages of product usage: buying, opening package, cooking, and eating a dehydrated vegetable product (Schifferstein et al., 2013).

the senses during various stages of making and consuming a coffee beverage: water heating, jar handling, cup preparation, and drinking. In many cases the importance and satisfaction ratings differed between experience moments, but followed the same trends. Importance and satisfaction for vision were rated high during the entire experience (7–9 on a scale ranging from 0 to 10), while hearing ratings remained moderate and stable during the entire experience (5–7). Smell ratings were moderate-low during water heating (4–6) and very high at the other three stages (8–9). Touch and taste were rated moderate at the beginning of the experience (5–7) and then progressively increased. Over the five senses, the highest ratings were given to the sense of taste after drinking a cup of coffee (>9). Participants who consumed coffee for sensory enjoyment produced higher ratings for smell and taste than participants who consumed coffee to feel energized, while ratings for the other sensory modalities were similar in both groups.

In the domain of personal care products Churchill, Meyners, Griffiths, and Bailey (2009) distinguished between four different stages of usage in the experience of shampoo during washing of the hair: how the product felt on the hands before being applied to hair, how the product felt on the hands and hair during application (including the lather), how the hair felt after product usage while the hair was still wet, and how the hair felt when it was dry. In their study, each stage of usage involved its own list of 11, 14, or 16 perceptual sensory attributes. Only six sensory attributes were shared among the four stages: greasy, oily, silky, smooth, soft, and sticky. In addition, four items were rated to assess overall liking over the four stages. The results revealed a cross-modal effect of olfaction on touch perception: different odors differentially affected the perception of the texture characteristics of a product (shampoo) and also the perception of the substrate on which it was used (hair). An identical base shampoo product with different fragrances elicited different ratings for the perceived stickiness of the product or the hair in all four stages of usage. In addition, the different fragrances affected the perceived richness and thickness of the shampoo product before being applied to hair, and the perceived runniness and tackiness of the shampoo during application.

In all of the studies just mentioned, consumers evaluated products while they were going through the different stages of using them. It is also possible to follow consumer experiences in those cases where the consecutive stages of the experience are not strictly defined beforehand. This approach has been used recently in a project investigating the adoption of insects as food for human consumption, where the investigators wanted to keep close track of any emotional changes in the consumers' psychological adoption processes (Ortiz, 2014). The observations were recorded in the form of an emotional timeline with added remarks for each participant and each dish. Fig. 15.3 shows an example of such a timeline, following a participant who takes a sample of grasshoppers and pulverizes them, before using it in a recipe. On this timeline, you can observe evaluations of various sensory aspects (smell, texture, appearance), associations with other products (coffee, a paste), the personal meaning of the activity (therapy), and a wealth of emotions that are elicited (dislike, hope, satisfaction, disgust, worry). Qualitative analysis of such timelines can be helpful in obtaining a detailed account of the interrelationships between sensory perceptions, associative meanings, personal concerns, and emotional responses.

Researching consumer experiences with fast-moving consumer goods (FMCGs), such as foods and personal care products, may be laborious but it is usually easy to organize, because consecutive usage stages follow each other rapidly. In contrast, research on consumer durables may involve stages of use that are divided over much longer time periods. This may require longitudinal, ethnographic studies following individuals over the course of several weeks after the purchase of a single product (Karapanos, Zimmerman, Forlizzi, & Martens, 2009). Alternatively, however, participants may be asked, within a single contact, to recall the most salient experiences that they had within a given time period, and estimate temporal details regarding the recalled experiences (Karapanos, Martens, & Hassenzahl, 2012). In the latter case, researchers need to rely on the retrospective elicitation of users' experiences from memory, which limits the data's validity. Another approach to investigate behaviors that may occur at any point in time involves alarming participants at multiple points in time through mobile devices, in order to sample their experiences and to answer specific questions (Hektner, Schmidt, & Csikszentmihalyi, 2007). In the food realm, these methods may be applied to research effects that follow food consumption, such as the perceived satiety of food components (eg, Blundell & Macdiarmid, 1997; Hulshof, de Graaf, & Weststrate, 1993), the occurrence of boredom with specific products (Köster & Mojet, 2007), or in monitoring the evolution of image perception for food categories or brands over time (eg, Schifferstein, Candel, & van Trijp, 1998; Verbeke, 2001).

The current section has provided an overview of approaches used to study perception in various stages of people's interactions with (food) products. Next, I will focus more on the outcomes of such studies, by attempting to create an overview of the roles of the sensory modalities in the different stages. Since the number of empirical studies on some of the stages is limited, this overview also makes use of personal observations and anecdotal evidence. As a consequence, the overview is likely to be incomplete, and the reader should feel free to add any missing observations to the framework provided.

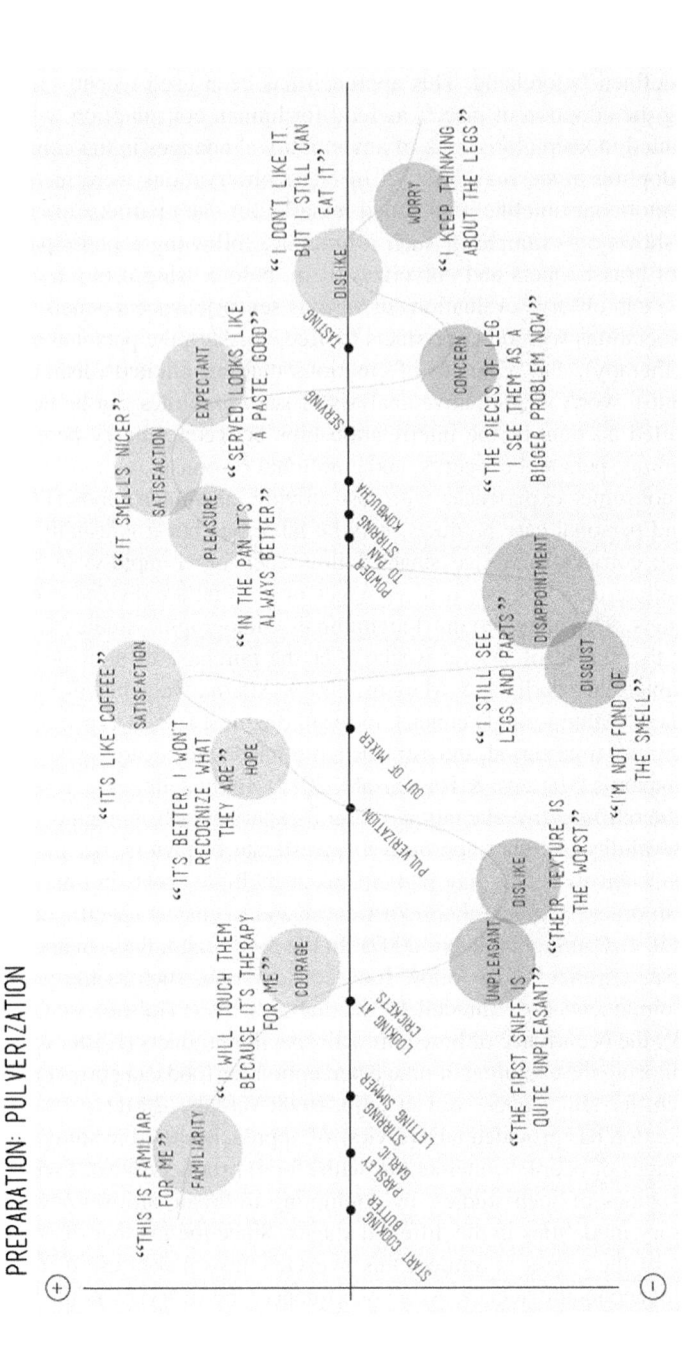

Figure 15.3 A timeline documenting a participant's cooking experience involving the pulverization of grasshoppers, showing a number of sensory, aesthetic, cognitive, and emotional associations (Ortiz, 2014).

3 Findings concerning the role of sensory perception in different interactions with food products

In this overview, a distinction is made between three major stages in interacting with foods: the packaged product, the unpacked product, and after consumption.

3.1 Stage 1: The packaged product

Many food products have been processed before they are sold to consumers: They have been cleaned, washed, cut, dried, fermented, heated, or chilled, and so on. After processing, most products are packaged. The package serves to protect and stabilize the contents, and facilitates handling during trading and distribution. In addition, it provides consumers with product information, and can play a role in facilitating its use and consumption (eg, Dekker, 2011).

1. Buying
   The buying outlet creates a context for product selection, which partly determines how the product itself is experienced during consumption (Wheatley & Chiu, 1977). The position of the product on the shelf (eye height) and the lighting can bring a product to the shopper's attention. In addition, the type of lighting can directly affect the perception and experienced character of the product (Barbut, 2001; Suk, Park, & Kim, 2012). Packaging shape and color play an important role on retail shelves, because consumers who move down long store aisles first see category facings from a distance and at an angle, and start processing the larger visual elements well before they can process finer details or read text (eg, Garber et al., 2008). Packaging should help in making the product stand out from its competitors on the shelves.

   During the buying process, packaging design helps to identify the category and brand to which the product belongs and to confer meaning or reinforce existing associations to the product. The package provides information about the producer, brand, origin, quantity, and nutritional value. The design of the exact packaging characteristics is critical, because it can suggest a certain identity for the contents that may enhance or interfere with its identification and evaluation (Cardello, Maller, Masor, Dubose, & Edelman, 1985; Piqueras-Fiszman & Spence, 2011). In addition, the package communicates concerning the properties of the contents and is likely to activate consumer expectations. If designed sensibly, the package can enhance the way in which consumers experience the content (eg, Becker, van Rompay, Schifferstein, & Galetzka, 2011; Spence & Piqueras-Fiszman, 2012).

   Given the description above, it will be no surprise that the sense of vision is generally judged to be most important during product purchase (Fenko et al., 2010; Schifferstein et al., 2013). Vision plays a crucial role in locating and obtaining information about the product. In addition, many food packages allow consumers only to inspect their contents visually through a transparent material, if at all. In some cases, however, consumers may also gain an impression of the weight or shape of ingredients through touch (eg, the contents of a flexible sachet). Although food packaging generally blocks most direct sensory perception, this may be partly compensated for by displaying lively pictures that stimulate mental imagery in the blocked senses. For instance, Schifferstein et al. (2013) reported that imagining the taste of the final, cooked product was quite important for consumers' product evaluations during the buying stage (see Fig. 15.2), even though the package of their dehydrated product did not allow for any direct taste perception.

2. Transporting and storage

    When packaged foods are transported in a cart or basket, carried in a bag, or stored at home, the product's shape, size, and weight are probably most important for allowing convenient handling. In addition, the packaging material's characteristics that determine the pleasantness of the feel and its slipperiness, its fragility, or its robustness will have an impact on how the product is experienced during transportation. If the product needs to be chilled or frozen for storage, the change in temperature will affect its feel and slipperiness. The majority of these aspects depend greatly on the sense of touch.

3. Opening the package

    The opening of a package mostly involves a mechanical interaction in which the hands and tactual feedback play an important part, under the guidance of the interpretation of visual cues on the packaging. In addition, the mechanical interaction will probably produce some sounds that give feedback on the type of packaging material and the user's actions. After the seal of the package has been broken, the smell of the product may be perceived for the first time, and if the product can be consumed right out of the package, the taste can also be sampled.

    The ensemble of visual (Becker et al., 2011; Mizutani et al., 2010), tactual (Krishna & Morrin, 2007; McDaniel & Baker, 1977), and auditory (Brown, 1958) stimuli that are available and are produced during the opening of the package are likely to affect how its content will be perceived. In this case, the perception in one modality may evoke expectations for what will be perceived in another sensory modality (Cardello & Sawyer, 1992; Schifferstein, 2001; Schifferstein & Cleiren, 2005), and thereby affect perception in that modality.

    However, the physical properties of the package can also have a direct effect on sensory perception. For instance, the shape of a container might affect flavor release (Hummel, Delwiche, Schmidt, & Huttenbrink, 2003; Spence & Wan, 2015) and the migration of compounds from the packaging material into the food may produce off-flavors (Janssens, Diekema, Reitsma, & Linssen, 1995). In addition, the shape and size of a container (Raghubir & Krishna, 1999; Wansink, 1996; Wansink & Van Ittersum, 2003) and the magnitude of the opening in the container (Farleigh, Shepherd, & Wharf, 1990; Greenfield, Maples, & Wills, 1983; Greenfield, Smith, & Wills, 1984) affect the amount that is consumed.

3.2 Stage 2: The unpacked product

In many cases, food products are not consumed directly from their package, but are prepared and served before they are eaten. Oftentimes, people interact already quite intensely with food products before they eat them and they tend to use all of their senses during these interactions (see Table 15.1).

1. Preparation and cooking

    First of all, people use their senses to check the quality of the ingredients they use. For instance, they check the color, texture, smell, and taste of ingredients (eg, vegetables, herbs, and cheese) to determine whether they are unripe, ripe, overripe, or spoiled. They also look and feel to determine whether the ingredients have the desired characteristics, whether they are clean, need to be washed or brushed, or whether some parts need to be removed.

    During the process of cutting they use vision and touch to determine the ideal shape of the ingredients and they receive textural feedback from the resistance of the knife and the sounds produced during the slicing of ingredients. When they mix ingredients, they can see the colors of multiple sources merging into a homogeneous blend, they may well smell the formation of a harmonious flavor blend, and they can taste the mixture to check whether ingredients were mixed in the right proportions. If mixing involves whipping air into the blend, the cook can monitor the look and feel of the creation of the foam and may perceive changes in sound during whipping.

Table 15.1 Involvement of the senses in preparing and cooking food: some examples

	Vision	Touch			Audition	Smell	Taste
		Shape and size	Texture	Temperature			
Check quality of ingredients							
Ripening status	Color		Softness			Intensity	Sweetness, sourness
Spoilage	Brown spots		Variety of textures		Crispness	Off-flavors	
Cleanliness	Presence of dirt		Presence of dirt				
Check progress of process							
Remove parts	Visual difference between skin, flesh, and core		Texture differences between skin, flesh, and core			Flavor release	
Cutting, slicing	Regularity of pieces	Regularity of pieces					
Blending, marinating	Homogeneous color					Flavor release	
Creating a foam	Thickness		Thickness		Pitch of blender sounds	Check flavors	Check taste
Heating treatments: Cooking, baking, frying	Browning, bubbles of cooking process		Softness, stickiness	Warmness	Bubbling and sizzling of cooking process	Cooking and baking flavors	Check taste

A cooking, baking, or frying process may be monitored by seeing and hearing bubbles of evaporating water, feeling temperature changes, evaluating texture changes by sticking a fork in the food, smelling the baking flavors, and tasting a small portion of the dish. Hence, food preparation and cooking is a real multisensory activity, in which all senses are used consecutively or simultaneously.

2. Serving

In a restaurant setting, during the act of serving, waiters may perceive the colors, shapes, textures, weight, smells, and perhaps also some sounds of the foods they are about to serve. In addition, if a client orders a dish en flambé, this will obviously add extra visual, auditory and olfactory sensations to the meal experience. Most attention, however, may be directed towards how the components of a dish are presented on a plate. Contemporary plating is primarily directed at creating an impressive visual image, where colors and shapes create a balanced, interesting whole. Recently, investigators have started to tackle this topic, by investigating the principles of visual organization (Michel, Woods, Neuhäuser, Landgraf, & Spence, 2015; Zellner, Lankford, Ambrose, & Locher, 2010) and relationships with the visual aesthetics of paintings (Deroy, Michel, Piqueras-Fiszman, & Spence, 2014; Spence, Piqueras-Fiszman, Michel, & Deroy, 2014).

3. Eating

People engage all their senses when they eat. Since this is the stage where they actually put the food into the mouth, their primary focus of attention may be on their chemosensory perception (see chapter: Attention and Flavor Binding, on this point): the aromas they smell orthonasally just before the food enters their mouth, the taste perceived when the food is in the oral cavity, and the flavors that reach the olfactory epithelium via the retronasal pathway. People also feel the rough, sticky, or slippery surface of the product on their tongue and they feel the thickness, hardness, and stiffness of the product mass in their mouth when they masticate (see chapter: Oral-Somatosensory Contributions to Flavor Perception and the Appreciation of Food and Drink). In addition, they hear the crunching, crackling, crispy sounds while they bite, and possibly the soft smacking and slurping sounds while they chew and swallow (Spence, 2015).

However, it is not just the sensory impressions of the food itself that determine the eating experience. The context in which people consume food also has a major effect on how a dish is perceived (Meiselman, 2008). Factors that influence their exact experience include the appropriateness of the food in that setting (Piqueras-Fiszman & Jaeger, 2014a, 2014b), the characteristics of the location and its atmosphere (Bell, Meiselman, Pierson, & Reeve, 1994; Cardello, Bell, & Kramer, 1996; Spence & Piqueras-Fiszman, 2014), the presence of others at the table (de Castro, 1994; Feunekes, de Graaf, & van Staveren, 1995), etiquette rules that apply, the type of furniture, the table arrangement, and the tableware used (Schifferstein, 2009; Spence, Harrar, & Piqueras-Fiszman, 2012; Stummerer & Hablesreiter, 2013). Hence, the eating experience can be influenced by many contextual variables that are extrinsic to the food itself.

3.3 Stage 3: After consumption

After consuming food, people may still perceive some flavors in their mouth, either as a lingering aftertaste, or due to gasses escaping from the stomach, for instance during belching. In addition, they may perceive some postingestive effects of eating. Through enteroception they may perceive the fullness of their stomach and intestines, giving them an indication of their degree of satiety (Bellisle, Drewnowski, Anderson, Westerterp-Plantenga, & Martin, 2012; Hulshof et al., 1993). In addition, they may feel depleted, because the body needs energy to digest the food, or they may feel refreshed or energetic after consuming substances that provide energy (eg, sugar) or mental stimulation (eg, caffeine). However, these postingestive effects do not make use of the exteroceptive senses that we focused on above.

The food-related activities that people perform after consumption will probably be focused on storing any leftovers, or on disposing food waste and packaging materials. Alternatively, in some instances the remaining packaging materials may find another use (eg, using a jar as a flower pot, creating decorations out of tea bags, and so on). Involvement of the senses in this case depends highly on the type of activities people engage in.

4 Conclusions

Although the exact role of the senses in the interaction with food products is highly dependent on the particular food involved, the different stages of the interaction, and the exact nature of the activities involved, we can also discern a general interaction pattern that describes how people interact with foods. It starts out with a first exploration at a distance (vision), followed by closer inspection and active engagement with the product (touch), through to a more intimate contact with the product that ultimately involves ingestion (taste and retronasal smell), possibly followed by postingestive effects. The orthonasal smell may already be involved in an early stage if it can be perceived from a distance (eg, freshly baked bread), but in other cases it partakes first during the opening of the package. Hence, each food experience will involve a highly dynamic and varied repertoire of sensations. Therefore, in the design process for food products and their packaging it is important to take into account the possible variations in sensory impressions that may be generated, and it may be helpful to develop a scenario of sensory encounters that describes the different stages the consumer goes through while interacting with a product (eg, MacDonald, 2002).

If a producer would like to convey a certain message or stress a certain product feature, various product aspects can be specifically tuned in order to communicate this aspect through multiple senses (Schifferstein, 2011). For instance, Nikolaidou (2011) created a soup package that conveyed naturalness through multiple sensory channels, while remaining in line with the principal values of the company brand. She developed an interaction scenario describing how the sensory modalities were stimulated during purchase and product use (see Fig. 15.4). In this case, requiring consumers to tear the package from the display in the shelf mimicked the harvesting of produce in the field. Similar to products from nature, the package had a rich texture and allowed the consumer to smell the contents of the soup. Furthermore, the crispy sounds from the packaging material enhanced the sensory richness of the experience, which she found was an important characteristic when it came to simulating natural experiences.

In conclusion, this overview of the involvement of the senses in the interaction with food products shows that all of the senses can potentially play a role in our interaction with food products. Although people are highly aware of the important role of taste (4.86), smell (4.21), and vision (4.20) in their experience of foods, they rate touch as somewhat less important (3.13), and audition as the least important (1.74) (from Schifferstein, 2006; all ratings on a 5-point scale from 1 = very unimportant

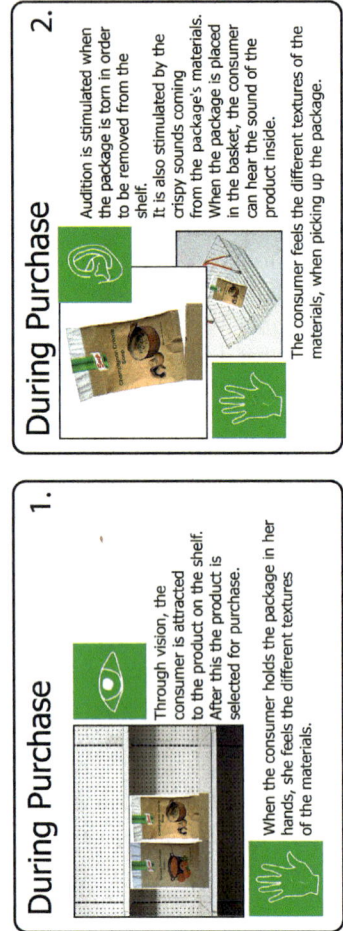

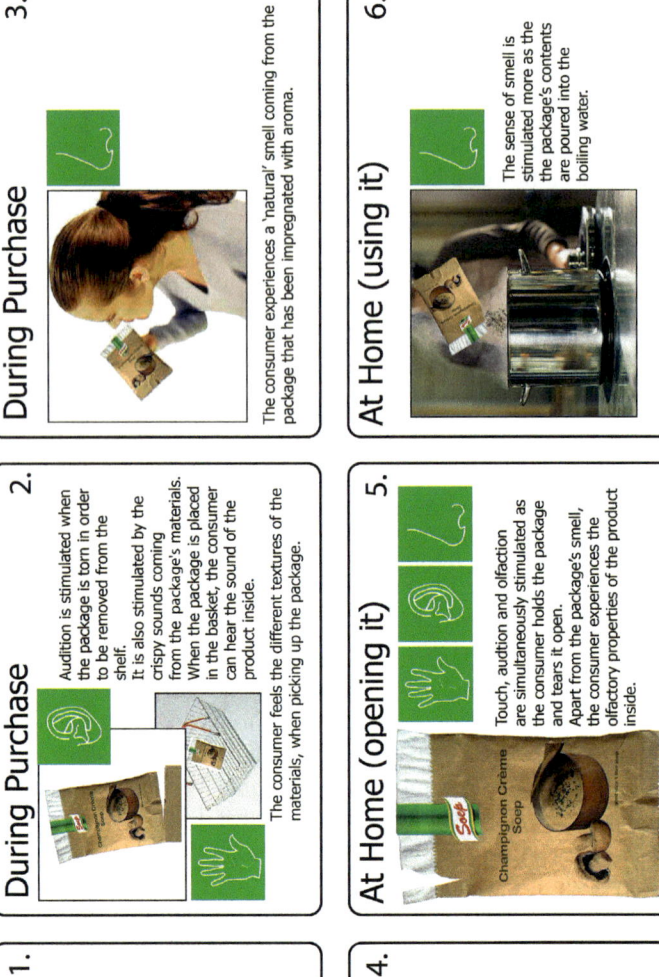

Figure 15.4 Sensory interaction scenario for a soup package conveying naturalness (adapted from Nikolaidou, 2011).

to 5 = very important). Nonetheless, several examples have been described in which the roles of touch and audition are evident, including the opening of the package, the preparation of the food, and eventual food consumption. Hence, the careful monitoring and evaluation of sensory inputs during the interactions that people have with food products is an important tool for unraveling the richness of their sensory experiences.

Probably, more future studies on food experiences will take into account both the complexity of the experience (including its multisensory perception, aesthetics, meaning, and emotions) and its dynamic character. This will yield a richer and more complete understanding of the subjective food experiences. In-depth knowledge of these interactions provides an important resource when it comes to trying to understand the consumers' responses to current products, but also for overcoming consumption barriers for uncommon products (eg, Deroy, Reade, & Spence, 2015) and for developing new market offerings (Lundahl, 2012; Moskowitz, Saguy, & Straus, 2009). I hope that the overview of methodologies provided here will help to stimulate research that takes into account all the subtle dynamics of our everyday sensory experiences that are inherently multisensory.

References

Barbut, S. (2001). Effect of illumination source on the appearance of fresh meat cuts. *Meat Science*, 29, 181–191.

Becker, L., van Rompay, T. J. L., Schifferstein, H. N. J., & Galetzka, M. (2011). Tough package, strong taste: The influence of packaging design on taste impressions and product evaluations. *Food Quality and Preference*, 22, 17–23.

Bell, R., Meiselman, H. L., Pierson, B. J., & Reeve, W. G. (1994). Effects of adding an Italian theme to a restaurant on the perceived ethnicity, acceptability, and selection of foods. *Appetite*, 22, 11–24.

Bellisle, F., Drewnowski, A., Anderson, G. H., Westerterp-Plantenga, M., & Martin, C. K. (2012). Sweetness, satiation, and satiety. *The Journal of Nutrition*, 142, 1149S–1154S.

Blundell, J. E., & Macdiarmid, J. I. (1997). Fat as a risk factor for overconsumption: Satiation, satiety, and patterns of eating. *Journal of the American Dietetic Association*, 97, S63–S69.

Brakus, J. J., Schmitt, B. H., & Zarantonello, L. (2009). Brand experience: What is it? How is it measured? Does it affect loyalty? *Journal of Marketing*, 73, 52–68.

Brown, R. L. (1958). Wrapper influence on the perception of freshness in bread. *Journal of Applied Psychology*, 42, 257–260.

Burke, R. R., Harlam, B. A., Kahn, B. E., & Lodish, L. (1992). Comparing dynamic consumer choice in real and computer-simulated environments. *Journal of Consumer Research*, 19, 71–82.

Cardello, A. V., Bell, R., & Kramer, F. M. (1996). Attitudes of consumers toward military and other institutional foods. *Food Quality and Preference*, 7, 7–20.

Cardello, A. V., & Sawyer, F. M. (1992). Effects of disconfirmed consumer expectations on food acceptability. *Journal of Sensory Studies*, 7, 253–277.

Cardello, A. V., Maller, O., Masor, H. B., Dubose, C., & Edelman, B. (1985). Role of consumer expectancies in the acceptance of novel foods. *Journal of Food Science*, 50, 1707–1714. 1718.

Churchill, A., Meyners, M., Griffiths, L., & Bailey, P. (2009). The cross-modal effect of fragrance in shampoo: Modifying the perceived feel of both product and hair during and after washing. *Food Quality and Preference*, 20, 320–328.

de Castro, J. M. (1994). Family and friends produce greater social facilitation of food intake than other companions. *Physiology & Behavior*, *56*, 445–455.

Dekker, M. (2011). Food packaging design. In A. R. Linnemann, C. G. P. H. Schroën, & M. A. J. S. van Boekel (Eds.), *Food product design; an integrated approach* (pp. 197–205). Wageningen: Wageningen Academic Publishers.

Deroy, O., Michel, C., Piqueras-Fiszman, B., & Spence, C. (2014). The plating manifesto (I): From decoration to creation. *Flavour*, *3*, 6.

Deroy, O., Reade, B., & Spence, C. (2015). The insectivore's dilemma, and how to take the West out of it. *Food Quality and Preference*, *44*, 44–55.

Desmet, P. M. A., & Hekkert, P. (2007). Framework of product experience. *International Journal of Design*, *1*, 57–66.

Farleigh, C. A., Shepherd, R., & Wharf, S. G. (1990). The effect of manipulation of salt pot hole size on table salt use. *Food Quality and Preference*, *2*, 13–20.

Fenko, A., Schifferstein, H. N. J., & Hekkert, P. (2010). Shifts in sensory dominance between various stages of user-product interactions. *Applied Ergonomics*, *41*, 34–40.

Fenko, A., Schifferstein, H. N. J., Huang, T. C., & Hekkert, P. (2009). What makes products fresh: The smell or the colour? *Food Quality and Preference*, *20*, 372–379.

Feunekes, G. I. J., de Graaf, C., & van Staveren, W. A. (1995). Social facilitation of food intake is mediated by meal duration. *Physiology & Behavior*, *58*, 551–558.

Garber, L. L., Hyatt, E. M., & Boya, U. O. (2008). The mediating effects of the appearance of nondurable consumer goods and their packaging on consumer behavior. In H. N. J. Schifferstein & P. Hekkert (Eds.), *Product experience* (pp. 581–602). Amsterdam: Elsevier.

Greenfield, H., Maples, J., & Wills, R. B. (1983). Salting of food: A function of hole size and location of shakers. *Nature*, *301*, 331–332.

Greenfield, H., Smith, A. M., & Wills, R. B. (1984). Influence of multi-holed shakers on salting on food. *Human Nutrition: Applied Nutrition*, *38*, 199–201.

Hekkert, P., & Schifferstein, H. N. J. (2008). Introducing product experience. In H. N. J. Schifferstein & P. Hekkert (Eds.), *Product experience* (pp. 1–8). Amsterdam: Elsevier.

Hektner, J. M., Schmidt, J. A., & Csikszentmihalyi, M. (2007). *Experience sampling method: Measuring the quality of everyday life*. Thousand Oaks, CA: Sage.

Hulshof, T., de Graaf, C., & Weststrate, J. A. (1993). The effects of preloads varying in physical state and fat content on satiety and energy intake. *Appetite*, *21*, 273–286.

Hummel, T., Delwiche, J. T., Schmidt, C., & Huttenbrink, K. B. (2003). Effects of the form of glasses on the perception of wine flavors: A study in untrained subjects. *Appetite*, *41*, 197–202.

Janssens, J. L., Diekema, N. W., Reitsma, J. C., & Linssen, J. P. (1995). Taste interaction of styrene/ethylbenzene mixtures in an oil-in-water emulsion. *Food Additives and Contaminants*, *12*, 203–209.

Karapanos, E., Martens, J. -B., & Hassenzahl, M. (2012). Reconstructing experiences with iScale. *International Journal of Human-Computer Studies*, *70*, 849–865.

Karapanos, E., Zimmerman, J., Forlizzi, J., & Martens, J.-B. (2009). *User experience over time: An initial framework*. Paper presented at the Twenty-Seventh Annual SIGCHI Conference on Human Factors in Computing Systems-CHI '09, Boston.

Köster, E. P., & Mojet, J. (2007). Boredom and the reasons why some new food products fail. In H. MacFie (Ed.), *Consumer-led food product development* (pp. 262–280). Cambridge, UK: Woodhead.

Krishna, A., & Morrin, M. (2007). Does touch affect taste? The perceptual transfer of product container haptic cues. *Journal of Consumer Research*, *34*, 807–818.

Labbe, D., Ferrage, A., Rytz, A., Pace, J., & Martin, N. (2015). Pleasantness, emotions and perceptions induced by coffee beverage experience depend on the consumption motivation (hedonic or utilitarian). *Food Quality and Preference*, *44*, 56–61.

Larson-Powers, M., & Pangborn, R. M. (1978). Paired comparison and time-intensity measurements of the sensory properties of beverages and gelatins containing sucrose or synthetic sweeteners. *Journal of Food Science, 43*, 41–46.

Law, E.L.C., Roto, V., Hassenzahl, M., Vermeeren, A.P.O.S., & Kort, J. (2009). Understanding, scoping and defining User eXperience: A survey approach. *Paper presented at the Human Factors in Computing Systems Conference, CHI '09, Boston, MA.*

Lawless, H. T., & Heymann, H. (1998). *Sensory evaluation of food: Principles and practices.* New York, NY: Chapman & Hall.

Lundahl, D. (2012). *Breakthrough food product innovation through emotions research.* San Diego, CA: Academic Press.

MacDonald, A. (2002). The scenario of sensory encounter: Cultural factors in sensory-aesthetic experience. In W. S. Green & P. W. Jordan (Eds.), *Pleasure with products: Beyond usability* (pp. 113–123). London: Taylor & Francis.

McDaniel, C., & Baker, R. C. (1977). Convenience food packaging and the perception of product quality: What does "hard-to-open" mean to consumers? *Journal of Marketing, 41*(4), 57–58.

Meilgaard, M. C., Civille, G. V., & Carr, B. T. (1991). *Sensory evaluation techniques* (2nd ed.). Boca Raton, FL: CRC.

Meiselman, H. L. (2008). Experiencing food products within a physical and social context. In H. N. J. Schifferstein & P. Hekkert (Eds.), *Product experience* (pp. 559–580). Amsterdam: Elsevier.

Michel, C., Woods, A. T., Neuhäuser, M., Landgraf, A., & Spence, C. (2015). Rotating plates: Online study demonstrates the importance of orientation in the plating of food. *Food Quality and Preference, 44*, 194–202.

Mizutani, N., Okamoto, M., Yamaguchi, Y., Kusakabe, Y., Dan, I., & Yamanaka, T. (2010). Package images modulate flavor perception for orange juice. *Food Quality and Preference, 21*, 867–872.

Moskowitz, H. R., Saguy, I. S., & Straus, T. (Eds.), (2009). *An integrated approach to new food product development.* Boca Raton, FL: CRC Press.

Nikolaidou, I. (2011). Communicating naturalness through packaging design. In P. M. A. Desmet & H. N. J. Schifferstein (Eds.), *From floating wheelchairs to mobile car parks* (pp. 74–79). The Hague: Eleven International.

Ortiz, M.A. (2014). *Welcoming insects at the table: Designing a new experience of eating insects* (Master thesis). Delft: Delft University of Technology.

Özcan, E., & van Egmond, R. (2012). Basic semantics of product sounds. *International Journal of Design, 6*, 41–54.

Pineau, N., Schlich, P., Cordelle, S., Mathonnière, C., Issanchou, S., Imbert, A., et al. (2009). Temporal dominance of sensations: Construction of the TDS curves and comparison with time-intensity. *Food Quality and Preference, 20*, 450–455.

Piqueras-Fiszman, B., & Jaeger, S. R. (2014a). Emotion responses under evoked consumption contexts: A focus on the consumers' frequency of product consumption and the stability of responses. *Food Quality and Preference, 35*, 24–31.

Piqueras-Fiszman, B., & Jaeger, S. R. (2014b). The impact of evoked consumption contexts and appropriateness on emotion responses. *Food Quality and Preference, 32*, 277–288.

Piqueras-Fiszman, B., & Spence, C. (2011). Crossmodal correspondences in product packaging. Assessing color–flavor correspondences for potato chips (crisps). *Appetite, 57*, 753–757.

Raghubir, P., & Krishna, A. (1999). Vital dimensions in volume perception: Can the eye fool the stomach? *Journal of Marketing Research, 36*, 313–326.

Schifferstein, H. N. J. (2001). Effects of product beliefs on product perception and liking. In L. Frewer, E. Risvik, & H. Schifferstein (Eds.), *Food, people and society: A European perspective of consumers' food choices* (pp. 73–96). Berlin: Springer Verlag.

Schifferstein, H. N. J. (2006). The relative importance of sensory modalities in product usage: A study of self-reports. *Acta Psychologica, 121*, 41–64.

Schifferstein, H. N. J. (2009). The drinking experience: Cup or content? *Food Quality and Preference, 20*, 268–276.

Schifferstein, H. N. J. (2011). Multi sensory design. *Paper presented at the DESIRE'11 Conference on Creativity and Innovation in Design, Eindhoven, the Netherlands.*

Schifferstein, H. N. J., & Cleiren, M. P. H. D. (2005). Capturing product experiences: A split-modality approach. *Acta Psychologica, 118*, 293–318.

Schifferstein, H. N. J., Candel, M. J. J. M., & van Trijp, H. C. M. (1998). A comprehensive approach to image research: An illustration for fresh meat products in the Netherlands. *Tijdschrift voor Sociaal-wetenschappelijk onderzoek van de Landbouw, 13*(3), 163–175.

Schifferstein, H. N. J., Fenko, A., Desmet, P. M. A., Labbe, D., & Martin, N. (2013). Influence of package design on the dynamics of multisensory and emotional food experience. *Food Quality and Preference, 27*, 18–25.

Spence, C. (2015). Eating with our ears: Assessing the importance of the sounds of consumption on our perception and enjoyment of multisensory flavour experiences. *Flavour, 4*, 3.

Spence, C., Harrar, V., & Piqueras-Fiszman, B. (2012). Assessing the impact of the tableware and other contextual variables on multisensory flavour perception. *Flavour, 1*, 7.

Spence, C., & Piqueras-Fiszman, B. (2012). The multisensory packaging of beverages. In M. G. Kontominas (Ed.), *Food packaging: Procedures, management and trends* (pp. 187–233). Hauppauge, NY: Nova.

Spence, C., Piqueras-Fiszman, B., Michel, C., & Deroy, O. (2014). Plating manifesto (II): The art and science of plating. *Flavour, 3*, 4.

Spence, C., & Piqueras-Fiszman, B. (2014). *The perfect meal: The multisensory science of food and dining.* Chichester, UK: Wiley-Blackwell.

Spence, C., & Wan, X. (2015). Beverage perception and consumption: The influence of the container on the perception of the contents. *Food Quality and Preference, 39*, 131–140.

Stummerer, S., & Hablesreiter, M. (2013). *Eat design.* Vienna: Metroverlag.

Suk, H.-J., Park, G.-L., & Kim, Y. (2012). Bon appétit! An investigation about the best and worst color combinations of lighting and food. *Journal of Literature and Art Studies, 2*, 559–566.

van Ittersum, K., Pennings, J. M. E., Wansink, B., & van Trijp, H. C. M. (2007). The validity of attribute-importance measurement: A review. *Journal of Business Research, 60*, 1177–1190.

Verbeke, W. (2001). Beliefs, attitude and behaviour towards fresh meat revisited after the Belgian dioxin crisis. *Food Quality and Preference, 12*, 489–498.

Verlegh, P. W. J., Schifferstein, H. N. J., & Wittink, D. R. (2002). Range and number-of-levels effects in derived and stated measures of attribute importance. *Marketing Letters, 13*, 41–52.

Vyas, D., & van der Veer, G.C. (2006). Experience as meaning: Some underlying concepts and implications for design. *Paper presented at the 13th European conference on cognitive ergonomics ECCE '06: Trust and control in complex sociotechnical systems,* New York.

Wansink, B. (1996). Can package size accelerate usage volume? *Journal of Marketing, 60*(3), 1–14.

Wansink, B., & Van Ittersum, K. (2003). Bottoms up! The influence of elongation on pouring and consumption volume. *Journal of Consumer Research, 30*, 455–463.

Wheatley, J. J., & Chiu, J. S. Y. (1977). The effects of price, store image, and product and respondent characteristics on perceptions of quality. *Journal of Marketing Research, 14*, 181–186.

Zellner, D. A., Lankford, M., Ambrose, L., & Locher, P. (2010). Art on the plate: Effect of balance and color on attractiveness of, willingness to try and liking for food. *Food Quality and Preference, 21*, 575–578.

Sensory Branding: Using Brand, Pack, and Product Sensory Characteristics to Deliver a Compelling Brand Message

David M.H. Thomson
MMR Research Worldwide, Wallingford, Oxfordshire, United Kingdom

1 Taking a multisensory approach to branding

We humans interface with the world around us through our senses. The flow of neural activity triggered by objects when they impinge upon our sensory receptors, causes complex representations to come-to-mind. These representations imbue the stimulating objects with identity and meaning and engender affective reactions that influence our immediate and future behavior towards the object and determine how we remember it.

Our relationship with the world around us is fundamentally multisensory, in so far as all of our senses are stimulated simultaneously almost all of the time. Consequently, the representations that come-to-mind are invariably the product of multiple sensory inputs. For example, the flavor perceived when eating something does not arise solely from smell and taste but also from inputs from all of the other senses too. That is why it is relatively easy to modify perceived flavor by coloring particular foods differently using odorless and tasteless food dyes whilst all other factors remain constant (Spence, 2015c). Similarly, the flavor or texture of foods can be intensified when, for example, the diner hears complementary sounds at the moment of consumption. Such demonstrations are sometimes castigated as gimmickry rather than serious science. However, they are actually rather powerful demonstrations of the multisensory nature of all perceptual phenomena.

Although product developers and sensory scientists usually consider foods from a multisensory perspective (appearance, aroma, flavor, texture, and aftertaste), sound is often neglected, even although it has a profound effect on perceived texture and sometimes on flavor as well (Spence, 2015a). Branding is implemented primarily through visual and auditory inputs and sometimes via texture. However, until recently, smell and most especially taste have been all but ignored as inputs to branding. Most packaging developers focus on the form (shape and fundamental nature) and functionality of the pack but give scant attention to the associated sensory characteristics (Spence & Gallace, 2011; Spence & Wang, 2015). All of these are missed opportunities as far as branding is concerned (Spence, 2015b)!

Brands are essentially bundles of conceptual associations that have been created by design or evolved by happenstance. Either way, these conceptual associations deliver emotional outcomes that may be positively or negatively rewarding for the individual (Thomson, 2016). The more positive the conceptual association, the greater the value attached by the individual to the brand and the more it is revered. Knowingly or otherwise, people buy brands for the emotional outcomes (and hence the reward) that they deliver and not just the functional benefits derived from the product.

Sensory branding involves taking a multisensory approach to designing and optimizing brands so that they deliver the optimal and consequently the most valued, motivating and revered combination of emotional outcomes. In practice, this means using all five of the principal sensory channels as "routes" for delivering this optimum emotional impact. More recently, the scope of sensory branding has been widened to embrace the idea that sensory characteristics intrinsic to the associated packaging and product may also be used to reinforce or modify branding (see Spence, in press). This brings with it the notion of brand–pack and brand–product consonance (or dissonance), where the sensory profiles of pack and brand are adjusted or otherwise modified to reinforce or alter the conceptual profile (bundle of conceptual associations) of the brand, so as to maximize delivery of rewarding emotional outcomes. Sometimes, through lack of awareness or neglect, the sensory profiles of pack and product may end-up failing to reinforce or even detracting from the brand. This effect is known as brand–packaging or brand–product dissonance. Failure to optimize the sensory characteristics of pack and product as a means of reinforcing branding constitutes a significant loss of opportunity. However, to allow packaging and product, by virtue of their sensory profiles, to detract from the brand would be nothing short of negligence. In today's crowded marketplace, there is no room for either of these shortcomings. To stand out from the crowd, products require powerfully motivating brands. Anything less and the product will be lost in a sea of mediocrity, only catching the attention of consumers when discounted, with all that this implies for profitability. When launching a new product, it is essential that brand–pack and brand–product dissonance should, at the very least, be minimized. Ideally, all three components should be highly consonant. This really matters because new products generally face the challenge of displacing existing products from consumers' repertoires, either by being functionally better in some way or other or by delivering a more rewarding spectrum of emotional outcomes. Because this is a significant challenge, it is possibly one of the key reasons why most new products and almost all new brands fail (Nielsen, 2014a,b). One of the principal causes is failure to minimize brand–pack–product dissonance and maximize consonance. Sensory branding provides a means of addressing this.

Several books have been written on what amounts to sensory branding. Martin Lindstrom's *Brand Sense: Sensory Secrets Behind the Stuff We Buy* (Lindstrom, 2005) was one of the first. Lindstrom famously describes how Singapore Airlines apparently used a particular fragrance (Stefan Floridian Waters) across every aspect of the passenger experience (cabin and lounge atmosphere, wipes, soaps, and personal fragrances), so that it became as much a part of the airline's brand signature as its logo, its livery, and the fabric patterns used on its soft furnishings and staff attire. Although certainly not his original idea, Lindstrom deserves much credit for bringing sensory branding to the attention of the wider marketing and sensory communities. And where Lindstrom led the way, many, many others have followed (see Hultén, 2011; Hultén et al., 2009; Krishna, 2010, 2012, 2013).

Interesting as the Singapore Airlines example might be, such anecdotes are largely avoided in this chapter. Instead, emphasis is placed on the science of sensory branding, most particularly in exploring the mechanisms through which sensory characteristics impact on branding. This requires us to delve into and otherwise consider the fundamental nature of sensory characteristics, pleasantness, and conceptual association. The *Duality of Reward Hypothesis* is introduced as a model for understanding how reward motivates human behavior and how sensory branding may deliver reward via the mechanisms proposed within this hypothesis.

2 A quick word about mental constructions

One of the first things to establish when working in this area of science is that there are few facts, at least as we normally think of them. Indeed we cannot be definite about the nature of any of the mental representations mentioned herein, or even that they should exist at all. This might seem surprising to those amongst us who imagine that we exist in a world of neatly ordered certainties but the functioning of the mind holds at least as many uncertainties as quantum physics and possibly more.

The phenomenological uncertainties associated with the mind and with mental processing led to "behaviorism"; a philosophy championed by Watson (1925) and later by Skinner (1976) amongst others. Behaviorism acknowledges the outward aspects of stimuli and the consequent behavior but dismisses the inward experiential aspects. Towards the end of the 20th century, the pendulum swung towards the opposite extreme with the advent of "cognitive psychology" (Posner, 1978); an alternative to behaviorism that gives cognizance to the notion of the mind, thought, and internal mental processes. From my perspective, it is important to draw the reader's attention to these two schools of psychology, and to behaviorism in particular, if only to make the point that much of what follows in this chapter should be construed as hypotheses and reasoned opinion rather than fact. Bearing this profound caveat in mind, let us now consider the various types of representation that seem to come-to-mind when humans interact with objects.

3 Sensory characteristics

Laymen most commonly allude to sensory characteristics in terms of the "taste" of foods and beverages or the "smell" and "feel" associated with homecare and personal care products and so forth. Within normal consumption or usage scenarios most people typically give little or no immediate thought to the sensory characteristics of a product unless, of course, it deviates from their expectation; for example, the chunk of chocolate that they've just slipped into their mouth is more or less bitter than anticipated, it's softer or harder than expected, or it delivers more or less immediate pleasure than they thought it would do. Such deviations make us take notice and cause us to be more analytical. Such disconfirmed expectations are rarely positively valenced, either (Piqueras-Fiszman & Spence, 2015).

It is envisaged that the capacity to perceive sensory characteristics may have evolved to allow animals to identify substances likely to be either harmful or beneficial. Although this faculty would probably have been fairly crude at first, those members of a species who were better at identifying the harmful from amongst the beneficial (and vice versa) would have been more likely to survive and so evolution would probably have selected on this basis. By way of example, some of the most toxic substances that occur in nature happen to contain substances (alkaloids) that trigger the perception of bitterness. Whilst this may have arisen by coincidence, those animals that developed the faculties that allow them to detect these compounds, to perceive bitterness and then to experience unpleasantness would have had a better chance (increased probability) of survival. Similarly with acidity and the link between perceived sourness and the experience of unpleasantness. Acid is, of course, a product of bacterial degradation of foods, with all that this implies. Conversely, sodium chloride triggers the perception of saltiness. Salt is necessary for osmotic regulation and sodium ions (Na^+) are fundamental to the generation of action potentials in neurons and sensory receptor cells. Those animals that developed the capacity to detect and perceive saltiness and to experience pleasantness in so doing, would again have had a better chance of survival. The same principle applies to sugar and sweetness. Sugar has two very important properties: It happens to trigger the perception of sweetness and it is also a safe and readily accessible source of energy for metabolic function. This includes fueling the fight-or-flight response which obviously facilitates survival in the face of danger. In spite of recent bad publicity, sugar undeniably plays a fundamental role in animal metabolism. We're attracted to sweetness because, when perceived at particular levels, it delivers pleasure (which we perceive because it often provides needed calories) (see chapters: Flavor Liking and Pleasure of food in the brain). Experiencing pleasure is rewarding so we're motivated to consume sweet-tasting foods and drinks in order to obtain the associated reward.

Germane to this is the capacity to experience some degree of pleasantness or unpleasantness with particular sensory characteristics. Experiencing pleasantness or unpleasantness triggers certain behaviors (ie, expectoration or continued consumption in the foregoing examples). This alerts us to the possibility that pleasantness or unpleasantness, when experienced fairly immediately upon interaction with a stimulus, may have evolved as a behavior-controlling mechanism.

4 Pleasantness and reward

As mentioned above, some animals evolved the capacity to create other types of mental constructions, experienced consciously as degrees of pleasantness or unpleasantness, and to associate these so-called "affective constructions" with particular sensory characteristics. It is the capacity to experience some degree of pleasantness or unpleasantness simultaneously with, or possibly even ahead of (Zajonc, 1980), the associated sensory characteristics that determines the degree to which an individual is attracted to or repulsed by the stimulating object in question.

There are many other examples of human behaviors that are mediated by sensory pleasure. Pain is very directly associated with tissue damage. It goes without saying that pain is generally unpleasant so we're motivated to withdraw from or otherwise deal with the cause of pain immediately. As a consequence, we generally avoid further tissue damage. Pleasure also influences some other primal behaviors that are fundamental to the continuance of the species, such as motivating us to participate in sexual intercourse, motivating us to drink when thirsty, motivating us to eat when hungry, and motivating us to stop eating when satiated (Hetherington & Havermans, 2013).

In all of the foregoing examples, pleasure or displeasure is experienced immediately upon sensory stimulation. It exerts an instantaneous, unsubtle, real-time influence upon us, thereby motivating us to behave in particular ways and crucially, to do so right now! Finer and more subtle control may occur via alliesthesia, defined as the dependence of the experience of pleasure or displeasure on the internal milieu of the organism. Consequently, a stimulus capable of ameliorating the state of the internal milieu may be perceived as pleasant whereas a stimulus that upsets the internal milieu may be unpleasant (Cabanac, 2009). Alliesthesia is thought to be involved in the fine control of appetite.

Since deriving pleasure and minimizing displeasure is rewarding, it would seem most apt to describe this form of sensorially triggered, immediate reward as *Reward via Immediate Sensory Pleasure* (Thomson, 2016). The principle of *Reward via immediate sensory pleasure* is summarized schematically in the top part of Fig. 16.1.

Presumably, evolution would have selected on the basis of this reward mechanism because the behavioral consequences it motivates made us more successful in terms of living long enough and being healthy enough to procreate and thereafter surviving long enough until our offspring also reach sexual maturity, so that the cycle of life may be repeated (Darwin, 1859).

Reward is difficult to define phenomenologically but it does have some fairly obvious features. For example, it's generally recognized that reward is something that happens in the mind. However, any relationships that may exist between the psychological phenomenon and the corresponding physiological phenomena are likely to be subtle and complex (Kringelbach & Berridge, 2009).

Reward may be positive or negative. Positive reward motivates us to begin and to continue doing whatever it is that's triggering the reward. It also encourages us, through learning, to repeat this behavior at the next appropriate opportunity. Negative reward motivates us to attenuate or stop the triggering behavior and to avoid repeating it. The greater the reward the greater the motivation to either start, continue, attenuate or stop the reward-triggering behavior. This leads to the rather more profound generalization that behavior is driven by reward.

The straightforwardness of the relationship between sensory characteristics and pleasure is readily apparent, especially with foods and beverages. One of the key challenges often handed down to product developers is to reformulate existing products so that they are "liked more" or to develop new products that exceed a specified "liking threshold." "Liked more" in this context usually means significantly preferred when the revised and original formulations are compared directly in a consumer test or the

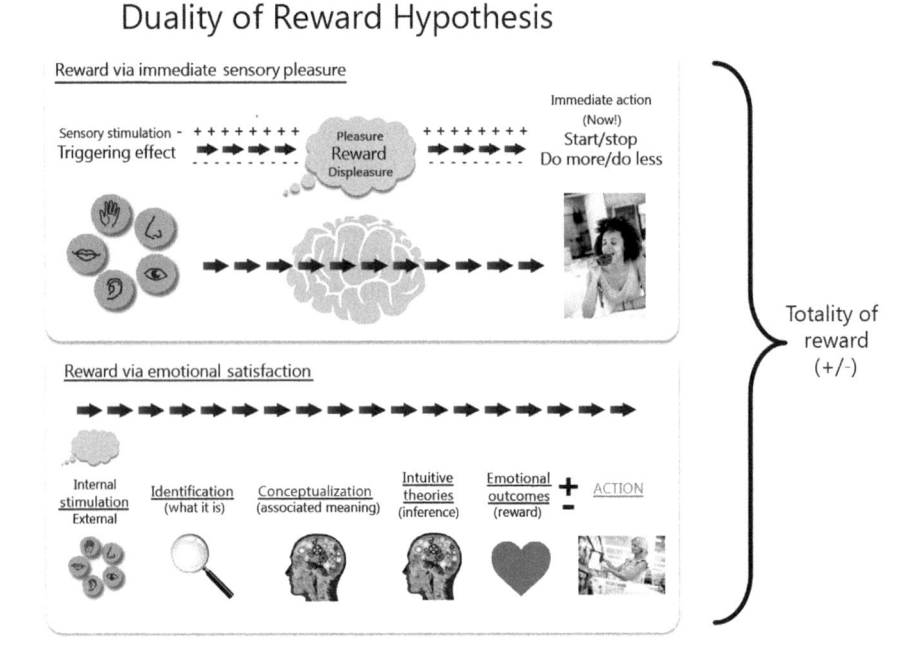

Figure 16.1 Schematic representation of the "Duality of Reward Hypothesis."

former achieves a significantly higher mean liking rating on a hedonic scale than the latter. Exceeding a specified "liking threshold" usually means that the mean liking rating achieved by the new product exceeds some specified (but often rather arbitrary) minimum, known as an action standard. The process through which this is achieved is called sensory optimization (or sensory-hedonic optimization). In practice, this usually means that the product developer will manipulate the sensory profile of the product by changing the ingredients and/or the process, in such a way that the consequent experience of "liking" is as great as possible for the largest possible proportion of target consumers.

Over the past 20 years or so, research tools have been developed which facilitate this outcome with astonishing reliability (although I am personally aware of many commercial examples that demonstrate this unequivocally, none unfortunately can possibly be published). This is indicative perhaps of the straightforwardness of the relationship between sensory characteristics and *"Reward via immediate sensory pleasure."* However, there's a problem! Developing products that are highly "liked" may achieve success in prelaunch preference tests, they may exceed a specified action standard and thereby trigger launch into the market and they may be enjoyed at first. However, in spite of this, it transpires that "liking" is actually a rather poor predictor of longer-term product adoption.

This should not be surprising! Anecdotally, we don't always buy what we like nor do we always like what we buy. If we revert to the general view that behavior (long-term purchase and consumption behavior in this case) is driven by reward, then this suggests that the form or aspect of reward captured when we establish product preference or measure liking in a product test is different in some way or other from the form of reward that motivates longer-term purchase and consumption behavior. This doesn't mean that conducting preference tests or measuring liking of branded or unbranded products in consumer research is invalid but it does suggest that this form of research may be failing to capture certain aspects of reward that are crucial in motivating longer-term affiliation to a product. In some respects, instantaneous product liking is analogous to the immediate sexual attraction that might drive infatuation whereas long-term product adoption is more akin to a deep, meaningful, loving and lasting relationship. Whilst sexual attraction may have a significant role in a lasting relationship, it requires a lot more than the pleasure of sex to bond two people together in the longer term. The same seems to be true of our relationship with products.

Working on the premise that preference as captured in a product test and liking as measured on an hedonic scale are both likely to embrace *Reward via immediate sensory pleasure*, this leads to the inescapable conclusion that longer-term consumption behavior may be motivated by additional aspects of reward that may be phenomenologically discrete from *Reward via immediate sensory pleasure*. This is an important realization because it leads to the further notion that the sensory characteristics of a product may also be involved in delivering these other aspects of reward. It's the relationship between sensory characteristics and these other aspects of reward that is the basis of sensory branding. In order to explore these other aspects of reward and through doing so, the fundamental scientific basis of sensory branding, we need to think more expansively about the other types of representation that may come to mind when people interact with everyday things such as brands, products, packaging, "objects" in general, places, people, events, and so forth.

5 Conceptual associations

All things carry with them what might be described as associated "conceptual meaning." The nature of this associated meaning can be readily understood by imagining a square panel of purple color (for example), printed on a sheet of paper or displayed on a screen. Depending on the resolution of the printer or the screen, this colored panel might be characterized using descriptors such as purpleness, blueness, redness, lightness/darkness, roughness/smoothness, glossiness/mattness, etc. These are the sensory characteristics of the colored panel and collectively these sensory characteristics constitute the sensory profile of this purple panel. Most lay people would have no difficulty in identifying and describing these sensory characteristics with little or no prompting. If the same people were asked to elaborate on what the color purple meant to them, what it made them think about or how it made them feel, most would struggle to understand the question and why it's being asked, never mind answer it! However, it is readily determined that the color purple in some cultures is associated with concepts such as confident, modern, genuine, powerful, classy, sophisticated, etc. (Gains, 2013; Thomson, 2016) and that this associated meaning determines how we feel about things that are colored purple and what they mean to us. These associations are otherwise known as latent associations (Stafford-Clark, 1965), implicit associations (Greenwald, Klinger, & Schuh, 1995) or, in the author's preferred terminology, conceptual associations (Thomson, 2010).

By analogy with sensory profiling, the various conceptualizations associated with an object constitute its conceptual profile (see Thomson & Crocker, 2015, for a brief review). In effect, this means that the panel of purple color mentioned previously has both a sensory profile and a conceptual profile. Indeed, all "things," including brands, products, packaging, "objects" in general, places, people, events, etc., have a sensory profile and a conceptual profile and both determine how the "thing" in question seems to us, what it means to us and how it influences us. For example, in encountering someone for the first time, we're influenced immediately by how they look (height, body shape, face shape, skin, eye and hair color, etc.), the sound of their voice and even how they smell (fragrance worn or body odor), as well as the sensory characteristics of their clothes, shoes, spectacles, and so forth. Knowingly or otherwise, we instantaneously form an opinion about this individual based on the conceptualizations implicitly associated with these sensory characteristics and this determines how we feel about them and how we might react to them in the first place. We may, of course, modify our opinion of the individual should we get to know them rather better but those initial impressions, formed on the basis of sensory-driven conceptual associations, can sometimes be difficult to change. Unfortunately, such conceptual associations are also the basis of rather uglier human traits such as bias and prejudice (Greenwald et al., 1995).

Referring back to the color purple for a moment, it's evident from the example presented above that there seems to be a direct causal relationship between sensory characteristics and conceptual associations. Empirical evidence has been presented elsewhere (Thomson, Crocker, & Marketo, 2010) to demonstrate that modest variations in the sensory profiles of products within a particular category (eg, milk chocolate) can have a direct influence on the conceptual profile of the object in question. This naturally extends to the notion that sensory characteristics per se also have conceptual associations. Anecdotally, bitterness has sometimes been associated with masculinity, aggression, and harshness, whereas sweetness might well be associated with care, nurture, and even love. This implies that variations in the sensory characteristics of any object could lead to changes in the associated conceptual profiles. This hypothesis is germane to the fundamental scientific basis of sensory branding. However, in order to develop this idea, we must first consider the mechanism through which conceptual associations might influence behavior.

6 Linking conceptualization to reward

A possible mechanism through which conceptualizations trigger reward and thus behavior is proposed in the lower part of Fig. 16.1. Fundamental to this mechanism is the idea that conceptual associations trigger emotional outcomes and that these emotional outcomes may be positively or negatively rewarding to some extent, as determined by the nature of the emotional outcome, the psyche of the individual concerned, the individual's lifetime of learned experiences, their mental and physical state at the moment of triggering, and the context in which the triggering event occurs.

Triggering events may be "internal" or "external." Internal triggering is where we recall things from memory, such as laughing and joking with a particular friend, which leads to happiness (emotional outcome). Recall may be deliberate (via cognitive thought processing) or the memory may come-to-mind automatically. External triggering occurs when we interact with "objects" (eg, brands, products, advertisements, books, music, art, colors, etc.), with other living things (eg, people, pets, etc.) or when we participate in "events" (eg, social gatherings, work, and other routine aspects of daily life), all of which can lead to an ever-changing array of emotional outcomes (and consequently, an ever-changing reward profile). This, in turn, can modify our behavior on a moment-by-moment basis.

Some aspects of emotional outcome may be experienced there and then, as feelings of happiness, eagerness or contentment, for example, or as feelings of sadness, anxiety, uncertainty, etc. Such obvious feelings may be immediately and obviously rewarding and consequently, may have an immediate and obvious influence upon our behavior. Other aspects of reward may not be experienced directly but indirectly as an urge to start, continue or stop doing something without necessarily knowing why. Whether apparent or otherwise, reward might also impact sometime after the triggering event, experienced perhaps as an inexplicable glow of satisfaction or as unattributable pangs of guilt or self-doubt. Feedback mechanisms allow the various outcomes to be factored into subsequent reward predictions. As emotional outcome is the proposed vehicle of reward in such cases, the author has labeled this aspect of reward *Reward via emotional outcome*.

Since *Reward via emotional outcome* is characteristic of the most highly evolved animals, especially humans, and *Reward via immediate sensory pleasure* is more widespread across the lower orders, it's tempting to assume that the latter would have evolved earlier than the former. However, even if this were the case, the two are hardly independent. If something delivers great *Reward via immediate sensory pleasure*, the chances are that it may also deliver *Reward via emotional outcome* and also that one may augment the other (just think about eating chocolate). Conversely, few people experience much in the way of immediate sensory pleasure when they taste beer for the first time. However, beer has the capacity to deliver positive emotional outcomes because of the social context in which it is normally consumed and also because of the intoxicating effects of alcohol. Consequently, many of us eventually learn to derive pleasure from the sensory characteristics of beer! There may also be tensions between the two aspects of reward. For example, we may be induced to eat chocolate, because the sensory characteristics deliver positive *Reward via immediate sensory pleasure* but we are finally dissuaded from doing so by the prospect of negative *Reward via emotional outcome*, driven by longer-term concerns about obesity or diabetes. There may also be tensions within *Reward via emotional outcome*. Someone may seek the calming and relaxing effect that eating chocolate can deliver but this may be in conflict with the negativity associated with others' perceptions of indulgence. The consequences of this offsetting of reward will differ across individuals, depending on

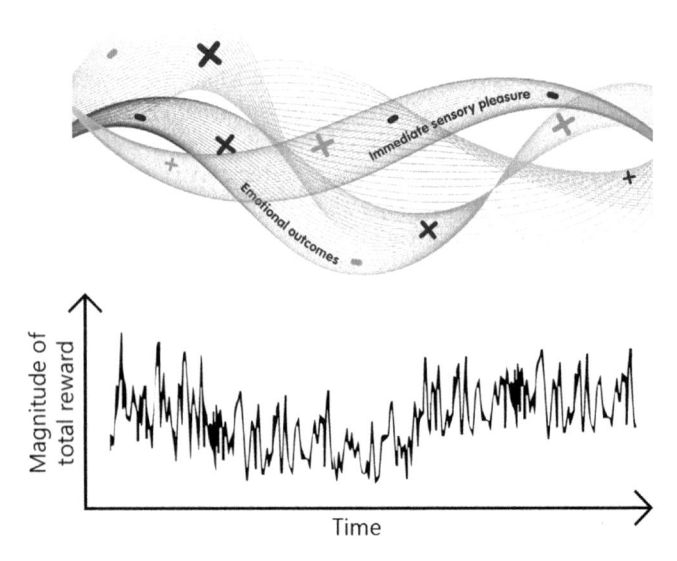

Figure 16.2 Schematic representing constant integration of Reward via immediate sensory pleasure and Reward via emotional outcome to yield Totality of Reward. Based on the "Duality of Reward Hypothesis."

the relative potency of the various aspects of reward to that individual. It may also differ within the same individual depending on the context in which the offsetting takes place so we may, for example, choose to eat chocolate in one set of circumstances but an apple or a banana in other seemingly similar circumstances.

It is proposed that all aspects of *Reward via immediate sensory pleasure* and *Reward via emotional outcome*, whether experienced or not, whether immediate or subsequent and whether positive or negative, integrate to deliver *Totality of reward* and it's this integral that finally influences our behavior (Fig. 16.2). The unremitting process of integration, much of which will occur below the level of conscious awareness (Ellis, 1995; Ellis & Newton, 2010), leads to constant fluctuations in *Totality of reward* and hence in motivation, as the events of life impinge upon us on a moment by moment basis. Since the integration of two aspects of reward is proposed, this theory is referred to (by the author, at least) as the *Duality of Reward Hypothesis* (Thomson, 2016).

Totality of reward is a hypothetical construction in so far as the theory proposes that integration of the two aspects of reward should occur below the level of conscious awareness and that the integral is never actually experienced. This implies that it is fundamentally impossible for the individual concerned to access the *Totality of reward* by thinking about it, with all that this implies for market research and brand, pack, and product development. However, postulating that sensory characteristics drive conceptual associations which trigger the emotional outcomes that deliver reward that finally influences behavior, gives an initial clue as to how sensory branding might function. To complete this understanding, it's necessary to first consider the fundamental nature of brands and branding.

7 The fundamental nature of brands

As mentioned previously, brands are essentially bundles of conceptual associations. The qualitative nature of these conceptualizations and their strength of association with the host brand, are the essence of the brand's conceptual profile. This raises the question as to how particular conceptualizations come to be associated with a particular brand. To understand this, we need to go back to the origins of branding.

The word "brand" derives from the practice of burning a mark into the hide of cattle or horses, to identify the animal as belonging to an individual, ranch, or business. This process was (and sometimes still is) conducted using a red hot branding iron. The mark seared into the animal's hide is the "brand mark" or simply the "brand." In today's marketing parlance this might be thought of as a brand logo.

Around the middle of the 19th century, when grocery retailing began to develop, shop owners would often label their in-house products with the name of their business so that consumers could relate the quality of the product (most particularly the claim that it was "pure" and not adulterated) to its source, in the hope of building confidence in the fidelity of the product and thereby encouraging repeat purchasing. This was certainly the case with John Walker & Sons of Kilmarnock, Scotland, whose "house" blends of Scotch whisky were much appreciated by the local populace (Hughes, 2005). The name "John Walker & Sons" conferred the product with an identity in the same way as branding cattle. It wasn't very long before people were asking for "Walker's Kilmarnock Whiskies" by name and so the brand began to evolve. In the first place, "Walker's Kilmarnock Whiskies" was largely associated with fidelity and consistent delivery of a particular style and quality of Scotch whisky. This distinguished them from other lesser Scotch whiskies of more dubious quality and less pleasing taste. Extolling these particular virtues was the general theme of John Walker & Sons' marketing strategy for many decades to come and so the brand grew in popularity across Scotland, across the United Kingdom, across the (then) British Empire and across the United States. A truly international brand was in gestation!

In 1908 John Walker & Sons introduced the now iconic striding man as their logo and rebranded their whiskies as Johnnie Walker. By today's standards, linking the name "Walker" to a walking figure might seem somewhat simplistic and naïve but these were less sophisticated times and it proved to be a masterstroke! In adopting the new logo and brand name, Johnnie Walker was effectively grafting the style, personality and other human-like traits of the striding man onto their brand. In the author's terminology, the conceptualizations associated with the striding man became subsumed into the Johnnie Walker brand and eventually came to define it, at least in part. Over the ensuing century and more, Johnnie Walker and its successors (now Diageo) have developed the look, style, and meaning of the striding man with great success. At the time of writing, Johnnie Walker is the most widely distributed and largest-selling brand of Scotch whisky and the second largest whisky brand in the World.

Johnnie Walker's striding man is one of the very first and best examples of personality branding. Although personality branding is still very relevant, it has become recognized that considering brands solely in terms of human-like personality traits may be constraining and much too narrow. By design or by default, the character of most successful brands usually extends well beyond human-like traits to encompass an array of other emotional, abstract and functional conceptualizations. This led to the notion that brands have conceptual profiles (Thomson & Crocker, 2015) rather than personality profiles (Aaker, 1997).

Other conceptual associations may arise, at least in part, by virtue of the words, slogans, straplines, narratives, and other verbal devices that marketers and advertisers use to tell us what their products can do for us (functional branding) and most particularly these days, how they'll make us feel (emotional branding). Johnnie Walker initially used "Born 1820—Still going Strong!" Now the message is "Keep walking." Such messages may be direct and obvious or subliminal. We may choose to (or try to) ignore or otherwise discount them but well-crafted marketing and advertising messages can have an insidious effect upon us, making us think in particular ways about brands and products without us necessarily realizing it, and thereby imbuing the brand with additional conceptual associations and reinforcing or otherwise modifying those conceptualizations that exist already.

The next section describes how multisensory branding extends branding beyond narratives and personalities to include the full panoply of mental representation engendered by sensory inputs from the brand per se but also from the associated packaging and product.

8 Extending sensory branding to pack and product

As originally conceived by Lindstrom (2005) amongst others, taking a multisensory approach extends branding beyond vision and sound to embrace supplementary inputs from touch, smell, and taste (Spence & Gallace, 2011). Assuming, of course, that the conceptualizations associated with these supplementary sensory characteristics are consonant with the conceptual profile of the brand, sensory branding should augment the brand so that it delivers the desired emotional outcomes more forcefully, thereby rendering interaction with the brand more rewarding for those who are engaged by its message.

More recently, the concept of sensory branding has been extended to reinforcement of the brand message via the sensory characteristics of the associated packaging and product. Although we're more used to thinking about foods and beverages in terms of their sensory characteristics, as with all "objects," they also have associated conceptual profiles. The conceptual profile of an unbranded product (ie, a product devoid of its associated branding) arises from two influences (Thomson & Crocker, 2015; Thomson, 2016): *Category Effect* – which relates to the fundamental nature of the product (ie, milk chocolate, Scotch whisky, air freshener, shampoo, skin cream, etc.). *Sensory-Specific Effect* – which derives from the sensory characteristics that define a particular product and differentiate it from other similar products within the same category (eg, the sensory characteristics that differentiate Cadbury, Galaxy/Dove, Hershey, Suchard, etc., milk chocolates). Relatively small sensory differences can sometimes have quite a profound influence upon the associated conceptual profile of the product (Thomson et al., 2010), and consequently upon the brand. This raises some very interesting questions about sensory optimization. Should products be optimized to deliver maximum *reward via immediate sensory pleasure*, as is the norm currently, or should they be optimized to maximize overlap of brand and product conceptual profiles (known as brand–product consonance), so as to reinforce the emotional outcomes implicit in the brand message? Bearing in mind that one of the primary purposes of branding is to deliver emotional outcomes that are rewarding for those people who are amenable to the brand message, it could be argued that in the case of revered brands the latter optimization strategy is likely to be the more profitable route in every sense. Often as not, the two optima will be markedly different. It's understandably difficult for people involved in product optimization to accept that it might be better in the longer term to aim for a lower mean liking rating in a product test in order to garner a greater degree of *Reward via emotional outcome*, especially since the latter is fundamentally unmeasurable. However, if the success rate of new product development (NPD) is to improve, radical "out-of-the-box" thinking is essential. Brand–product dissonance, caused by optimization of products for immediate pleasure rather than longer-term emotional outcomes, is considered by this author at least, to be one of the primary causes of new product failure.

Essentially the same rationale applies to the nature or form of the packaging (rather than what's printed upon it). The fundamental nature of the pack (ie, whether bottle, can or pouch, etc.) is analogous to the "category effect" mentioned above. Other factors such as the aspect ratio (short and stubby versus long and narrow), shape (curvy versus sharp and angular), feel (squashy or firm when squeezed, hard or soft to touch, smooth or textured finish), opacity (transparent, translucent or opaque), look (shiny or glossy), color, and the sound it makes when handled (eg, the different sounds made by foil, greaseproof paper, cellophane, etc.), effectively constitute the sensory profile of the pack (Spence, in press). Consequently, the conceptual profile of the pack is determined jointly by the nature or form of the pack (category effect) and its particular sensory characteristics (sensory-specific effect).

Packaging plays an interesting role in the mix. Rationally, it might be imagined that packaging should be optimally functional in every respect, including visual impact and on-shelf standout. However, packaging must also act in support of the brand and otherwise, it must also be consonant with the conceptual profile of the product category. Sometimes this means a trade-off between functionality and aesthetics. For example, consumers who buy into the premium quality wine and Single Malt Scotch whisky (SMSW) categories, generally expect that the closure should be a cork rather than a screw cap, even although the latter is much more practical. This expectation arises because the conceptualizations associated with a screw cap are, for whatever reason, dissonant with the conceptual profiles of premium wines and Single Malts. Consequently, the conceptual profile of the pack, as determined by its form and its sensory profile, should be "designed" to overlap with the conceptual profile of the brand to the greatest possible extent. The word "design" is used very deliberately in this regard, as a means of introducing the idea of "designed conceptual profiles" for packaging. Designers tend to think of packaging in terms of physical form but the design focus should arguably be on the conceptual profile, with the form of the pack and the other physical elements from which the pack's sensory profile arises, being the means through which the "designed conceptual profile" is achieved.

Conducting conceptual profiling on design prototypes, to determine whether or not the design strategy is delivering the desired conceptual profile is simple, reliable and relatively inexpensive (Thomson & Crocker, 2015). Achieving maximum alignment between brand and pack conceptual profiles is highly desirable because the pack conceptual profile will augment, reinforce and otherwise support the brand message. This is known as brand–pack conceptual consonance. When the pack conceptual profile is misaligned with that of the brand (brand–pack conceptual dissonance) it undermines and otherwise dilutes the brand message, usually to the detriment of the branded product as a whole. As with product, there are tensions between two different optimization objectives (optimum functionality versus optimum brand–pack consonance) and it can often be very difficult to reconcile the two.

9 "The Matrix"

Product developers are, in effect, governed by the needs of three different "masters"; creating maximum functionality, achieving the optimum array of emotional outcomes to deliver maximum reward, and maximizing liking. These three requirements should be delivered across the three primary elements of the product; that is, the branding, the packaging, and the product. This idea is captured in "the matrix" (see Fig. 16.3).

The primary purpose of "the matrix" (Thomson, 2010) is to highlight to branding, packaging, and product developers, the necessity to simultaneously consider the needs of their "three masters" in finding an optimum. It also serves to remind them that the final product (ie, the branded entity) will have the greatest possible chance of longer-term success when the conceptual profiles of pack and product are consonant with that of the brand ("fit-to-brand").

It makes no sense whatsoever that the brand, the packaging, and the product should be developed separately from each other, as is often the case currently. A more rational approach would be to identify the key conceptual elements of the brand (assuming of course that the brand has primacy) using an agreed and prioritized lexicon of emotional and abstract conceptual terms (effectively the conceptual profile of the brand or the new product concept) and issue this as a brief to the packaging and product developers. In the case of packaging, this should be designed and developed to deliver the best possible conceptual overlap with the brand, whilst minimizing functional compromise (in that order of priority). Likewise with the product, it too should be optimized so that its conceptual profile has maximum overlap with the conceptual profile of the brand whilst minimizing the impact in terms of measured liking.

There may be other circumstances where the conceptual profile of the product should have priority over that of the brand. This may occur when the "taste" of the product is very distinctive and instantly recognizable. Products such as Cadbury's Dairy Milk, Red Bull, Marmite, and Guinness come to mind immediately, although there are many more worldwide. In each case, the "taste" (sensory profile) of the

Matrix – template for branded product diagnostics

Anticipated functionality = based on functional conceptualizations which may/may not reflect actual functionality

Emotional outcomes = emotional outcomes arising from functional, emotional and abstract conceptualizations

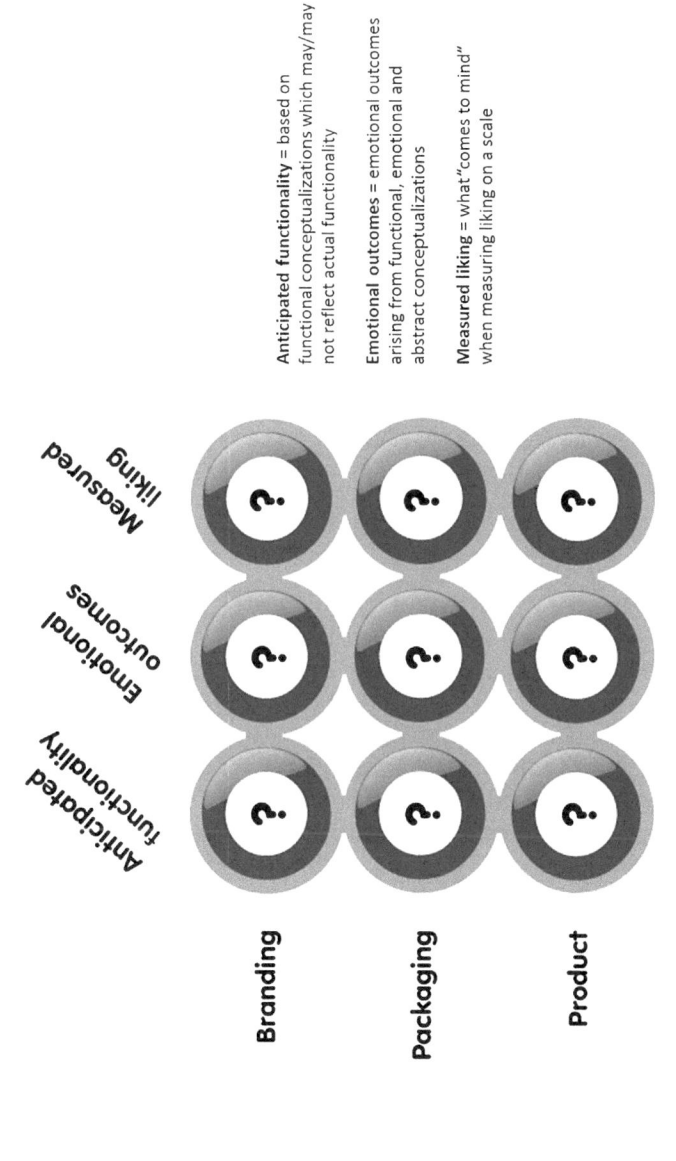

Measured liking = what "comes to mind" when measuring liking on a scale

Figure 16.3 "The Matrix" brand evaluation tool as used to explore consonance across the primary elements of branded goods.
Source: © MMR Research Worldwide Ltd. 2016.

product and the brand are as one because the brand is instantaneously recognized from the "taste." Since the sensory profile of the product is immutable with such products, by virtue of the rationale put forward in this chapter, the conceptual profile of the product (which arises from its sensory profile) rather than from the brand, becomes defining. In practice, this means that advertising and all other aspects of marketing should focus on delivering a brand conceptual profile that is consonant with that of the product. Guinness' association with rugby union and its legendary *surfer* advertising campaign are classic executions of this particular strategy.

Make no mistake, aligning branding, packaging, and product across the three touch-points of functional promise, emotional outcome, and liking, can be very challenging. Again, this is the sort of radical "out-of-the-box" thinking that's required to break the cycle of failure in NPD. The first step is to recognize the need to conduct "matrix analyses" and to factor this into brand, packaging, and product development briefs. The second step for those responsible for brand health and development is to use the matrix as a means of deconstructing and analyzing their branded goods. Try to write a short narrative for each of the nine cells in the matrix for a brand that you're either responsible for or simply a brand that you interact with regularly! This is an exercise that the author conducts every year with final-year undergraduate students at the University of Reading in the UK. The shear depth of analyses that can be achieved via the matrix, even with students who have no past experience with brand analyses, never ceases to surprise me.

10 Sensory signatures

Although the author professes no particular knowledge of Cadbury's Dairy Milk, it's quite well-known across the industry that any product that bears this particular brand name, whether it's chocolate confectionery, hot or cold drinks, cakes or ice cream, must have what might be described as *Cadbury-ness*; that is, the key sensory characteristics that define Cadbury's Dairy Milk. This is a first class example of a "sensory signature," where the brand is immediately and unmistakeably recognizable from the "taste" of the product. Kettle Chips is another good example, where the tough but fracture-able texture and the oily look and mouthfeel are immediately recognizable as Kettle. Again, the author has no particular knowledge of Kettle's product development strategy but it does seem highly unlikely that such a sensory profile would have garnered best-in-category liking ratings, at least in the early life of the product, before the brand became established and the sensory signature became inextricably associated with the brand. However, the success of the brand worldwide speaks volumes for this strategy.

When Red Bull was first launched, many industry commentators questioned the rationale of launching a product with such an unusual and highly polarizing sensory profile. It seems unlikely that the liquid would have performed well in any unbranded carbonated beverage benchmarking test! However, this rather unique sensory profile made several very bold proclamations on behalf of Red Bull: "I'm different from the others!," "I'm my own man/woman!," "I'm not fruity and I'm not a cola—I'm something completely different!," "I'm like nothing that you've tried before!," "I'm different and because of that I'll do something different for you or to you!," "I'm not for kids!," "You may not like me but that's your problem!," "You may not like me but you want me!," and so on. All of these proclamations fitted very well with other aspects of the brand message and the target market, especially in the early years when the brand was more closely associated with a somewhat irreverent rave culture (Buchholz, 2008). These were (and still are) emotional outcomes that, knowingly or otherwise, Red Bull's target consumers seek. Nobody should be in any doubt that the sensory profile of the drink is the sensory signature of the Red Bull brand. Whether by luck or happenstance, the success of this strategy is unquestionable. Red Bull has since strayed into cola (with limited success) and more recently into exotic flavors. Although the rationale for doing so is easy to imagine, delivering completely different flavors devoid of the Red Bull sensory signature, could diminish the brand message. Whilst the marketers at Red Bull have a tremendous reputation for innovation and success, only time will tell whether or not they've got it right on this occasion.

There are many examples across the laundry, homecare, and personal care product categories of fragrance-based sensory signatures. However, rather surprisingly, this asset is seriously underutilized across the food and beverage categories. Giving a product or a range of products a distinct and recognizable sensory signature and making sure that the associated conceptual profile is aligned with that of the brand, is a very potent branding tool.

11 Consonance and "fit-to-brand"

Direct ratings of "fit-to-brand," where the respondent is asked to quantify the degree of fit between product and concept (or brand), often form part of standardized concept evaluation protocols. The object, of course, is to gauge consonance between the conceptual profiles of these two key elements. Somewhat surprisingly, this idea rarely extends to fit between the form of packaging and brand, in spite of the fact that the visual characteristics of the pack form are often perceived in advance of other elements of the branded entity, thereby creating an initial impression that may be difficult to modify subsequently.

Whilst the aim of measuring "fit-to-brand" is laudable, it's questionable whether or not research participants have the capacity to make this particular determination, since so much of what influences "fit" is likely to occur below the level of conscious awareness. This means that it's doubtful whether or not a high-fidelity representation of "fit-to-brand" could actually come-to-mind when research participants are asked to quantify it directly, thereby rendering the metric of questionable validity. Moreover, this metric is often highly correlated with liking ratings, suggesting that whatever it is that comes to mind may be confounded by liking.

Doubts about direct measurement led to the idea of a derived index of "fit-to-brand" obtained by quantifying consonance between the conceptual profile of the brand and that of the corresponding product. As discussed above, all "objects" have a conceptual profile (and a sensory profile) and this extends to branding, packaging, and product. By capturing and quantifying their conceptual profiles using an appropriate lexicon of abstract and emotion-related conceptual terms, it's possible to plot brand versus pack or brand versus product, calculate the correlation coefficient and use this as a derived index of fit-to-brand. (This process has been implemented very successfully by MMR Research Worldwide on behalf of many of its clients.)

Practical implementation of conceptual profiling is described in detail by Thomson and Crocker (2015). The example presented below of a derived index of "fit-to-brand" for two brands of SMSW, is extracted in brief from Thomson (2015). A short description of the SMSW category may be helpful at this stage.

"Scotch whisky" is a geographical indication that describes a particular and very popular style of whisk(e)y that can only be made and matured in Scotland. "Single Malt" alludes to one of the two processes through which Scotch whisky may be produced. "Malt" refers to the fact that the whisky is prepared using only malted barley as a substrate. Malting is the process whereby the barley is soaked, causing the moisture content inside the grain to increase to the point where it begins to germinate. During the process of germination, the grain produces enzymes which convert starch stored in the endosperm into sugars which are subsequently fermented into alcohol. The germination process must be halted at the point where enzyme potential is at a maximum whilst enzymatic conversion of starch into sugar is at a minimum. (Conversion happens at a later stage in the process.) This is achieved by drying the grain using warm air or peat smoke. The latter process leaves residual phenolic compounds on the barley grains which impart a distinctly smoky/medicinal character (known as "peatiness") upon the subsequent Scotch whisky (key brands are Ardbeg, Bowmore, Caol Ila, Lagavulin, and Laphroaig). This is one of the principal manufacturing variables in SMSW production. The other is maturation. Scotch whisky must be matured in oak casks. Normally, these are ex-bourbon casks from Kentucky or ex-sherry casks from Spain. The former imparts a vanilla, sweet, confectionery, biscuit-like character on the Scotch whisky (eg, standard expressions of Glenfiddich, The Glenlivet, and Glenmorangie, amongst many others) whilst the latter imparts a winey, dark fruit character (eg, Macallan Sherry Wood, and Aberlour A'bunadh).

Lagavulin 16 years old is a "peated" SMSW where the spirit is matured using a relatively high proportion of ex-sherry casks, which yields a complex, rich but quite mellow smoky character. Indeed, it has a distinct sensory signature that is highly revered amongst aficionados of peated Scotch whiskies. The Glenlivet 12 years old is an "unpeated" SMSW that is matured using a high proportion of fresh ex-bourbon casks, giving it a distinct bourbon-like character that is devoid of any hints of smokiness. It is one of the largest-selling brands of SMSW. Whilst the sensory profiles of these two SMSWs have some features in common, they are also very obviously and distinctly different.

As described earlier, the conceptual profile of an unbranded product derives from two principal sources; the "category effect" (SMSW in this case) and the "sensory-specific effect" (ie, aspects of the sensory profile that are unique to each of the SMSWs). Since both brands are SMSWs, the "category effect" will not differentiate the conceptual profiles of the two brands but the profound sensory differences most certainly will do. Fig. 16.4 shows the hierarchy of conceptual associations for the two unbranded SMSWs.

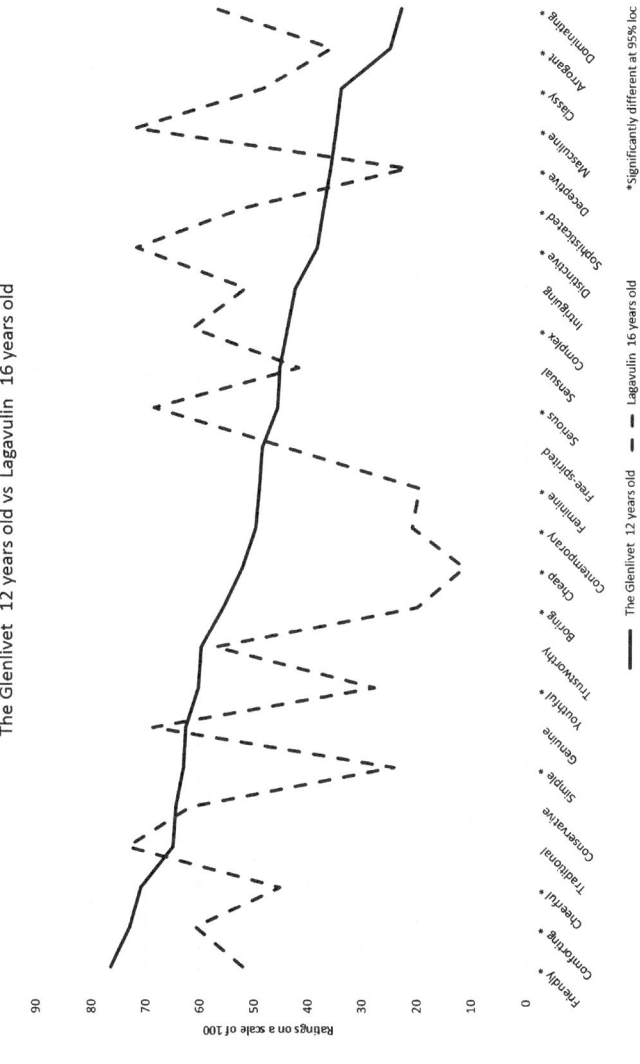

Figure 16.4 Summaries of the unbranded product conceptual profiles of The Glenlivet and Lagavulin.

In brief, The Glenlivet (unbranded) is conceptualized as being more *friendly, comforting*, and *cheerful* whereas Lagavulin (unbranded) is more *traditional, masculine, distinctive, genuine*, and *serious*. This alone suggests that these two SMSWs have the potential to deliver two very different emotional outcomes.

There are more than 100 SMSW distilleries. "Single" refers to the fact that all of the whisky is sourced from one particular distillery (eg, Glenfiddich Distillery, Macallan Distillery, Lagavulin Distillery, etc.). Effectively each distillery is a brand. One of the major problems facing the SMSW category as a whole is that almost all of these 100+ "brands" essentially communicate the same brand message. The conceptual profiles of almost all credible SMSW brands are characteristically *comforting, classy, distinctive, genuine, sophisticated, traditional, trustworthy* but not *arrogant, boring, or cheap* – known as the "Key 10"). In essence, the conceptual profile of the category as a whole is very strong but further differentiation amongst brands is relatively modest.

When the conceptual profile of the Lagavulin brand (not shown) is plotted against the conceptual profile of the corresponding unbranded liquid (Fig. 16.5), the two are very obviously correlated, especially across the "Key 10" conceptual associations.

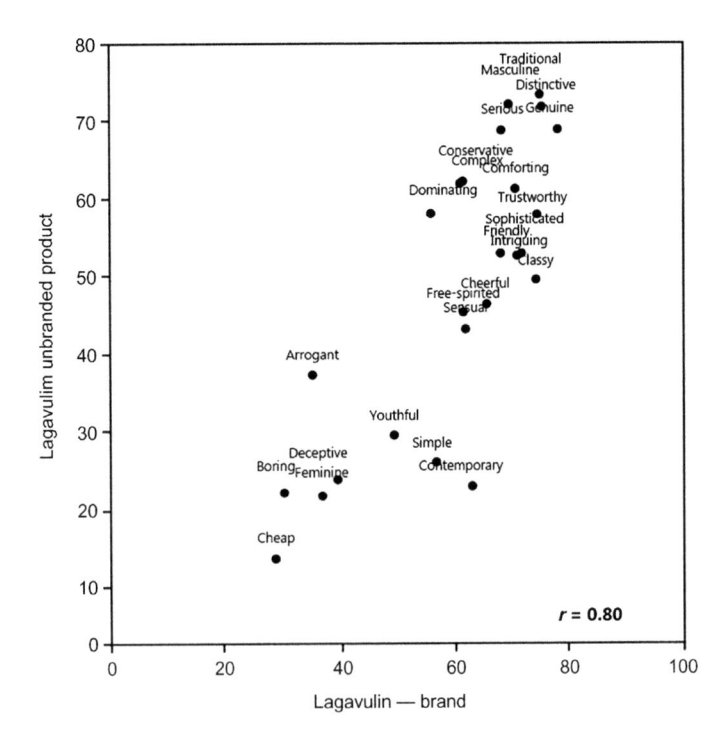

Figure 16.5 Lagavulin: biplot of conceptual profiles of product versus brand.

This suggests a high degree of brand–product consonance and consequently, excellent "fit-to-brand." Conversely, the biplot for The Glenlivet (Fig. 16.6) shows no obvious correlation and consequently, very poor "fit to brand." Although The Glenlivet is widely distributed globally, it is one of the cheaper brands in the category and it is often discounted. Lagavulin sells at a significant premium over The Glenlivet and it is almost never discounted. This level of brand–product dissonance is rather unexpected, bearing in mind that The Glenlivet is a category-leading brand. One possible explanation might be that consumers are attracted to the SMSW category as a whole and it is the conceptualizations associated with the category rather than with specific brands that deliver rewarding emotional outcomes. If so, this reduces the brand to little more than a tag to label SMSWs that may have been enjoyed or otherwise and consequently as an aid to future selection. This takes us all the way back to the genesis of branding and suggests that many SMSW brands have failed to evolve. (This is ironic since many of the iconic Blended Scotch whisky brands were amongst the first real brands.) In many respects, Lagavulin is the epitome of sensory branding and The Glenlivet is the antithesis.

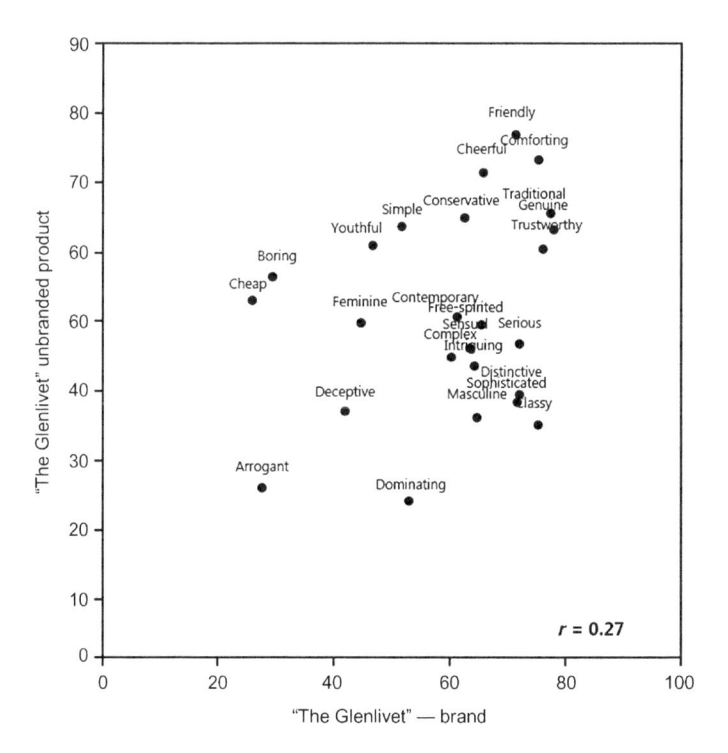

Figure 16.6 The Glenlivet: biplot of conceptual profiles of product versus brand.

12 Conclusion

As initially conceived, sensory branding was about taking a more extensive, multisensory approach to branding, so as to include touch, smell, and taste in addition to vision and audition in brand sensory profiles. This chapter further extends the remit of sensory branding, to include the sensory characteristics associated with pack and product in presenting, delivering and reinforcing the brand message. Germane to this process is the notion that the three primary elements of branded goods, the brand, the pack, and the product, each has a sensory profile and a conceptual profile and that the former plays a critical role in determining the latter. This, in turn, means that manipulating the sensory characteristics of brand, pack, and product provides a means of influencing the brand message and the emotional outcomes it delivers. This places sensory characteristics at the true interface of brand, pack, and product.

Marketing needs to become more scientific. In marketing circles, science is often thought of as an impediment; something that restricts and otherwise stymies creativity. Nothing could and should be further from the truth. "Best practice" from a scientific point of view, advocates that scientists should create hypotheses which attempt to explain occurrences in the natural world. The *Duality of Reward Hypothesis* attempts to do this in terms of human behavior. Whilst it's only a hypothesis, it does provide a framework for understanding and exploring *why people do the things that they do* and it has led to some exciting and useful methodological developments (eg, conceptual profiling) for brand and product development. We hope to extend this further by investigating how conceptual associations might be used to predict the emotional outcomes that drive reward.

Taking a multisensory approach across the three primary touch points of all branded goods (brand, pack, and product) to deliver a compelling brand message is a much underutilized resource in the marketer's toolkit. By adopting more radical, "out-of-the-box" thinking, underwritten by "good science," perhaps the curse of failure that has dogged new product and new brand development for so many decades, might finally be broken.

References

Aaker, J. L. (1997). Dimensions of brand personality. *Journal of Marketing Research, 34*, 347–356.

Buchholz, S. (2008). *When brands get wings: Red bull's secret to marketing success.* Southampton, UK: University of Southampton.

Cabanac, M. (2009). *The fifth influence: Or, the dialectics of pleasure.* iUniverse.com.2223

Darwin, C. (1859). *On the origin of the species by means of natural selection.* London, UK: John Murray.

Ellis, R. D. (1995). *Questioning consciousness: The interplay of imagery, cognition and emotion in the human brain.* Amsterdam: John Benjamins.

Ellis, R. D., & Newton, N. (2010). *How the mind uses the brain (to move the body and image the universe).* Chicago, IL: Open Court.

Gains, N. P. (2013). *Brand esSense*. London, UK: Kogan Page.
Greenwald, A. G., Klinger, M. R., & Schuh, E. S. (1995). Activation by marginally perceptible ("subliminal") stimuli: Dissociation of unconscious from conscious cognition. *Journal of Experimental Psychology: General, 124*, 22–42.
Hetherington, M., & Havermans, R. C. (2013). Sensory-specific satiation and satiety. In J. E. Blundell & F. Bellisle (Eds.), *Satiation, satiety and the control of food intake: Theory and practice* (pp. 253–269). Cambridge, UK: Woodhead.
Hughes, J. (2005). *Still going strong: A history of scotch whisky advertising*. Stroud, UK: Tempus.
Hultén, B. (2011). Sensory marketing: The multi-sensory brand-experience concept. *European Business Review, 23*, 256–273.
Hultén, B., Broweus, N., & van Dijk, M. (2009). *Sensory marketing*. Basingstoke, UK: Palgrave Macmillan.
Kringelbach, M. L., & Berridge, K. C. (2009). Towards a functional neuroanatomy of pleasure and happiness. *Trends in Cognitive Sciences, 13*, 479–487.
Krishna, A. (Ed.). (2010). *Sensory marketing: Research on the sensuality of products*. London: Routledge.
Krishna, A. (2012). An integrative review of sensory marketing: Engaging the senses to affect perception, judgment and behavior. *Journal of Consumer Psychology, 22*, 332–351.
Krishna, A. (2013). *Customer sense: How the 5 senses influence buying behaviour*. New York, NY: Palgrave Macmillan.
Lindstrom, M. (2005). *Brand sense*. New York, NY: Simon & Schuster.
Nielsen (2014a). *Nielsen breakthrough innovation report* (June 2014—US edition).
Nielsen (2014b). *Nielsen breakthrough innovation report* (South East Asia—1st edition).
Piqueras-Fiszman, B., & Spence, C. (2015). Sensory expectations based on product-extrinsic food cues: An interdisciplinary review of the empirical evidence and theoretical accounts. *Food Quality & Preference, 40*, 165–179.
Posner, M. I. (1978). *Chronometric explorations of mind*. Hillsdale, NJ: Erlbaum.
Skinner, B. F. (1976). *About behaviorism*. New York, NY: Vintage Books.
Spence, C. (2015a). Eating with our ears: Assessing the importance of the sounds of consumption to our perception and enjoyment of multisensory flavour experiences. *Flavour, 4*, 3.
Spence, C. (2015b). Leading the consumer by the nose: On the commercialization of olfactory-design for the food & beverage sector. *Flavour, 4*, 31.
Spence, C. (2015c). On the psychological impact of food colour. *Flavour, 4*, 21.
Spence, C. (in press). Multisensory packaging design: Colour, shape, texture, sound, and smell. In M. Chen & P. Burgess (Eds.), *Integrating the packaging and product experience: A route to consumer satisfaction*. Oxford, UK: Elsevier.
Spence, C., & Gallace, A. (2011). Multisensory design: Reaching out to touch the consumer. *Psychology & Marketing, 28*, 267–308.
Spence, C., & Wang, Q. (J.) (2015). Sonic expectations: On the sounds of opening and pouring. *Flavour, 4*, 35.
Stafford-Clark, D. (1965). *What Freud really said*. New York, NY: Schocken Books.
Thomson, D. M. H. (2010). Going beyond liking: Measuring emotional and conceptual profiles to make better new products. In S. R. Jaeger & H. J. H. MacFie (Eds.), *Consumer-driven innovation in food and personal care products* (pp. 219–274). Cambridge, UK: Woodhead.
Thomson, D. M. H. (2015). Expedited procedures for conceptual profiling of brands, products and packaging. In J. Delarue, B. Lawlor, & M. Rogeaux (Eds.), *Rapid sensory profiling techniques and related methods: Applications in new product development and consumer research* (pp. 91–118). Cambridge, UK: Woodhead.

Thomson, D. M. H. (2016). Emotions and conceptualisations. In H. Meiselman (Ed.), *Emotion measurement*. Cambridge, UK: Woodhead.

Thomson, D. M. H., & Crocker, C. (2015). Application of conceptual profiling in brand, packaging and product development. *Food Quality & Preference, 40*, 343–353.

Thomson, D. M. H., Crocker, C., & Marketo, C. G. (2010). Linking sensory characteristics to emotions: An example using dark chocolate. *Food Quality & Preference, 21*, 1117–1125.

Watson, J. B. (1925). *Behaviorism*. New York, NY: W.W. Norton & Company.

Zajonc, R. B. (1980). Feeling and thinking: Preferences need no inferences. *American Psychologist, 35*(2), 151–175.

Index

Note: Page numbers followed by "*f*" and "*t*" refer to figures and tables, respectively.

A

Affective priming, 139–140
Affordance, 139
Alanine-Valine-Isoleucine (AVI), 196
Alliesthesia, 317
AMY1, 189
Amygdala, multisensory interaction in, 216, 225, 237
Analytic attitudes, role in flavor perception, 169–170
Androstenone, 191
Anosmia, 29, 191
Anticipatory cues, 2
Aromas, specificity of ANS activity in, 256–258, 257*f*
Artificial food color, 118–119
Astringency, 60–62, 188–189
Attentional accounts, of flavor binding, 24–29
   attentional capture, 24–27
   attentional channel, 27–29
Audition, in flavor perception, 2–4, 15. *See also* Sound
Autonomic nervous system (ANS), 7
   activity, 249–251, 254*t*–255*t*
     measures of, 253
     patterns of, 253–264, 256*f*
     temporal dynamics of, 258–260, 259*f*, 260*f*
   responses to flavors, 249
     emotions, 251–252, 252*f*
     measurements in real consumer world, 263–264
     measures in consumer research, applications of, 260–263, 261*f*, 262*f*
     specificity, 251–252, 252*f*, 256–258, 257*f*

B

Background music, 93–96
Background noise, 89–93, 91*f*
Behavior(ism), 315
   food color on, psychological effects of, 116–119
     artificial/natural color, 118–119
     color variety, 117
     off coloring, 117
β-ionone, 192
Beverage preparation, 82–84
Bitterness, 195–197
   TAS2R16, 196–197
   TAS2R38, 196
Boar taint, 191
Bottom-up priming, 145–147, 146*f*
Brain
   and bodily reactions, 6–7
   hungry, 239–240
   obese, 240
   principles, of eating, 212–213
   reward processing in, 223–227
     hedonic network, mapping, 224–225
     pleasure cycle, 225–227
   scanning. *See* Magnetic resonance imaging
Branding, 2, 241, 313
   conceptual associations, 319–320
   consonance, 329–333, 332*f*, 333*f*
   extending to pack and product, 324–326
   fit-to-brand, 329–333, 332*f*, 333*f*
   fundamental nature of, 323–324
   linking conceptualization to reward, 320–322, 322*f*
   matrix, 326–328
   mental constructions, 315
   multisensory approach to, 313–315
   pleasantness, 316–319

Branding (*Continued*)
  reward, 316–319
  sensory characteristics, 315–316
  sensory signature, 328–329
Brand–packaging consonance, 314
Brand–packaging dissonance, 314
Brand–product consonance, 314
Brand–product dissonance, 314

C

Carbonation, 60–62
  sound of, 88
l-Carvone, 192
CD36 (cluster of differentiation 36), 194–195
Central nervous system (CNS), 249–251
Change-blindness, 171–172
Chemesthesis, 15, 189–190
Chromosomes, 185–186
Cineole, 192
Cis-3-hexen-1-ol, 192
Cognitive psychology, 315
Color
  on behavior, psychological effects of, 116–119
    artificial/natural color, 118–119
    color variety, 117
    off coloring, 117
  brands and, 115–116
  of containers, 278–280
    colored cans, 280
    colored cups and mugs, 278–279, 279f
    colored glasses, 278
  in flavor perception, 4, 29, 107
  marketing, 119
  names and, 115–116
  psychological effects of, 108–115
    flavor identity, 108–110, 109f, 110f
    taste/flavor intensity, 110–113, 112f
  psychological effects of, individual differences in, 119–124
    cross-cultural differences, 120
    developmental differences, 121
    expertise, 121–122
    genetic differences, 122–124
Color–flavor associations, cross-cultural, 120f
Conceptual associations, 319–320
Conceptual priming, 139

Conditioned stimulus–unconditioned stimulus (CS–US) pairing, 160–161
Congruency, 162
  and oral referral, 49–50
Consonance, 329–333, 332f, 333f
Consummatory cues, 2
Consumption behavior, drinking vessels and, 284–287, 286f
Containers, color of, 278–280
  colored cans, 280
  colored cups and mugs, 278–279, 279f
  colored glasses, 278
Copy number variation (CNV), 187, 189–190
Creaminess, 188–189
Crispness, 64–65, 68–69
  sound of, 86–88, 87f
Crunchy, 70–71, 88–89
Cue(s)
  anticipatory, 2
  consummatory, 2
  exteroceptive, 2
  interoceptive, 2

D

DNA (deoxyribonucleic acid), 185–186, 187f
Drinking receptacle, 269
  and consumption behavior, 284–287, 286f
  future research directions of, 288–289, 289f
  marketing practice, implications for, 287–288
  sensation transference, 269, 277–284
    color of containers, 278–280
    texture of drinking vessel, 280–283, 281f, 282t, 283f
    weight of drinking vessels, 284
  shape of, 270–277
    expectations regarding type/flavor of drink, 270–272, 270f
    wine glass, 272–277, 273f, 276f
Drinks, specificity of ANS activity in, 256–258, 257f

E

Eating
  behavior, flavor memory influence on, 177–178
  brain principles of, 212–213

computational processing related to, 213–223
   food-related rewards, sensory perception of, 216–222
   hunger and hedonia, 215–216
   hunger–homeostasis relationship, 215–216
   identity, 222–223
   intensity, 222–223
   quality, 222–223
Ecologically relevant measurement, 170
Evaluative conditioning, 157–160
Evaluative learning, models of, 160–161
Expectations, 133, 236–237
   receptacle regarding type/flavor of drink, 270–272, 270f
   sensory, setting, 108–115
Experiences, 236–237
Explicit memory, 169–180
Exteroceptive cues, 2

F

Fattiness, 60–62
Fatty Acid Translocase. *See* CD36 (cluster of differentiation 36)
Fight-or-flight response, 316
Fit-to-brand, 329–333, 332f, 333f
Flavor, 1–2, 15. *See also* Taste
   adaptive significance of, 163
   binding. *See* Flavor binding
   definition of, 2, 59, 134f, 135–137, 140
   expectations, 133, 236–237
   experiences, 236–237
   hedonics and memorability of, 5
   identity, 108–110, 109f, 110f
   intensity, 110–113, 112f
   learning, and memory over lifetime, 175–177
   memory. *See* Memory
   modalities. *See* Flavor modalities
   multisensory perception. *See* Multisensory flavor perception
   neuroscience of, 235
   object, 2
   priming, multisensory. *See* Priming
   recollection versus novelty detection, 178–179
   sensory complexity of, 136f
   system, 39f

Flavor binding, 2, 16–22, 27, 50–51, 161–162
   attentional accounts of, 24–29
   broader perspective, 21–22
   capacity to perceive parts and wholes, 20–21
   characteristics of, 22
   interactions between senses, 18–19
   perceived location, 19–20
   preattentive accounts of, 22–24
   retronasal olfaction, modality/content dissociation in, 16–18
   temporal properties, 20
Flavor–consequence learning, 157–160
Flavor–flavor learning, 157–159
Flavor liking, 155
   development of, 155–156
   evaluative conditioning, 157–160
   evaluative learning, models of, 160–161
Flavor modalities, 38–40
   gustation, 38, 40t
   olfaction, 38, 40t
   somesthesis, 38–39, 40t
Food
   color. *See* Color
   intake, functional outcome of, 198–201, 201f
   likes/dislikes determination, oral-somatosensation role in, 68–69
   preparation, 82–84
   specificity of ANS activity in, 256–258, 257f
   texture, 60, 65–69, 66f, 87–88, 188–189
"Food 386", 115–116

G

Genetics, 185–188
GNAT3, 193
GPR120, 194–195
Guaiacol, 192
Gustation, 25–26, 38, 40t

H

Hedonia, 211
   hunger and, 215–216
Hedonic network, 5–6, 212f
   mapping, 224–225
Homeostasis–hunger relationship, 215–216

Hunger
  and homeostasis, relationship between, 215–216
  and hedonia, relationship between, 215–216
Hungry brain, 239–240

I

Identity, 222–223
Implicit memory, 170–178
Inosine monophosphate (IMP), 194
Intensity, 222–223
Interoceptive cues, 2
Isobutyric acid, 192
Isovaleric acid, 192

L

Labeled Magnitude Scale, 111–112
Labeling, 242–243
Learning, 211, 226–227
  evaluative learning, models of, 160–161
  flavor–consequence, 157–160
  flavor–flavor, 157–159
  and memory over lifetime, 175–177
Lexical Decision Task (LDT), 142
Liking, 211, 213, 225–226
  threshold, 317–318
Linkage disequilibrium (LD), 186–187, 192
Localization, 25, 27–29

M

Magnetic resonance imaging, 140–141
Marketing color, 119
Mediodorsal thalamus, multisensory interactions in, 237
Medium tasters, 6
Memory, 5, 169
  distortion, 173–174
  explicit, 169–180
  forms of, 170–175
  implicit, 170–178
  influence on eating behavior, 177–178
  odor-emotional, 223
  over lifetime, flavor learning and, 175–177
  real-world, 178–180
Mental constructions, 315
Mere exposure effect, 155–156, 270–272
Messenger RNA (mRNA), 186

Minor allele frequency (MAF), 186–187
Misfit theory of spontaneous conscious odor perception (MITSCOP), 171–172
Modal attention, 25–26
Molecular biology, 185–188
Monosodium glutamate (MSG), 158–159, 193–194, 238, 242–243
Mouthfeel, 2–3, 62, 221–222
  definition of, 62
Multisensory, definition of, 134–135, 134*f*
Multisensory flavor perception, 60–62
  individual differences in, 185
Multisensory flavor priming. *See* Priming
Multisensory integration, 1–4
  perceptual evidence of, 44
Multisensory interactions, perceptual evidence of, 42–44
Multisensory neural perception, in neural circuits, 237–240

N

Natural food color, 118–119
Negative reward, 317
Neural substrates, 67–68
Neurogastronomy, 235
Neuroscience of flavor, 235
  branding, 241
  flavor expectations, 236–237
  flavor experiences, 236–237
  labeling, 242–243
  neural circuits, 237–240
  pricing, 241–242
Nontasters, 6
Novelty detection versus flavor recollection, 178–179
Nucleus accumbens (NAc), multisensory interaction in, 225–226
Nucleus of the solitary tract (NST), 41–42
  multisensory interaction in, 219

O

Obese brain, 240
Odor, 4. *See also* Taste
  anosmias, 190–192
  fingerprints, 218
  mechanisms of binding of, 161–162
  memory, 171–172
  objects, 171–172

priming, 143–145
rancid, 144
recognition, 171–172
referral, retronasal, 45–49, 47*f*, 48*f*
and viscosity, interaction between, 63–65
Odor-emotional memory, 223
Odor-enhancement effect, 19
Off coloring, 117
Oleogustus, 194–195
Olfaction, 38, 40*t*, 135, 190
orthonasal, 15, 17
retronasal, 15–19
Olfactory location illusion, 162
Omnivore's Paradox, 156
OR4N5, 191–192
OR6A2, 191–192
OR7D4, 191
Oral referral, 1–2, 44–51, 137
conditions for, 49–50
congruency, 49–50
spatial synchrony, 50
temporal synchrony, 50
retronasal odor referral, 45–49, 47*f*, 48*f*
sensory and neural mechanisms, 50–51
taste referral, 45
Oral-somatosensation, 2–3, 19–20, 62–67
mouthfeel, 62
neural substrates, 67–68
oral stereognosis, 63
role in expected and experienced satiety, 69, 70*f*
role in food likes/dislikes determination, 68–69
temperature, 65
texture, 65–67, 66*f*
viscosity, 63–65
Oral stereognosis, 63
Orbitofrontal cortex (OFC)
caudomedial, 239–240
lateral, 239–240
mid-anterior, 225
multisensory interaction in, 6, 216, 220–221, 223–226, 235–242
Orthonasal olfaction, 15, 17

P

Packaging
extending sensory branding to, 324–326
sounds, 84–85

Parchment skin illusion, 86
Pavlovian conditioning, 160–161
Perceived location, 19–20
Perceptual priming, 139
Phenotypes, 187–188
Phenylthiocarbamide (PTC), 196, 198–199
Pleasantness, 316–319
Pleasure, 317
cycle, 212*f*, 225–227
Positive reward, 317
Preference, 5–6
Pricing, 241–242
Priming
affective, 139–140
bottom-up, 145–147, 146*f*
conceptual, 139
definition of, 133, 137–140, 138*f*
flavor, 143–145
multisensory flavor, 133
odor, 143–145
perceptual, 139
repetition, 138
situated perception and behavior, conceptual framework for, 147–148, 148*f*
taxonomy of, 140–143, 141*t*
top-down, 145–147, 146*f*
Product
emotions, 250–251
extending sensory branding to, 324–326
Proline-Alanine-Valine (PAV), 196
6-*n*-Propylthiouracil (PROP), 189, 195–196, 198–199
Proteins, 185–186
Proustian phenomenon, 223
Provencal rose paradox, 5
Psychological effects, of color, 108–115
on behavior, 116–119
artificial/natural color, 118–119
color variety, 117
off coloring, 117
flavor identity, 108–110, 109*f*, 110*f*
individual differences in, 119–124
cross-cultural differences, 120
developmental differences, 121
expertise, 121–122
genetic differences, 122–124
taste/flavor intensity, 110–113, 112*f*

Q
Quality, 222–223
  fusion, 44

R
Receptacle
  appropriateness, 7
  and consumption behavior, 284–287, 286f
  future research directions of, 288–289, 289f
  marketing practice, implications for, 287–288
  sensation transference, 269, 277–284
    color of containers, 278–280
    texture of drinking vessel, 280–283, 281f, 282t, 283f
    weight of drinking vessels, 284
  shape of, 270–277
    expectations regarding type/flavor of drink, 270–272, 270f
    wine glass, 272–277, 273f, 276f
Repetition priming, 138
Retronasal odor referral, 45–49, 47f, 48f
Retronasal olfaction, 15, 19, 236
  modality/content dissociation in, 16–18
Reward, 316–319
  food-related, sensory perception of, 212–213, 215–222
    mouthfeel, 221–222
    smell, 216–221
    sound, 221
    taste, 219–221
    vision, 221
  hypothesis, duality of, 318–319, 318f, 321–322
  linking conceptualization to, 320–322, 322f
  negative, 317
  positive, 317
  processing, in brain, 223–227
    hedonic network, mapping, 224–225
    pleasure cycle, 225–227
RNA (ribonucleic acid)
  messenger, 186
  transfer, 186

S
Saltiness, 197–198
Satiation, 213
Satiety, 213
  expected and experienced satiety, oral-somatosensation role in, 69, 70f
  sensory specific, 261–262
SCNN1B, 198
Sensation transference, 7, 269, 277–284
  color of containers, 278–280
    colored cans, 280
    colored cups and mugs, 278–279, 279f
    colored glasses, 278
  texture of drinking vessel, 280–283, 281f, 282t, 283f
  weight of drinking vessels, 284
Senses. *See also* Smell; Taste; Vision, in flavor perception
  interactions between, 18–19
  role in user–product interactions, 297
    after consumption, 306–307
    dynamics of sensory perception, research approaches to, 298–301, 299f, 300f, 302f
    packaged product, 303–304
    unpacked product, 304–306
Sensory branding. *See* Branding
Sensory characteristics, 315–316
Sensory dominance, 121, 239
Sensory inputs, pathways and convergence of, 40–42, 41f
Sensory optimization, 317–318
Sensory perception, 1–4
  impact of, 7–8
Sensory perception, 133
Sensory signature, 328–329
Sensory specific satiety, 261–262
Sentence completion technique, 18
Single nucleotide polymorphisms (SNPs), 186–187
  in CD36, 195
  in *OR4N5*, 191–192
  in *OR6A2*, 191–192
  in *OR7D4*, 191
  in *TAS1R1*, 194
  in *TAS1R3*, 193–194
  in *TAS2R1*, 191–192
  in *TAS2R3*, 197
  in *TAS2R4*, 197
  in *TAS2R5*, 197
  in *TAS2R16*, 196–197
  in *TAS2R19*, 197
  in *TRPA1*, 189–190

Situated Inference Model, 138–139
Skin conductance, 253, 256–258, 259f, 260–264, 261f, 262f
Smell, 20, 216–219. *See also* Senses
 in flavor, multisensory convergence of, 220–221
Somatic marker hypothesis, 250
Somatosensation, oral. *See* Oral-somatosensation
Somatosensory stimulation outside the mouth, contribution of, 69, 71f, 72f
Somesthesis, 38–39, 40t
Sonic seasoning, 96–99, 97f
Sound. *See also* Audition, in flavor perception
 background music, 93–96
 background noise, 89–93, 91f
 of carbonation, 88
 of crispness, 86–88, 87f
 in flavor perception, 2–4, 81, 221
 of food and beverage preparation, 82–84
 of packaging, 84–85
 sonic seasoning, 96–99, 97f
Sourness, 198
Spatial attention, 25
Spatial coincidence, and oral referral, 50
Static pattern rotation, illusion of, 136f
Supertasters, 6
Sweetness, 192–193
Synthetic attitudes, role in flavor perception, 169–170

T

Tactile acuity, 6
Tactile sensations, 188–190
 role in taste referral, 46–49
Tactile stimulation, of oral cavity, 64
TAS1R1, 194
TAS1R2, 193
TAS1R3, 193–194
TAS2R1, 191–192
TAS2R3, 197
TAS2R4, 197
TAS2R5, 197
TAS2R16, 196–197, 199–200
TAS2R19, 197
TAS2R38, 196, 199–200, 201f
TAS2R44. *See* TAS2R16
TAS2R48. *See* TAS2R19

Taste, 1–2, 15, 219–220. *See also* Flavor; Odor; Senses
 aversion, 159–160
 bitterness, 195–197
 in flavor, multisensory convergence of, 220–221
 individual differences in, 6
 intensity, 110–113, 112f
 mechanisms of binding of, 161–162
 oleogustus, 194–195
 referral, 45
 saltiness, 197–198
 sourness, 198
 sweetness, 192–193
 umami, 193–194
 and viscosity, interaction between, 63–65
Taste-enhancement effect, 18–19
Taster(s)
 medium, 6
 nontasters, 6
 status, 6
 supertasters, 6
Temperature–taste interaction, 65
Temporal properties, of flavor binding, 20
Temporal synchrony, and oral referral, 50
Texture
 of drinking vessel, 280–283, 281f, 282t, 283f
 flavor, 60, 65–69, 66f, 87–88, 188–189
Thermal-taste illusion, 65
Tip-of-the-nose phenomenon, 223
Top-down priming, 145–147, 146f
Transfer RNA (tRNA), 186
Trichloroanisole (TCA), 192
Trimethylamine, 192
TRPA1, 189–190
TRPV1, 198

U

Umami, 193–194
Unisensory cortex, in multisensory flavor perception, 237–238

V

Ventral pallidum (VP)
 multisensory interaction in, 225–226
Ventriloquism, 64–65

Ventroposterior medial nucleus of the thalamus (VPM$_{PC}$), 41–42
Viscosity, 63–65
  interaction with taste/odor perception, 63–65
Vision, in flavor perception, 2–4, 15, 221. *See also* Senses
Visual sensory dominance, 6

W

Wanting, 158–159, 211, 213, 225–226
Weight, 70–71
  of drinking vessels, 284
Wine glass, shape of, 272–274, 273*f*
  evidence of influence, 274–276, 276*f*
  physical versus psychological influence of, 276–277

Edwards Brothers Inc.
Ann Arbor MI. USA
December 21, 2017